ADVANCES IN CONSTRUCTION, REAL ESTATE, INFRASTRUCTURE AND PROJECT MANAGEMENT

This volume presents a curated selection of high-impact, cutting-edge research papers from ICCRIP-2024, covering a broad spectrum of themes across construction, real estate, infrastructure, and project management. It captures innovative methodologies, technological advancements, and interdisciplinary solutions relevant to today's complex built environment challenges. Contributions span both academic and industry perspectives, offering actionable insights and strategic frameworks. The book serves as a dynamic platform for knowledge exchange among scholars, practitioners, and decision-makers. It is a vital resource for those seeking to understand emerging trends and future directions in these sectors. Academicians, researchers, and professionals alike will find this compilation highly valuable.

ADVANCES IN CONSTRUCTION, REAL ESTATE, INFRASTRUCTURE AND PROJECT MANAGEMENT

Edited by
Anil Kashyap
Sushma S. Kulkarni
Rajni Kant Rajhans
Jolita Kruopienê
Shashank B. S.

CRC Press
Taylor & Francis Group
Boca Raton London New York

CRC Press is an imprint of the
Taylor & Francis Group, an **informa** business

First edition published 2026
by CRC Press
4 Park Square, Milton Park, Abingdon, Oxon, OX14 4RN

and by CRC Press
2385 NW Executive Center Drive, Suite 320, Boca Raton FL 33431

British Library Cataloguing-in-Publication Data
A catalogue record for this book is available from the British Library

ISBN: 978-1-041-13432-9 (hbk)
ISBN: 978-1-041-13433-6 (pbk)
ISBN: 978-1-003-66981-4 (ebk)

DOI: 10.1201/9781003669814

Typeset in Times New Roman
by Aditiinfosystems

Contents

Advances in Construction, Real Estate, Infrastructure and Project Management – Anil Kashyap et al. (eds)
© 2026 Taylor & Francis Group, London, ISBN 978-1-041-13433-6

List of Figures

Advances in Construction, Real Estate, Infrastructure and Project Management – Anil Kashyap et al. (eds)
© 2026 Taylor & Francis Group, London, ISBN 978-1-041-13433-6

List of Tables

Advances in Construction, Real Estate, Infrastructure and Project Management – Anil Kashyap et al. (eds)
© 2026 Taylor & Francis Group, London, ISBN 978-1-041-13433-6

Conference Organizing Committee

Conference Chief Patron

Dr. Anil Kashyap
President & Chancellor,
NICMAR University, Pune

Conference Patron

Dr. Sushma S. Kulkarni
Vice Chancellor,
NICMAR University, Pune

Conference Convener

Dr. Rajni Kant Rajhans
Dean-Research & Development,
NICMAR University, Pune

Conference Committee Members

- **Prof. Abhishek Shrivas** – Assistant Professor
- **Prof. Aritra Halder** – Assistant Professor
- **Dr. Ashish Rastogi** – Assistant Professor
- **Dr. Deepak M.D.** – Assistant Professor
- **Dr. Dhaarna** – Assistant Professor
- **Dr. Hindavi Pavan Tikate** – Assistant Professor
- **Dr. Komal Raosaheb Handore** – Assistant Professor
- **Dr. K. Tulasi Vigneswara Rao** – Assistant Professor
- **Dr. Nilesh Agarchand Patil** – Assistant Professor
- **Dr. Pradeepta Kumar Samanta** – Associate Professor
- **Prof. Sameer Jain**-Assistant Professor
- **Dr. Shashank B.S.** – Assistant Professor
- **Dr. Sudarsan J. S.** – Assistant Professor
- **Dr. Sudarshan Dattatraya Kore** – Assistant Professor

Acknowledgment

The successful organization of the **8th International Conference on Construction, Real Estate, Infrastructure, and Project Management (ICCRIP 2024) at NICMAR University, Pune**, would not have been possible without the collective efforts of numerous individuals and organizations. We extend our sincere gratitude to all contributors who played a pivotal role in making this event a grand success.

We express our heartfelt appreciation to **NICMAR University's leadership, faculty, and administrative staff** for their unwavering support in organizing this prestigious conference. Their commitment to fostering academic excellence and industry collaboration has been instrumental in the continued success of ICCRIP.

A special note of thanks to all **authors and researchers** whose high-quality contributions have enriched the proceedings, *Advances in Construction, Real Estate, Infrastructure, and Project Management*. The conference received an overwhelming response, and the insightful research papers presented have significantly added to the body of knowledge in construction, real estate, infrastructure, and project management.

We are deeply grateful to our esteemed **keynote speakers, session chairs, panelists, and reviewers**, whose expertise and valuable feedback ensured the high academic standards of the conference. Their guidance and critical evaluations have helped enhance the quality of research shared during ICCRIP 2024.

We extend our sincere appreciation to our **industry partners, sponsors, and collaborating institutions** for their generous support and active participation. Their engagement has strengthened the link between academia and industry, fostering meaningful discussions on real-world challenges and solutions.

Our gratitude also goes to the **organizing committee, technical committee, and student volunteers,** who worked tirelessly behind the scenes to ensure the seamless execution of the event. Their dedication, meticulous planning, and teamwork have been instrumental in making ICCRIP 2024 a resounding success.

Finally, we thank all **participants, attendees, and delegates** who joined us from across the globe. Your active engagement, thought-provoking discussions, and enthusiasm have contributed immensely to the success of this conference.

We look forward to continued collaboration and knowledge-sharing in the years to come and hope that the insights gained from ICCRIP 2024 will inspire future research and innovations in the built environment.

ICCRIP 2024
NICMAR University, Pune

Preface

The 8th International Conference on Construction, Real Estate, Infrastructure, and Project Management (ICCRIP 2024), organized by NICMAR University, Pune, on August 23–24, 2024, served as a premier platform for knowledge exchange and industry-academic collaboration. Continuing its legacy of fostering innovation and research in the built environment, ICCRIP 2024 featured insightful discussions across a wide spectrum of emerging challenges and advancements in the CRIP sectors.

The conference welcomed over 1,000 participants, including distinguished academicians, industry leaders, researchers, and professionals from across the globe. With a rigorous selection process, 250 research papers were accepted for presentation, reflecting cutting-edge developments in areas such as risk analysis, construction waste minimization, ESG compliance, lean construction, BIM integration, sustainable materials, and infrastructure project management.

The proceedings, titled *Advances in Construction, Real Estate, Infrastructure, and Project Management*, encapsulate these pioneering research contributions. Some of the key topics covered in this volume include:

- **Risk Management & Project Performance**: Risk analysis using FMEA, smart city project management, success factors in TQM, and claims in the Indian construction industry.
- **Sustainability & Green Construction**: ESG impact on infrastructure companies, green concrete materials, biochar's role in sustainable construction, and carbon market comparisons.
- **Technological Innovations**: Integration of BIM and Last Planner System (LPS), VR/AR in construction safety, and IGBC-compliant software models.
- **Infrastructure & Real Estate Trends**: Relationship between infrastructure development and real estate demand, valuation parameters of real estate, and feasibility of prefabricated container houses in India.
- **Emerging Challenges & Future Outlook**: Predicting climate change impact on coastal structures, labor productivity in construction profitability, and delays in road construction projects.

We extend our gratitude to all authors, reviewers, and organizing committee members whose efforts have made this volume possible. It is our sincere hope that the research presented in this proceeding will contribute to the continued evolution of construction, real estate, infrastructure, and project management, inspiring further advancements in these crucial fields.

Editors

Advances in Construction, Real Estate, Infrastructure and Project Management – Anil Kashyap et al. (eds)
© 2026 Taylor & Francis Group, London, ISBN 978-1-041-13433-6

Foreword

The rapid growth of the construction, real estate, and infrastructure sectors in India and across the world has introduced new challenges and opportunities for industry professionals and researchers alike. As technology, sustainability, and market dynamics continue to evolve, interdisciplinary collaborations and knowledge-sharing platforms like ICCRIP have become vital in shaping the future of the built environment.

ICCRIP 2024 has once again demonstrated the importance of bringing together industry experts and academic scholars to address pressing concerns and explore innovative solutions. This year's conference has delved into critical issues, from infrastructure risk analysis and labor productivity to sustainable materials and digital transformation. The volume *Advances in Construction, Real Estate, Infrastructure, and Project Management* presents ground breaking studies that offer valuable insights into these key areas.

In particular, topics such as ESG compliance in the engineering, procurement, and construction (EPC) industry, the feasibility of prefabricated container houses, transport corridor megaproject enablers and barriers, and climate change predictions using deep learning highlight the multifaceted nature of construction research today. By addressing both theoretical advancements and practical applications, these papers offer a well-rounded perspective on the future of the industry.

NICMAR University, Pune, has played a pivotal role in advancing research and training in these sectors. ICCRIP 2024 is yet another testament to its dedication to bridging academia and industry, fostering innovation, and equipping professionals with the knowledge to navigate an ever-changing landscape.

As you explore the insights presented in this volume, we encourage you to apply these findings to your own work and to continue engaging in collaborative research efforts. May this collection serve as a catalyst for further innovation and excellence in construction, real estate, infrastructure, and project management.

Editors

Advances in Construction, Real Estate, Infrastructure and Project Management – Anil Kashyap et al. (eds)
© 2026 Taylor & Francis Group, London, ISBN 978-1-041-13433-6

About the Editors

Prof. (Dr.) Anil Kashyap is the first President and Chancellor of NICMAR University Pune and Hyderabad. He has 30 years of experience in practice, academia and research – both in India and abroad. In his leadership roles, Dr. Kashyap has been Head of UWE Bristol School of Geography and Environmental Management, Deputy Head of Coventry University School of Energy, Environment and Construction, and Founding Professor and Director of RICS School of Real Estate before joining NICMAR as the Director General in September 2021. His expertise includes Infrastructure development and finance, Urban Regeneration, and Real Estate Research. Dr. Kashyap holds a Bachelor's Degree in Civil Engineering with distinction from NIT Kurukshetra, Masters in Urban Planning from School of Planning & Architecture, New Delhi and a Ph.D. from University of Ulster, United Kingdom. Dr. Kashyap is an Alumnus of Harvard University with training in Higher Education Leadership and Change Management. He is a Fellow of the Higher Education Academy of the United Kingdom. Dr. Kashyap always had a passion for academia from early career days and was awarded the Ulster University Fellowship for his PhD. He has extensive experience of innovative curriculum design. Dr. Kashyap has published over 100 papers in peer reviewed journals, international and national conferences till date. He has supervised 5 PhDs and has been examiner for PhD students in India and the United Kingdom. He has also served as examiner for PhD students in India and abroad. Dr. Kashyap has undertaken research funded by research councils, charities and international institutions. Dr. Kashyap has been Chartered Member of RICS, Chartered Member of Royal Town Planning Institute London, Fellow of Higher Education Academy, United Kingdom and Fellow of Institute of Town Planners, India.

Prof. (Dr.) Mrs. Sushma Shekhar Kulkarni is the Vice Chancellor of NICMAR University, Pune, India, and Chair of GEDC. She was formally the Director at Rajarambapu Institute of Technology, Maharashtra, India. She has completed her Ph.D. in Civil Engineering (2003), M.E. in Construction Management (1993), from Shivaji University, Kolhapur, India, and B.E. in Civil Engineering (1987), from Visvesvaraya National Institute of Technology, Nagpur, India. She has been holding the post of Principal/Director for more than 16 years. She has transformed an affiliated Engineering college into an Autonomous Institute with excellent credentials. Institute has NAAC "A+" grade, and all its eligible programs {more than 60%} are accredited & re-accredited by NBA. She has successfully handled World Bank-funded Technical Education Quality Improvement program (TEQIP I & II) with a funding of Rs. 15 crores. She has 37 years of teaching & research experience in the Civil Engineering program. Her areas of interest are Concrete Technology, TQM, Quality Circles, Construction Management, Watershed Management, Outcome Based Education, and Women Empowerment. She has received more than 35 awards/honors for her outstanding academic work. She has contributed to more than 155+ research papers and, has two patents provisionally registered. She has delivered more than 140+ expert talks on diverse topics in India and outside India. Also, she worked as a panelist in workshops and webinars organized by IUCEE, IFEES, GEDC, etc at national and international levels. Under her guidance, 5 students have completed their Ph.D. and 7 are pursuing it. She has also guided more than 70 UG & PG students in their research projects.

Advances in Construction, Real Estate, Infrastructure and Project Management

Dr. Rajni Kant Rajhans is Dean-Research & Development at NICMAR University, Pune. Dr. Rajhans has earned academic credentials such as a B.Sc. Engineering (Electronics & Communication Engineering with Distinction), MBA (Finance) & Ph.D. (Management). He teaches Managerial Accounting, Financial Management, Project Appraisal, Real Estate Investment & Finance, Corporate Valuation, and other related courses. A state rank holder in the engineering entrance exam, and merit scholarship recipient in all four semesters of MBA, Dr. Rajhans was awarded the Teaching Excellence Award at NICMAR in 2019. Dr. Rajhans has research interests in Corporate Finance, Valuation, and Green Finance. He is an associate editor and editorial advisory board member at Emerging Markets Case Studies, Emerald Publication. He is the series editor of "How-to-Guides", a case-based business research method published by Sage Publishing, and editor of the Journal of Real Estate, Construction & Management (Sage Publishing).

Jolita Kruopienê with a Ph.D. in Environmental Engineering, is a professor and senior researcher at Kaunas University of Technology, Lithuania, working in the fields of chemical risk management, solid waste, environmental impact, and natural resources. Her activities cover research, lecturing, supervising MSc and PhD students, and providing consultancy and training to industries as well as authorities. She has experience as a researcher and coordinator in various international projects. She is editor-in-chief of Environmental Research, Engineering and Management Journal (ISSN 1392-1649. eISSN 2029-2139).

Dr. Shashank B. S., an IGBC-accredited faculty, is a distinguished academician and administrator with a doctoral degree in Civil Engineering. With over a decade of experience in teaching, research, and academic leadership, he is spearheading the undergraduate Civil Engineering program at NICMAR University. An expert in Concrete Technology, Dr. Shashank is deeply committed to embedding digitalization and sustainability into engineering education. His research interests encompass self-healing concrete, advancements in concrete technology, and alternative building materials. His contributions have been widely recognized through publications in reputed national and international journals and conferences. As an educator, he has mentored over 20 undergraduate and postgraduate theses, fostering innovative thinking and problem-solving among students.

Advances in Construction, Real Estate, Infrastructure and Project Management – Anil Kashyap et al. (eds)
© 2026 Taylor & Francis Group, London, ISBN 978-1-041-13433-6

1

Critical Analysis of Road Construction Delays in the Indian Context

Pratik Bhangale

P. G. Student,
Department of Civil Engineering,
Dr. Vishwanath Karad MIT World Peace University,
Pune

Pravin Minde*

Assistant Professor,
Department of Civil Engineering,
Dr. Vishwanath Karad MIT World Peace University,
Pune

■ **ABSTRACT:** The Ministry of Statistics and Program Implementation (MOSPI, India) released a report in November 2023, revealed that 845 infrastructure projects, each valued at ₹150 crore or more, were delayed, representing 46.1% of the total. This alarming statistic highlights the pervasive challenges in Indian road construction, a sector critical to national economic growth. This study investigates the root causes of road project delays and proposes actionable mitigation strategies. Drawing on extensive literature reviews, surveys, and interviews, the research utilises the Relative Importance Index (RII) and Statistical Package for the Social Sciences (SPSS) to rank delay factors systematically. The findings identify weather conditions, material shortages, and design changes as the most significant contributors to project delays. Additional factors, such as permitting and regulatory hurdles, labour shortages, financial constraints, and equipment failures, also play a role but to a lesser extent. The study emphasizes the need for modern solutions, including Building Information Modeling (BIM), prefabrication techniques, the use of advanced construction materials, and real-time project monitoring systems, to address these issues effectively. These strategies are designed to enhance project efficiency, reduce delays, and optimize resource allocation. By adopting such innovative approaches, project managers and policymakers can improve the reliability of construction schedules and support sustainable infrastructure development. The insights provided are not only critical for India but also hold relevance for other developing nations grappling with similar challenges in their road construction sectors. This study advances the knowledge of delay factors and identifies doable strategies to lessen their effects.

■ **KEYWORDS:** Construction delays, Indian road projects, Mitigation strategies, Relative importance index, Developing nations

*Corresponding author: pravin.minde@mitwpu.edu.in

DOI: 10.1201/9781003669814-1

1. INTRODUCTION

1.1 Background

The development of a nation's infrastructure critically enables its economic growth and development. It is considered the backbone of any nation, providing support for the nation's march ahead [2]. A nation's infrastructure must be well-equipped and designed to serve efficiently for smoother, uninterrupted connectivity. Since the earliest times, having good highways has been seen as crucial for a country's economic progress [13]. They help the country grow, work better, and make other industries like construction more productive [9]. Enhancing accessibility and mobility stands as the foremost objective of roads. Presently, emerging nations globally are emphasizing the enhancement and interconnection of their road infrastructures [7]. Road initiatives are being highlighted as the central agenda in their national financial plans, as a robust road network fosters economic advancement and overall national prosperity [15]. Timely project completion indicates efficiency, but the construction process often involves many unpredictable and variable factors. These factors include how parties perform, the availability of resources, the state of the environment, the participation of other parties, and contractual relationships. Nevertheless, it is not common for a project to be finished in the allotted time [3]. The paper focuses on identifying a research gap in the existing literature related to delay analysis in construction projects. It emphasizes the need for a more comprehensive study that maps the worldwide body of knowledge on the topic [5]. The study's novelty lies in its use of the science mapping method to identify gaps and current trends in the literature, offering a distinct and organized way of analyzing and visualizing large volumes of scientific documents. This approach contributes to a more comprehensive understanding of the research landscape on delay analysis in construction projects [16].

In recent years, many contractors, both small and large, have expressed concerns about the challenges of addressing delays in construction projects [10]. A key issue is their inability to pinpoint the critical causes of these delays during the construction process [8]. By ranking the significance of delay factors, project managers can better identify the most impactful issues and explore effective alternative solutions [4].

Previous studies have described the root causes of delays as specific situations or conditions that deviate from fundamental principles and are detailed enough to allow for corrective action. Building on this foundation, this research explores whether identifying root causes of delays significantly impacts project timelines and ranks the delay factors in road construction projects in India. The subsequent sections cover the literature review, research methodology, results and discussions, as well as conclusions and recommendations [4]. To evaluate delays in road construction projects, both the Relative Importance Index (RII) and statistical analysis using SPSS were employed. The RII method ranked delay factors based on their perceived significance, while SPSS provided a data-driven statistical comparison. Results from both methods were analyzed to identify patterns and disparities in rankings. This dual approach ensured comprehensive insights into the critical causes of delays.

1.2 Aim and Scope

This paper seeks to determine and examine the key causes of delays in Indian road construction projects, using the Relative Importance Index (RII) and SPSS for ranking, and to propose effective mitigation strategies. By exploring modern techniques such as (BIM), advanced materials, and real-time monitoring, the study offers actionable recommendations to minimize disruptions, enhance project efficiency, and support sustainable infrastructure development.

2. LITERATURE REVIEW

Table 1.1 provides a concise overview of various studies conducted in different countries to analyze construction industry practices. It highlights the sampling techniques, research methodologies, methods employed, and data collection strategies, emphasizing the widespread use of descriptive and statistical approaches with questionnaire-based surveys utilizing five-point Likert scales.

Table 1.1   Methodologies employed in earlier studies on delays on roads, construction, and highways

Source	Country / Context	Sampling (Dimensions, Tactics, Etc.)	Research Methodology	Research Method	Data Collection Strategy
[18]	Vietnam	Utilizing Spearman's rank correlation and indexes of frequency, severity, and importance	Descriptive and statistical	Questionnaire and desk-based investigation	The surveys were sent back. Five-point Likert scale in a questionnaire
[11]	Palestine	Random selection was used to choose the sample. The survey was not directed at clients, but rather at consultants and contractors. Of the 80 surveys, 64 were returned.	Descriptive and statistical	Questionnaire and desk-based investigation	The surveys were sent back. Five-point Likert scale in a questionnaire
[4]	Afghanistan	The sample was obtained from a random selection. 20 clients, 20 contractors, and 15 consultants make up the 60 stakeholders in the building industry	Descriptive and statistical	Questionnaire and desk-based investigation	The surveys were sent back. Five-point Likert scale in a questionnaire
[7]	Ghana	A survey was distributed to 125 respondents, comprising consultants, contractors, and owners. The 72 completed questionnaires were returned.	Descriptive and statistical	Questionnaire and desk-based investigation	The surveys were sent back. Five-point Likert scale in a questionnaire
[1]	Malaysia	The sample was chosen at random with a concentration on building projects. Contractors, consultants, developers, subcontractors, engineers, and architects made up the responders. 78 of the 450 responses were gathered for the survey using a questionnaire.	Descriptive and statistical	Questionnaire and desk-based investigation	The surveys were sent back. Five-point Likert scale in a questionnaire

Source: Compiled by authors

Table 1.3 presents the many ways in which researchers have ranked the reasons for building project delays using statistical techniques. Regression analysis, factor analysis, and are some of the techniques used to systematically identify and rank delay reasons.

Table 1.2 presents the many ways in which researchers have ranked the reasons for building project delays using statistical techniques. Regression analysis, factor analysis, and are some of the techniques used to systematically identify and rank delay reasons.

Table 1.2   Statistical methods to rank the causes of delays [2, 17].

Sr. No.	Quantitative Method	Description
1	Relative Importance Index method [RII]	A statistical method called RII is used to measure the relative value of different components in a dataset. It calculates a weighted average of respondents' ratings on the importance of different variables, aiding in prioritizing them based on their impact.
2	Frequency Index method, [FI] (%)	The Frequency Index method assesses variable importance by tallying their occurrence frequency within a dataset, offering a straightforward approach to identifying prevalent factors without relying on subjective judgments.
3	Severity Index method, [SI] [%]	The Severity Index method quantifies the impact of events or variables by assessing their severity or consequences, aiding in prioritizing based on the magnitude of their effects. It provides a structured approach to evaluate and rank the significance of different factors according to their potential impact.
4	Importance Index method [II]	The Importance Index method evaluates the significance of various factors by aggregating their ratings or weights, facilitating prioritization based on perceived importance. It provides a structured approach to rank factors according to their perceived value or relevance in a given context.
5	Relative Importance Weight method [RIW]	The Relative Importance Weight method assigns weights to factors based on their significance, enabling prioritization by their weighted impact, and streamlining decision-making processes.
6	Mean Score method [MS]	The Mean Score method assesses variable importance by calculating the average scores assigned to each factor, simplifying prioritization based on collective ratings. It offers a straightforward approach to ranking factors according to their perceived significance, utilizing the mean value as a measure of importance.
7	Rank correlation coefficient method [ρ]	The Rank Correlation Coefficient method evaluates variable importance by measuring the strength and direction of association between their ranks, facilitating prioritization based on their correlation. It provides insights into the relationships between factors, aiding in decision-making by identifying influential variables.
8	Weighted Opinion Average method [WA]	The Weighted Opinion Average method assesses variable importance by averaging ratings while considering their assigned weights, streamlining prioritization based on weighted impact. By integrating ratings and weights for a thorough assessment, it provides an organized method for ranking factors based on their perceived relevance.
9	Importance Weight method [IW]	The Importance Weight method assigns weights to factors based on their significance, aiding prioritization by their weighted impact, and simplifying decision-making processes. Given their differing degrees of effect, it provides an organized method for ranking things in order of perceived importance.

Source: Compiled by authors

Table 1.3 presents a comprehensive analysis of the major causes of project delays and outlines specific causes of delay, along with suggested mitigation measures from the perspectives of three key stakeholders: owners, contractors, and government authorities. It highlights how each stakeholder

Table 1.3 Causes of delays and mitigation measures

Sr. No.	Major causes	Specific causes of delay
1	Financial issues	Inadequate project funding
		Delayed payments to contractors
2	Design and planning	Incomplete or inaccurate project design
		Changes in design during construction
3	Contractual problems	Disputes over contract terms
4	Material supply	Delays in the delivery of materials
		Poor quality of materials
5	Equipment and machinery	Equipment breakdowns
6	Labor issues	Shortage of skilled labour
		Labour strikes or disputes
7	Environmental conditions	Adverse weather conditions (heavy rain, storms)
		Unexpected site conditions (e.g., soil issues)
8	Regulatory and legal	Delays in obtaining permits and licenses
		Land acquisition issues
9	Management inefficiencies	Poor project management and coordination
10	External factors	Political unrest or modifications to governmental regulations
		Impact of pandemics or health crises

Source: Compiled by authors

can address issues such as financial constraints, design and planning challenges, contractual disputes, material supply delays, equipment breakdowns, labour issues, environmental conditions, regulatory and legal obstacles, management inefficiencies, and external factors. The table serves as a structured framework to understand the role of each stakeholder in mitigating project delays and ensuring successful project delivery.

3. METHODOLOGY

Figure 1.1 shows an outline of a structured methodology for a research paper, focusing on the investigation of delays and risks in a specific context. It begins with a literature review, which provides a foundation by identifying existing studies, theories, and research gaps related to delays and risks. Following this, the identification of factors causing delays and risks is conducted, leveraging insights from the literature and potentially expert inputs. This step narrows the research focus to key variables of interest. Next, data collection is performed through a questionnaire survey, targeting relevant stakeholders or experts to gather primary data on the identified factors. For example, after being gathered, the data is analyzed using suitable statistical or qualitative methods to determine trends, correlations, and important discoveries. Finally, the research concludes with results and discussion, where the analyzed data is interpreted, compared with prior research, and contextualized to provide meaningful insights and recommendations. This stepwise approach ensures a logical flow, starting from theoretical foundations to empirical validation, aligning with standard research practices.

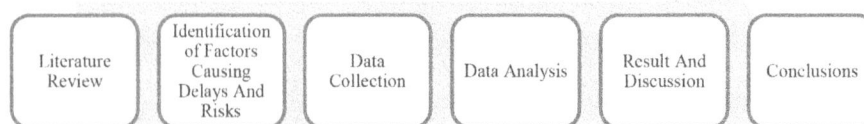

Fig. 1.1 Research methodology

Source: Authors

4. DATA COLLECTION

The questionnaire focused on ten critical areas contributing to delays, allowing for a comprehensive assessment of these issues. Responses varied based on individual perspectives, offering a range of viewpoints and insights into the multifaceted nature of construction project delays. To better illustrate the collected opinions, graphs were created to represent the responses. These visual aids effectively showcase the distribution of opinions and highlight trends and patterns in the feedback. The graphical representation enhances the understanding of diverse perspectives, making it easier to identify significant themes and commonalities in the stakeholders' views. Table 1.4 shows the summary of data from the respondents.

Table 1.4 Data from respondents

Respon-dents No.	Causes of Delay						
	Weather Conditions	Design Changes	Material Shortage	Permitting and Regulatory Approvals	Financial Problems	Equipment Failures	Labour Shortage
1	4	3	4	3	2	1	1
2	5	3	5	4	1	2	2
3	5	4	4	3	2	1	2
4	5	4	3	3	2	2	1
5	4	5	5	2	1	3	2
6	5	3	4	3	2	1	2
7	4	4	4	3	2	1	2
8	5	3	5	2	1	2	3
9	4	4	3	4	2	1	2
10	5	3	4	3	2	2	1
11	4	3	4	3	2	1	2
12	5	4	3	3	2	2	1
13	5	3	5	4	2	2	3
14	4	5	4	2	1	3	2
15	5	4	4	3	2	1	2

Respon-dents No.	Causes of Delay						
	Weather Conditions	Design Changes	Material Shortage	Permitting and Regulatory Approvals	Financial Problems	Equipment Failures	Labour Shortage
16	4	4	3	3	2	2	1
17	5	3	4	3	1	2	2
18	5	3	5	2	2	1	3
19	4	4	3	3	2	2	1
20	5	4	4	3	2	2	3
21	4	5	4	3	1	2	2
22	5	3	4	3	2	2	1
23	4	4	3	2	2	3	1
24	5	4	5	3	2	2	3
25	4	3	4	3	1	2	2
26	5	5	4	4	3	2	2
27	4	4	4	3	2	2	1
28	5	4	3	3	2	1	2
29	4	3	5	3	2	2	1
30	4	5	3	2	1	3	2
31	5	4	4	3	2	2	3
32	5	3	4	4	1	2	2
33	5	4	5	3	2	2	1
34	4	3	4	3	2	2	3
35	5	4	4	2	3	1	2
36	5	3	3	3	2	2	1
37	4	5	4	3	2	2	1
38	5	4	5	3	2	2	3
39	4	3	4	4	1	2	3
40	5	4	3	3	2	1	2
41	4	5	4	3	2	3	2
42	5	3	5	2	1	2	3
43	4	3	4	3	2	1	2
44	5	4	3	4	2	2	3
45	4	3	5	3	2	1	2
46	5	3	4	3	2	2	1
47	5	4	4	3	1	2	2
48	4	5	3	3	2	1	2
49	5	4	4	3	2	2	1
50	5	4	4	3	2	2	3
Sum	218	174	186	136	78	80	85
Total = 956							
Percentage	23	18	19	14	8	8	9

Source: Compiled by authors

Figure 1.2 illustrates the pie graph showing the key factors contributing to delays in road construction projects. Weather conditions are most significant, accounting for 23%of delays, followed by material shortage at 19%. Design changes make up 18%, while labour shortage contributes 14%. Labour shortage accounts for 9% and both financial problems and equipment failure contribute 8% each. These insights emphasize the diverse challenges faced during project execution.

Percentage

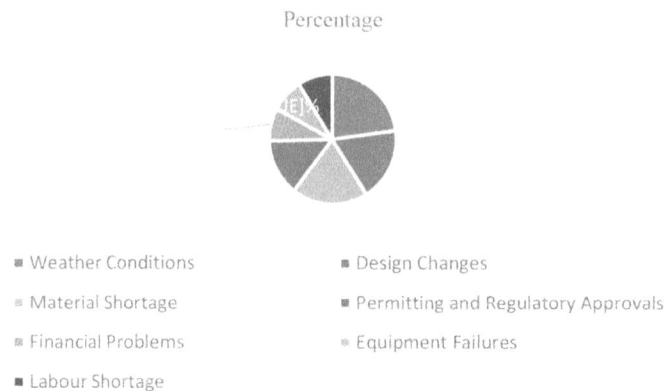

- Weather Conditions   ▪ Design Changes
- Material Shortage   ▪ Permitting and Regulatory Approvals
- Financial Problems   ▪ Equipment Failures
- Labour Shortage

Figure 1.2   Pie chart showing factors causing road construction projects to be delayed

Source: Authors

5. DATA ANALYSIS

The methodology incorporates data analysis using both the Relative Importance Index (RII) method and Statistical Package for the Social Sciences (SPSS) software, ensuring comprehensive insights into the collected data. The identified causes were grouped into four categories based on their Relative Importance Index (RII). Since it is challenging to propose solutions for every delay cause listed in the questionnaire, the focus was placed on suggesting measures for the most significant causes those that contribute the most to delays. SPSS (Statistical Package for the Social Sciences) is a robust statistical tool used for analyzing data, testing hypotheses, and generating comprehensive reports. For this study, SPSS can be applied to perform various statistical analyses.

5.1 Relative Importance Index approach (RII)

RII method was employed to identify the key causes of delay in construction projects. A 5-point Likert scale was used to determine the RII for each risk factor contributing to schedule delays in the Indian construction industry [12]. The data analysis, based on the responses from the questionnaire survey, followed the approach used in previous studies [14], which also applied the RII to evaluate the causes of delays. The Likert scale ranged from 1 (very low importance) to 5 (very high importance). The RII was then calculated as follows:

$$\mathbf{RII = \Sigma W/A \times N}$$

Here, W represents the total sum of the factors, A is the highest possible weight, and N refers to the total number of respondents for each variable.

The identified causes were grouped into five categories based on their Relative Importance Index (RII). Since it is challenging to propose solutions for every delay cause listed in the questionnaire;

the focus was placed on suggesting measures for the most significant causes—those that contribute the most to delays.

Table 1.5   Rank based on RII

IIR	≥ 0.750	0.700 to 0.749	0.600 to 0.699	0.400 to 0.599
Set	I	II	III	VI

Source: Compiled by authors

5.2 Prioritization of Delay Factors

Construction delay variables are ranked by RII in Table 1.6. Weather conditions are the most critical (RII = 0.792), followed by design changes and material shortages (RII = 0.744, 0.708). Permitting issues and equipment failures rank third, while financial and labour shortages rank lowest.

Table 1.6   Classification of delay causes using RII

Sr. No.	Causes of delay	RII	Rank
1	Weather conditions	0.792	1
2	Design changes	0.744	2
3	Material shortage	0.708	2
4	Permitting and regular approvals	0.632	3
5	Equipment Failures	0.604	3
6	Financial Problem	0.422	4
7	Labour shortage	0.412	4

Source: Compiled by authors

5.3 SPSS Approach

Statistical Package for the Social Sciences (SPSS) is a powerful tool widely used for analyzing and interpreting data in research. For ranking the causes of delays in construction projects, SPSS provides a systematic way to handle, process, and analyze survey or project data to identify the most significant factors. Table 1.7 presents delay causes ranked by their mean scores using SPSS. Weather conditions are the most significant delay factor (Mean = 4.58), followed by material shortages (Mean = 3.96) and design changes (Mean = 3.76). Permitting issues rank fourth, while labour shortages, financial problems, and equipment failures are the least impactful, ranked fifth to seventh, respectively.

Table 1.7   Delay causes ranked by SPSS

Sr no	Causes of delays	Mean	Rank
1	Weather conditions	4.58	1
2	Material shortage	3.96	2
3	Design changes	3.76	3
4	Permitting and regular approvals	2.98	4
5	Labour shortage	1.94	5
6	Financial Problem	1.82	6
7	Equipment Failures	1.80	7

Source: Compiled by authors

6. RESULT AND DISCUSSION

Construction project delays are frequent and can have a big impact on budgets and schedules. Effective project management requires the identification and comprehension of critical delay reasons. The Relative Importance Index (RII) and SPSS techniques are used in this discussion to rank the impact of delay factors on building projects. Weather conditions are the top delay factor, consistently ranked first with an RII of 0.792 and a mean score of 4.58, highlighting their significant influence on project timelines. Material shortages ranked second (RII = 0.708, Mean = 3.76), create delays by hindering the availability of essential resources. Design changes ranked third (RII = 0.744, Mean = 3.98), leading to rework and further delays. Permitting and regulatory approvals, ranked fourth (RII = 0.632, Mean = 2.98), can halt progress but are less impactful than the top three causes. Labour shortages (RII = 0.412, Mean = 1.94), financial problems (RII = 0.422, Mean = 1.82), and equipment failures (RII = 0.604, Mean = 1.80) are less significant but still contribute to delays. In conclusion, while weather, material shortages, and design changes are the most critical factors, addressing all delay causes is crucial for minimizing project disruptions and ensuring timely completion. Table 1.8 provides the Comparison by using RII and SPSS.

Table 1.8 Comparison by using RII and SPSS

Sr no	Causes of delays	RII	Mean by SPSS
1	Weather conditions	0.792	4.58
2	Design changes	0.744	3.76
3	Material shortage	0.708	3.96
4	Permitting and regular approvals	0.632	2.98
5	Equipment Failures	0.604	1.80
6	Financial Problem	0.422	1.82
7	Labour shortage	0.412	1.94

Source: Compiled by authors

7. CONCLUSION

Delays in road construction projects continue to challenge India's infrastructure development, with weather conditions, material shortages, and design changes emerging as the top causes. These factors disrupt timelines, inflate costs, and hinder economic progress. By leveraging the Relative Importance Index (RII) and SPSS methods, this study systematically ranks delay causes and highlights their multifaceted impacts. Weather conditions, the most critical factor, often stall projects due to unpredicted events like storms or heavy rains. Material shortages, ranked second, highlight inefficiencies in procurement and supply chain management. Design changes, third in rank, result in costly reworks and disrupted schedules. Permitting and regulatory approvals, although ranked lower, pose significant bureaucratic hurdles, while labour shortages, financial constraints, and equipment failures further aggravate project delays. Effective mitigation requires a multi-pronged approach. Stakeholders must adopt modern construction techniques such as BIM and prefabrication to streamline processes and reduce dependency on on-site conditions. Advanced materials and real-time monitoring systems can enhance adaptability to unpredictable challenges. Collaborative planning, improved resource allocation, and comprehensive risk assessments are critical to addressing these delays. The study underscores the necessity for proactive government policies to streamline permits and approvals, encourage skill development programs, and ensure consistent funding for projects. Contractors and project managers should also prioritize robust planning and transparent communication among all stakeholders. By addressing these delay factors holistically, stakeholders can mitigate risks, minimize disruptions, and improve the efficiency of road construction projects in India. This approach not only accelerates project delivery but also strengthens the foundation for sustainable economic growth.

REFERENCES

[1] Alaghbari, W., Razali, M., Kadir, A., Salim, A., & Ernawati (2007). The significant factors causing delay of building construction projects in malaysia. *Engineering, Construction and Architectural Management*, 14(2), 192–206. https://doi.org/10.1108/09699980710731308.

[2] Alfakhri, A. Y.Y., Ismail, A., & Khoiry, M. A. (2018). The effects of delays in road construction projects in tripoli, libya. *International Journal of Technology*, 9(4), 766–774. https://doi.org/10.14716/ijtech.v9i4.2219.

[3] Assaf, S. A., & Al-Hejji, S. (2006). Causes of delay in large construction projects. *International Journal of Project Management*, 24(4), 349–357. https://doi.org/10.1016/j.ijproman.2005.11.010.

[4] Aziz, R. F., & Abdel-Hakam, A. A. (2016a). Exploring delay causes of road construction projects in Egypt. *Alexandria Engineering Journal* 55(2), 1515–1539. https://doi.org/10.1016/j.aej.2016.03.006.

[5] Durdyev, S., & Hosseini, M. (2020). Causes of delays on construction projects: a comprehensive list. *International Journal of Managing Projects in Business*, 13(1), 20–46. https://doi.org/10.1108/IJMPB-09-2018-0178.

[6] Frimpong, Y., Oluwoye, J., & Crawford, L. (2003). Causes of delay and cost overruns in construction of groundwater projects in a developing countries; Ghana as a case study. *International Journal of Project Management*, 21(5), 321–326. https://doi.org/10.1016/S0263-7863(02)00055-8.

[7] Hussain, S., Zhu, F., Ali, Z., Aslam, H. D., & Hussain, A. (2018). Critical delaying factors: public sector building projects in Gilgit-Baltistan, Pakistan. *Buildings,* 8(1), 6. https://doi.org/10.3390/buildings8010006.

[8] Isramaulanan, A., & Yuliana, C. (2017). Analysis of factors cause delay project construction bridge in the city of Banjarmasin. *Tropical Wetland Journal*, 3(3), 24–30.

[9] Kayelle, E., Rabbani, E. K., & Macedo, M. (2023). Identifying the causes of delay using the analytic hierarchy process (AHP) method in Brazilian public road infrastructure projects. *Journal of Management and Sustainability*, 13(2), 45. https://doi.org/10.5539/jms.v13n2p45.

[10] Koushki, P. A., Al-Rashid, K., & Kartam, N. (2005). Delays and cost increases in the construction of private residential projects in Kuwait. *Construction Management and Economics*, 23(3), 285–294. https://doi.org/10.1080/0144619042000326710.

[11] Mahamid, I. (2022). Critical factors influencing the bid / No-Bid decision in the palestinian construction industry. *Engineering, Technology and Applied Science Research*, 12(1), 8096–8100. https://doi.org/10.48084/etasr.4538.

[12] Muneeswaran, G., Manoharan, P., Awoyera, P. O., & Adesina, A. (2020). A statistical approach to assess the schedule delays and risks in indian construction industry. *International Journal of Construction Management*, 20(5), 450–461. https://doi.org/10.1080/15623599.2018.1484991.

[13] Pai, S., Anand, N., Mittal, A., & Singh, I. (2020). Developing a framework for mitigation of project delays in roads and highways sector projects in india. *Journal of Management Research and Analysis,* 5(3), 357–366. https://doi.org/10.18231/2394-2770.2018.0057.

[14] Patil, S. K., Gupta, A. K., Desai, D. B., & Sajane, A. S. (n.d.). Causes of delay in indian transportation infrastructure projects. *IJRET: International Journal of Research in Engineering and Technology*. http://www.ijret.org.

[15] Rivera, L., Baguec, H., & Yeom, C. (2020). A study on causes of delay in road construction projects across 25 developing countries. *Infrastructures*, 5(10), 1–16. https://doi.org/10.3390/infrastructures5100084.

[16] Santoso, D. S., & Soeng, S. (2016). Analyzing delays of road construction projects in Cambodia: causes and effects. *Journal of Management in Engineering*, 32(6), 05016020. https://doi.org/10.1061/(asce)me.1943-5479.0000467.

[17] Venkatesh, P. K., & Venkatesan, V. (n.d.). Delays in construction projects: A review of causes, need & scope for further research. https://www.researchgate.net/publication/325381206.

[18] Yap, J. B. H., Goay, P. L., Woon, Y. B., & Skitmore, M. (2021). Revisiting critical delay factors for construction: analysing projects in malaysia. *Alexandria Engineering Journal*, 60(1), 1717–1729. https://doi.org/10.1016/j.aej.2020.11.021.

Advances in Construction, Real Estate, Infrastructure and Project Management – Anil Kashyap et al. (eds)
© 2026 Taylor & Francis Group, London, ISBN 978-1-041-13433-6

2

Geopathic Stress and Its Implications on Human Health, Pavement Distress and Road Safety: A Comprehensive Review

Shivani Shinde*,
Rohit Salgude, and Swathi Jamadar
Dr. Vishwanath Karad MIT World Peace University,
Pune, India

■ **ABSTRACT:** This research delves into the intricate relationship between geopathic stress, pavement distress, and road accidents, offering a comprehensive overview of their impact on both human health and infrastructure. Geopathic stress, steaming from the Earth's magnetic field fluctuations, has been linked to various physical and mental ailments in humans and can adversely affect the structural integrity of pavements and buildings. Through an analysis of existing literature and empirical studies, this paper elucidates the mechanisms through which geopathic stress influences human health, soil and pavement material properties, and road safety. It highlights the necessity of incorporating geopathic stress considerations into transportation planning and infrastructure development processes to mitigate accidents and preserve pavement quality. The findings underscore the importance of recognizing geopathic stress zones in sustainable development practices and call for concerted efforts to integrate this vital aspect into environmental impact assessments and infrastructure projects worldwide.

■ **KEYWORDS:** Geopathic stress, Human health, Soil properties, Built environment, Pavement, Road accident

1. INTRODUCTION

There is a magnetic field inherent to the earth [1]. The Earth acts as an electromagnet when it rotates on its axis, generating an electromagnetic field and currents of electricity within the melting metals that comprise its core that oscillates at a frequency of 7.83 Hz on average on the surface. This frequency is almost the same as the alpha range of human brainwaves. In tandem with this background magnetic field, life on Earth has developed [2].

---

*Corresponding author: shivanishinde273@gmail.com

DOI: 10.1201/9781003669814-2

The term "geopathic stress" describes the earth's inherent energy. Changes in the earth's magnetic field can lead to geopathic stress. These might be caused by man-made disturbances or natural disturbances like geological faults and subterranean water [1, 2].

The Greek term geo, which translates to "of the earth," and pathos, which signifies "suffering" or "disease," are the sources of the word "geopathic." The word "geopathic" literally means "from the earth suffering or disease." Numerous mammals and other creatures are believed to use the magnetic field's energy for migration and navigation. The environment that humans live in also includes electromagnetic radiation [1, 2]. The phenomenon is known as "geopathic stress" or "geopathic interference zones." "Location disturbance" or "disturbed zones" [3].

The energies of the Earth include gravitational force, water energy, geomagnetic energy, bioelectromagnetic (BEM) radiations, artificial radiations from electrical installations, and energy from subsurface streams. Three primary grids are linked to extremely high energies that have been determined to be significant to humans. The terms main, primary, and typical refer to these grids as shown in Figure 2.1 [4, 5]. Major: There are a total of 12 lines of intensity connected to these grid lines, which are spaced 16–20 meters apart. Principal: The grid lines have an intensity of 10 counts and are spaced 10–12 meters apart. Normal: The lines are spaced 5–6 meters apart and have a single intensity, which suggests that the region inside the grid is devoid of intensity lines [5].

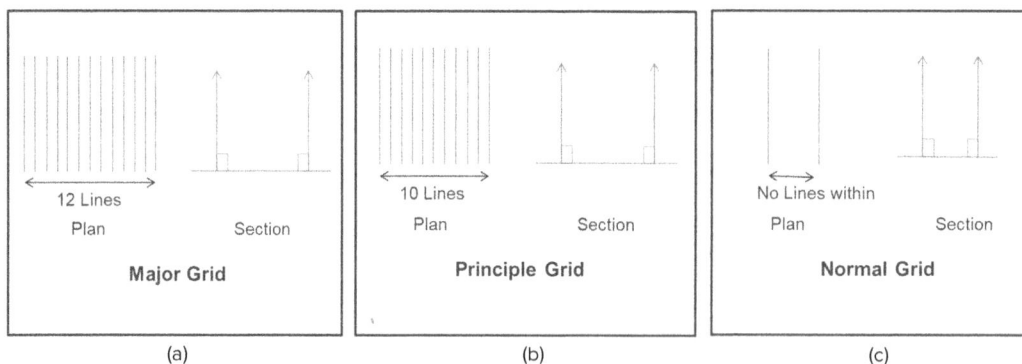

Figure 2.1 Bioelectromagnetic grids, (a) Major grid, (b) Principle grid, (c) Normal grid [47]

Source: Authors

Researchers who studied the undeniable effects of electromagnetic fields on living organisms included Dr. Hartmann, Prof. Endros, Dr. Palme, KE Lotz, and several others [5]. Dr. Hartmann's studies indicate that there are two parts to the geomagnetic field: the active sections are composed of lines that are 21 cm wide, and the neutral zone is between the lines. In the temperate zone, the grid is composed of these lines, which are spaced around 2.5 meters away from the East-West axis and 2.0 meters away across the North-South axis [5, 6].

According to Dr. Manfred Curry, The Currie grid, which is a diagonal grid to the cardinal directions or Hartmann network, is known to aid in cell proliferation. It is made up of energy lines that are found in the North East-South West and North West-South East directions at lengths of 3–4 meters and with a thickness of 40–50 cm [5, 6].

2. THE IMPACT OF GEOPATHIC STRESS

All living things, including people, animals, and plants, are susceptible to geopathic stress. It is mostly linked to sick building syndrome and may result in both mental and physical ailments. In addition, it could result in frequent mechanical breakdowns and pavement damage. Ancient methods such as dowsing, contemporary tools like the Vegetative Resonance Test (VRT) and Light Interference Technique (LIT), in addition to the Spinning Electric Vector Analyzer (SEVA) and gas discharge visualization (GDV) camera, are all capable of being utilized to identify energy from the beneath the surface of the earth at particular locations or in a built environment [7]. All living things are impacted by geopathic stress, which also a significant factor in the emergence of sick building syndrome in humans [7, 8]. All living things, including humans, are a component of the network that produces, receives, and emits electromagnetic fields (EMF) [8, 9]. As a result, the EMF has an impact on humans as exposure to geopathic stress may result in diminished performance, greater stress, decreased focus, a higher risk of accidents, and a higher frequency of sick days [7, 10].

2.1 On Human Body

Weak, slowly moving electrical brainwaves govern our mind, body, endocrine system, and immune system. These brainwaves crosstalk with other organs, including man-made and naturally occurring electromagnetic and radio waves, which may interact with electrical brainwaves to cause health risks [8]. Geopathic zones are areas on Earth that are known to cause health issues, according to Dharmadhikari et al. [11, 12]. Moreover, by impairing the body's subtle energy system (etheric body, meridians, chakras) and electrical system (coronary, brain, and muscles), geopathic stress also postpones the healing and recovery from an illness [1]. Long-term exposure to geopathic stress at work increases stress and changes a person's physiological state, causing headaches, elevated heart rate, and poor blood circulation. In addition, sleeping in geopathic stress zones may raise a person's risk of developing tachycardia and cancer, among other illnesses [13]. Geopathic stress is also known to be linked to poor sleep, which can cause daily tiredness and restlessness at night [14].

Numerous additional diseases have also been connected to geopathic stress, such as leukemia, Parkinson's disease, motor neuron diseases, multiple sclerosis, Crohn's disease, wasting and paralyzing diseases, endocrine disorders, other congenital genetic disorders, Down's syndrome, schizophrenia, and several other mental illnesses [2].

Human heart rates have been found to fluctuate between 10% and 15% in response to geopathic stress zones relative to normal zones. The heart rate change as a percentage of age is shown in Figure 2.2. (A rise in HR is indicated by a positive heart rate, and a decrease in HR is shown by a negative heart rate.) It alters blood pressure, skin resistance, body voltage, and the heart's regular operation [15–17].

The case studies of cancer and geopathic stress lines were examined, as Figure 2.3 illustrates. Grid lines are harmful to one's health, and crossing them beneath a bed might cause a patient to acquire various cancers [1, 10, 18].

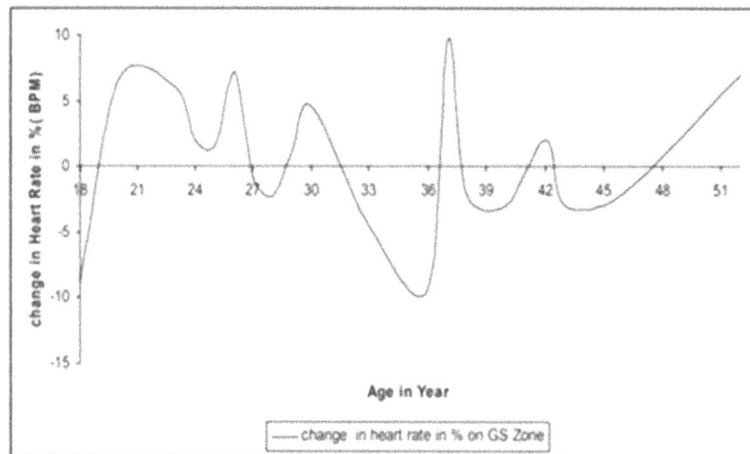

Figure 2.2   Heart rate variation as a percentage versus age [12]

Source: Authors

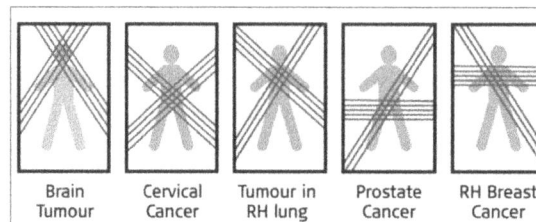

Figure 2.3   Geopathic stress lines and cancerm [54]

Source: Authors

2.2 On Built Environment

Below the earth's surface, ley lines can be found between large, hollow natural features like lakes, volcanoes, and caverns. In order to obtain their strength and channel their energy into the buildings, several spiritual constructions are constructed across these lines [6].

Numerous academics have released a great deal of work to help us understand how GS affects the built environment [1, 8, 11, 19–23]. Geopathic stress appears to affect all species of animals, plants, fungi, and microorganisms in addition to humans [11, 20, 24]. In a similar vein, it's critical to comprehend these natural energies, how they affect us, and how to account for them when designing the built environment in order to provide better outcomes and settings [6].

The movement of water veins may negatively impact individuals and structures because to the geopathic stress effect; one example of this is when walls begin to break. It may also detrimentally interfere with the transmission of sound [25]. It influences the rate of breakdown of machinery and equipment housed in geopathic stress zones as compared with machinery of the same kind that are maintained in normal zones [26].

In terms of architecture, it would be prudent to look for magnetic anomalies on a potential construction site because of the local earth's geology. House plans and electrical systems can be constructed so as to minimize the impact of powerful magnetic fields. Likewise, every metal structural element should be carefully addressed throughout design. The earth's natural direct magnetic field may be negatively impacted by walls and floors made of reinforced concrete. Preventing abnormalities of this kind becomes crucial, particularly in areas where people relax [5].

2.3 Geopathic Stress in Tunnel Engineering and Coal Mining

Tunnel engineering safety has long been a focus of scientific research to guarantee the safety and long-term viability of activities, particularly in coal mining. Geopathic stress in deep-buried tunnels during water inrush events is highlighted, and its effects on rock stability and deformation are examined [27, 28]. The effects of geopathic stress on tunnel stability investigated, with a focus on high ground stress, bias pressure, and load distribution [29]. Geophysical stress inversion was used to assess the stress distribution in alpine valleys, essential for predicting rock stability during excavation [30]. However, significant attention is given to horizontal geostress, especially in relation to how the fracture initiation point and failure surface direction of the rock mass are affected [31].

Geological stress is recognized as an important factor in contributing to tunnel settlement, alongside hydrophysical influences and bedrock instability. The role of high geological stress in accelerating tunnel deformation, particularly at the loess-silt interface, is highlighted. These constraints are found to amplify uneven distribution patterns and compromise the tunnel's overall structural integrity [32].

This stress is linked to failure mechanisms in coal–rock masses and the risk of mining disasters, such as rock bursts, emphasizing the importance of monitoring geopathic stress for safe mining operations [33]. And monitoring geopathic stress is deemed critical for ensuring safe mining operations [34].

Under condition of high geothermal temperatures and geotechnical stresses, thermal stresses of the rock mass cause significant changes in the morphology and extent of the plastic zone, which cannot be ignored and can lead to serious errors if they are not properly taken into account [35].

2.4 On Construction and Pavement Material

The characteristics of soil are affected by mild electromagnetic fields, which may be measured using ordinary equipment to estimate the density, water content, specific gravity, and liquid limit. Compared to non-stress zones, it was discovered that certain attributes are higher in stress zones [3, 36, 37]. Table 2.1 presents the impact of geopathic stress on the properties of soil.

The pavement on these soils was found to be significantly degraded as a result of wearing course, subbase, and subgrade failure [38]. Excessive rutting caused by an incorrect base mix led to the pavement's failure [39]. Numerous scholars have examined various factors that contribute to pavement deterioration, such as substandard building materials, inadequate drainage, unsatisfactory soil properties, excessive traffic volume, expansive subgrade soils, incorrect site supervision, and inadequate maintenance. Additionally, recommendations have been made to prevent pavement deterioration [12, 17, 19, 40, 41].

Table 2.1 Effect of geopathic stress on soil properties [42, 48]

Sr. No.	Soil Properties	Increases/ Decreases	By How many %
1	Moisture Content	Increases	3%
2	Specific Gravity	Increases	33%
3	Plastic Limit	Decreases	12%
4	Liquid limit	Increases	17%
5	Density	Increases	17%

Source: Compiled by authors

Road pavement distresses such as longitudinal cracking, corner cracks, spalling, patching, surface deterioration, joint condition, map cracking, and transverse cracking are impacted by the existence of geopathic stress. Several of the distresses are seen in the photos below, taken from Figures 2.4 to 2.8 [17, 44].

Figure 2.4 Pavement distress, (a) Corner crack, (b) Polished aggregate, (c) Pimping In action (d) Pot holes (e) Rut Depths [44]

Source: Authors

Cubes of concrete of grades M20 and M30 evaluated after 7, 14, and 28 days found that the compressive value of the concrete in stress zones was less than that of non-stress zones [17, 43]. Concrete subjected to geopathic stress exhibits a substantial reduction in compressive and flexural strength, with losses reaching 18.43% and 23.43%, respectively, after one year. Non-destructive testing (NDT) results on panels indicate progressive deterioration, suggesting a gradual weakening of the concrete over time [44]. Figure 2.5 depicts the variations in compressive and flexural strength in both the Geopathic Stress Zone (SZ) and the Non-Stress Zone (NSZ). Figure 2.5 depicts the

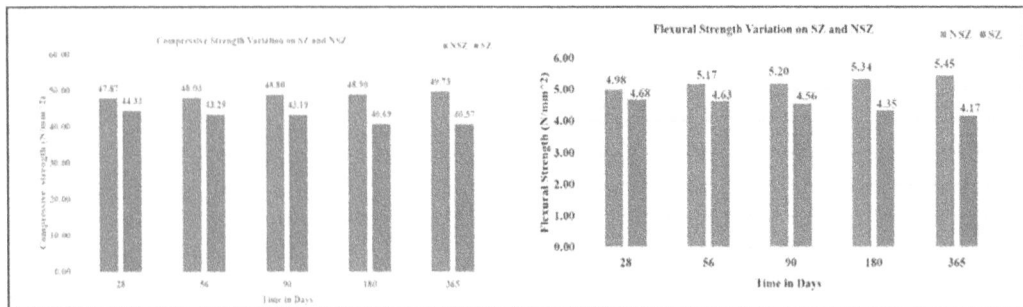

Figure 2.5 Compressive and flexural strength Variation in SZ and NSZ [32]

Source: Authors

variations in compressive and flexural strength in both the Geopathic Stress Zone (SZ) and the Non-Stress Zone (NSZ).

2.5 On Road Accidents

A crucial prerequisite of the transportation planning process is road safety. According to a 2019 estimate by the Ministry of Road Transport and Highways (MOR & TH), road accidents cost India 3.14% of its GDP [45].

The Geopathic Stress Study further demonstrates that drivers' response times lengthen in these areas, which ultimately results in collisions [17, 46–48]. Road accidents rise in tandem with the severity of pavement distress [17].

On the basis of this idea, a road accident prediction model has been developed for national highways with reference to pavement performance and road safety [49–54]. Regression models were created and automated and manual techniques were used to measure pavement distresses [55–60]. A few studies created accident prediction and reduction models based on a variety of variables, including population, vehicle type, traffic condition, volume of traffic, pedestrian traffic volume, traffic time, and precise road geometry design [46–48]. There aren't many prediction models that take geopathic stress into account as a crucial factor in road design [49–60]. Geopathic stress is proposed by researchers as a potential road design metric [55].

It is imperative that weak electromagnetic fields be considered when designing and analysing concrete pavement in order to minimize accidents in black areas and preserve pavement quality [55].

3. CONCLUSION

Geopathic stress must be included in the first survey for the design and planning of railways, roads, planning for towns, and other development of infrastructure. Thus, it is imperative to avoid developing residences or other infrastructure in a geopathic stress zone and to consider it an essential part of sustainable development and health and safety best practices.

According to the study, learning about natural building technology, architecture, and the effects of geo stress on individuals, animals, plants, and the environment from traditional and native sources can help to create new insights into the design of sustainable built environments.

According to the study, a deeper comprehension of geopathic stress zones by built environment professionals and their clients and private have a substantial, long-term influence on how built environment infrastructure development is organized, planned, constructed, run, and managed going forward. Additionally, it will advance economic growth, environmental safety, human health, and sustainable practices. The study concludes that for the planet's overall health, geopathic stress zones should be considered and incorporated into the assessment of environmental impacts and feasibility study for construction of infrastructure projects, such as the building of roads, bridges, housing, and industries, as well as the placement of machinery and plants.

Credit Authorship Contribution Statement

Shivani Shinde: Led the development of the review paper, including conceptualization, writing the original draft, and managing the overall structure, as well as performing critical reviewing and editing of the manuscript. Rohit Salgude: Contributed to the comprehensive literature review, synthesized and tabulated key insights, and assisted in identifying significant conclusions and trends within the reviewed studies. Swathi Jamadar: Supported in literature findings and contributed to the editing of the manuscript.

Declaration of Competing Interest

The authors declare that they have no known competing financial interests or personal relationships that could have appeared to influence the work reported in this paper.

ACKNOWLEDGMENTS

The authors gratefully acknowledge the invaluable contributions and insightful discussions from colleagues during the preparation of this review paper. We extend our thanks to the institutions that provided essential resources and support. We also acknowledge the extensive body of existing research, which served as a vital foundation for our investigation. Your expertise and encouragement have been integral to the development of this work.

REFERENCES

[1] Freshwater, D. (1997). Geopathic stress. *Complementary Complementary therapies in nursing and midwifery*, 3(6), 160–162. doi:https://doi.org/10.1016/S1353-6117(05)81003-0

[2] Creightmore, R. (2012). Geopathic stress. http://www. geomancygroup. org/geopathic, 2012 – safespace. net.nz

[3] Hacker, G. W., Eder, A., Augner, C., & Pauser, G. (2008). Geopathic stress zones and their influence on the human organism. *In Proceedings of the congress on Earth's Fields and Their Influence on Human Beings, Druskininkai, Lithuania* (pp. 8–17).

[4] Matanhelia, M. (1996). Geo bio energies in relation to human body and the built environment. Institute Of Environmental Design, Vallabh Vidyanagar.

[5] Sonal, A. S. (2012). Channelizing earth energies towards a healthy built environment. In *Proceedings of International Conference on Advances in Architecture and Civil Engineering*, 21, 862.

[6] Elrafie, N. S., Hassan, G. F., Abd Elrahman, A. S., & Elfayoumi, M. (2019). Assessing the effect of electromagnetic radiations on human beings in the built environment. *Journal of Engineering and Applied Science*, 66(4), 403–427.

[7] Tong, E. S., & Kong, C. K. (2021). An overview of impact of geopathic stress on environment and human health. *Progress in Drug Discovery & Biomedical Science*, 4(1). doi: 10.36877/pddbs.a0000174.

[8] Saunders, T. (2003). Health hazards and electromagnetic fields. Complementary Therapies in Nursing and Midwifery, 9(4), 191–197.

[9] Giannoulopoulou, L., Evangelou, A., Karkabounas, S., & Papamarinopoulos, S. (2018). The effects of geophysical anomalies on biology.*Journal of Scientific Exploration*, 32(3). doi:10.31275/2018.1295.

[10] Augner, C., Hacker, G. W., & Jekel, I. (2010). Geopathic stress zones: short-term effects on work performance and well-being?. *The Journal of Alternative and Complementary Medicine*, 16(6), 657–661. doi:10.1089/acm.2009.0499.

[11] Dharmadhikari, N. P., Rao, A. P., Pimplikar, S. S., Kharat, A. G., Aghav, S. D., Meshram, D. C., et al. (2010). Effect of geopathic stress on human heart rate and blood pressure. I*ndian Journal of Science and Technology*, 54–57.

[12] Dharmadhikari, N., Meshram, D. C., Kulkarni, S. D., Kharat, A. G., Sorate, R. R., Pimplikar, S. S., et al. (2012). Use of dowsing and geo-resistivity meter for detection of geopathic stress zone. *International Journal of Modern Engineering Research*, 1(2), 609–614.

[13] Manickam, S. (2018). Potential impact of geopathic radiation on environment and health. *Current World Environment*, 13, 25–30. doi:10.12944/CWE.13.Special-Issue1.0

[14] Dwivedi, R., Singh, A. P., & Appukuttan, R. (2024). Study of sleep quality in geopathic stresses areas and efficacy of enviromat in improving the same-a quantitative study., *International Journal of Biomedical Research & Practice,* 4(3), 1–5.

[15] Aghav, S., & Tambade, P. (2015). Investigating effects of Geopathic stress on health parameters in young healthy volunteers. *International Journal of Chemical and Physical Science,* 4, 28–34.

[16] Dharmadhikari, N. P., Meshram, D. C., Kulkarni, S. D., Kharat, A. G., & Pimplikar, S. S. (2011). Effect of geopathic stress zone on human body voltage and skin resistance.*Journal of Engineering and technology Research,* 3(8), 255–263.

[17] Salgude, R. R., Pimplikar, S. S., Kumbhar, P. S., & Shinde, K. D. (2022). Effect of geopathic stress on flexible pavement distresses and accidents. *Materials Today: Proceedings,* 65, 1498–1503.

[18] Gordon, R. (2013). The big four. Dulwich Health Society, London, UK.

[19] Croome, D. J. (1994). The effect of geopathic stress on building occupants. *Renewable Eenergy,* 5(5–8), 993–996. https://doi.org/10.1016/0960-1481(94)90122-8.

[20] Dubrov, A. P. (2008). Geopathic zones and oncological diseases. *In Proceedings of the Sixteenth BDA Congress on "Earth's Fields and Their Influence on Human Beings* (pp. 42–44).

[21] Kathe B (1989) Earth radiation. Word Asters Ltd.

[22] Storozuk, G. A. (2002). Geopathic zones and the iron stake method-no-2. A Dowser's Series. pp, 12–20.

[23] Von Pohl, G. F. (1993). Earth currents: causative factor of cancer and other diseases. Frech-Verlag, Germany.

[24] Hacker, G. W., Pawlak, E., Pauser, G., Tichy, G., Jell, H., Posch, G., et al. (2005). Biomedical evidence of influence of geopathic zones on the human body: scientifically traceable effects and ways of harmonization. *Forsch Komplementarmed Klass Naturheilkd,* 12(6), 315–327. doi: 10.1159/000088624.

[25] Choi, Y., Dobrucki, A. B., & Dobrucki, A. B. (2020). Forum Acusticum. Lyon, France. pp. 2607–2610, ff10.48465/fa.2020.0258ff. ffhal-03242418.

[26] Poddar, A., & Rana, S. (2014). Effect of geopathic stress and its correction on human body and machinery breakdown. *Medicine and Medical Sciences,* 1(3), 41–45.

[27] Hu, S., Wang, X., & Wang, E. (2024). Experimental study of true triaxial high pressure subcritical water impact fracturing. *Scientific Reports,* 14(1), 1150.

[28] Yang, J., Li, J., Zhu, H., Wu, H., Zhou, Z., & Li, J. (2024). A model study of water inrush in underground roadways. *International Journal of Geomechanics,* 24(12), 04024286.

[29] Li, J., Liu, Y., & Zhang, J. (2024). Simultaneous measurement of strain and displacement for railway tunnel lining safety monitoring. *Sensors,* 24(19), 6201.

[30] Tan, F., Guo, H., Pan, P., Wang, Z., Liu, X., & Zhou, Y. (2024). Integrated approach of predicting rock stability in high mountain valley underground caverns. *Underground Space,* 19, 317–341.

[31] Yi, Q., Shen, Z., Sun, G., Lin, S., & Luo, H. (2024). Comparative analysis of the stability of overlying rock mass for two types of lined rock caverns based on rock mass classification. *Applied Sciences,* 14(8), 3525.

[32] Wang, D., Zhao, X., Qiu, C., Guo, X., Du, Y., Li, X., et al. (2023). Experimental and numerical investigation on the damage mechanism of a loess–mudstone tunnel in cold regions. *Atmosphere* 14(9), 1391.

[33] Fu, Y., He, Y., & Li, C. (2023). Failure and acoustic emissions of coal–rock combinations with different dip angles in the shaqu no. 1 coal mine. *Advances in Civil Engineering,* 2023(1), 9969802.

[34] Sun, X., He, Y., Jin, T., Xie, J., Li, C., & Pang, J. (2023). Microseismic signal characteristics of the coal failure process under weak-energy and low-frequency disturbance. *Sustainability,* 15(19), 14387.

[35] Yin, Y., Jiang, H., Zhang, J., Lu, G., & Li, Q. (2024). The influence of rock thermal stress on the morphology and expansion pattern of the plastic zone in the surrounding rock of a deep-buried tunnel under high geothermal temperature conditions. *Applied Sciences,* 14(17), 7589.

[36] Salgude, R. R., Pailwan, P., Pimplikar, S., & Kolekar, D. (2024). Investigation of the effects from geopathic stress on the design thickness of flexible pavements. *World Journal of Engineering,* 22(1).

[37] Sorate, R. R., Dharmadhikari, A. K. N., & Bhagwat, S. (2013). Effect of geopathic stress zone on soil properties. *Elixir International Journal*, 12365–12367.

[38] Sorate, R. R., Zode, P. M., Hire, H. B., Kharat, A. G., Dharmadhikari, N. P., & Pimplikar, S. S. (2014). Geopathic stress: A threat to the built environment. *International Journal of Latest Technology in Engineering, Management & Applied Science*, 3, 30–32.

[39] Pimplikar, S. S., Kharat, A. G., Raval, P. M., & Dharmadhikari, N. P. (2006). Detection of geopathic stresses using light inference techniques. *New Building Materials & Construction World*, 11(9), 102–107.

[40] Dharmadhikari, N. P., Muthekar, V. V., Mahajan, C. S., Basavaiah, N., Kharat, A. G., Barde, S. I., et al. (2019). Vein width measurement of groundwater on earth's surface using semiconductor laser light and proton precession magnetometer. *ournal of Applied Geophysics*, 171, 103864.

[41] Gawand, A. B., & Gaikwad, S. B. To study the nature of geopathic stress caused by ground water using NaaV Meter and proton magnetometer. doi: http://dx.doi.org/10.20431/2347-3142.0603003.

[42] Chafekar, B. H., Jarad, G. P., Pimplikar, S. S., Dharmadhikari, N. P., Kharat, A. G., & Sorate, R. R. (2012). Effect of geopathic stress on pavement distresses.*Journal of Mechanical and Civil Engineering*, 1–8.

[43] Ambekar, S. S., & Bhilare, S. L. (2018). Effect of geopathic stress on concrete blocks. *International Journal of Scientific Research,* 7, 1392–1395.

[44] Kolekar, D. M., & Pimplikar, S. S. (2024). Investigation for influence of geopathic stress on structural behaviour of road elements. *Innovative Infrastructure Solutions*, 9(4), 106.

[45] Engineer, S. (2018). Ministry of road transport & highways government of India.

[46] Bradna, L. (2002). The influence of hydro pathogenic zones on drivers. Narendra Prakashan, 38–43.

[47] Kharat, A. G. (2000). Empirical and theoretical investigation on built environment. Ph. D Thesis, University of Pune.

[48] Sorate, R. R., Kharat, A. G., Shivshette, M., Desai, A., Nandgude, M., Ekal, P., & Sontakke, P. (2015). Geopathic stress: parameter for the occurrence of accidents. *The International Journal of Latest Technology in Engineering, Management & Applied Science*, 4(5), 1–4.

[49] Baek, J., Yoo, H. M., Lee, T. H., Park, Y. H., & Kim, M. I. (2014). Lessons from 20 years' experience of pavement management systems on national highways in Korea: focus on distress survey. http://worldcat.org/isbn/9780784478462

[50] Elghriany, A., Yi, P., Liu, P., & Yu, Q. (2016). Investigation of the effect of pavement roughness on crash rates for rigid pavement. *Journal of Transportation Safety & Security*, 8(2), 164–176. https://doi.org/10.1080/ 19439962.2015.1025458.

[51] Lee, J., Nam, B., & Abdel-Aty, M. (2015). Effects of pavement surface conditions on traffic crash severity. *Journal of Transportation Engineering*, 141(10), 04015020.

[52] Pimplikar, S. S., & Salgude, R. R. (2017). Road accident prediction models based on geopathic stress. *International Journal of Engineering Research in Mechanical and Civil Engineering*, 2456, 70–75.

[53] Serigos, P. A., Prozzi, J. A., de Fortier Smit, A., & Murphy, M. R. (2016). Evaluation of 3D automated systems for the measurement of pavement surface cracking. *Journal of Transportation Engineering*, 142(6), 05016003. https://doi.org/10.1061/(ASCE)TE.1943-546.0000841.

[54] Yao, Q., & Liu, S. (2014). Analysis of typical distress and cause on asphalt pavement for Liaoning highway. *Safe, Smart, and Sustainable Multimodal Transportation Systems* (pp. 919–924). https://doi.org/10.1061/ 9780784413623.088.

[55] Salgude, R. R., Pimplikar, S. S., Kolekar, D. M., Yadav, R. R., & Ghanekar, S. (2023). Investigation for effect of weak electric and magnetic fields at black spots on concrete pavement. *Materials Today: Proceedings*, 77, 805–812. https://doi.org/10.1016/j.matpr.2022.11.486.

[56] Pimplikar, S. S., Kharat, A. G., & Vaidya Sujata, G. S. (2005). A novel road design parameter. *International Conference START* (pp. 435–443).

[57] Agyemang, B., Abledu, G. K., & Semevoh, R. (2013). Regression analysis of road traffic accidents and population growth in Ghana. *International Journal of Business and Social Research*, 3(10), 41–47. doi:10.18533/ijbsr.v3i10.290.

[58] Alhasan, A., Nlenanya, I., Smadi, O., & MacKenzie, C. A. (2018). Impact of pavement surface condition on roadway departure crash risk in Iowa. *Infrastructures*, 3(2), 14. https://doi.org/10.3390/infrastructures3020014.

[59] Gupta, H., & Rokade, S. (2017). Development of crash prediction model using multiple regression analysis.*International Journal of Current Engineering and Scientific Research*, 4(6), 82–86.

[60] Sailaja, V., & Raju, S. S. (2015). Accident analysis on NH-18 by using regression model and its preventive measures. *International Journal of Scientific Research*, 4(4), 2467–2470.

Advances in Construction, Real Estate, Infrastructure and Project Management – Anil Kashyap et al. (eds)
© 2026 Taylor & Francis Group, London, ISBN 978-1-041-13433-6

3 Effective Implementation of Case-Based Learning in Teaching Construction Management Courses: The Case of Ethiopian Universities

Desalegn Girma Mengistu[1]

Associate Professor,
Department of Construction Technology and Management,
Hawassa University, Institute of Technology,
Hawassa, Ethiopia

Hawa Yesuf Ayele[2]

Lecturer,
Department of Construction Technology and Management,
Hawassa University, Institute of Technology,
Hawassa, Ethiopia

Firehiwot Birhanu Kibret[3]

Lecturer,
School of Teachers Education,
Hawassa University, College of Education,
Hawassa, Ethiopia

■ **ABSTRACT:** Project based and technical nature of construction management (CM) profession calls for effective integration of case-based learning (CBL) in teaching CM courses to equip the graduates with the required competence. The CBL can be integrated at institutional level in the curriculum and/or at course level by individual course instructor. This study majorly focusses on the course level implementation of CBL in the context of Ethiopian universities; mainly it addresses implementation level of the three dimensions of CBL: lesson plan, assessment method and implementation approach, and the associated challenges. In this study a mixed-method approach is used where structured questionnaire and semi structured interview were used in collecting the data to evaluate the implementation level of CBL dimensions and the challenges. In addition, CBL intervention was conducted on three selected CM courses. Finally, impact of the CBL intervention was evaluated using questionnaire in which 81 students were participated in giving feedback. Similarly, instructors of the selected courses have reflected on impact of the CBL. While frequency and descriptive statistics were used in evaluating the quantitative

[1]m.g.desalegn@gmail.com, m.g.desalegn@hu.edu.et; [2]hawa@hu.edu.et; [3]firehiwotb@hu.edu.et

DOI: 10.1201/9781003669814-3

data, the qualitative data was analyzed thematically focusing on specific objectives of the study. From the lesson plan perspective, allocating adequate duration to ensure appropriate balance between the teacher and students' activity and matching group size to the nature of intended activities were identified as areas of major improvement. The identified dominant assessment method is written examination and the dominant CBL implementation approach is group-based approach. CM practice needs teamwork in addition to individual competence. Dominantly using examination would not give the opportunity to assess individuals' and teams' concerted effort. Hence, the shortfall lies with the assessment methods not the approach. The major challenge identified was lack of instructors' exposure to real world practice.

■ **KEYWORDS:** Active learning method, Assessment method, Case-based learning, Construction management education, Ethiopia, Lesson plan

1. INTRODUCTION

The primary objective of academic institutes is equipping the graduates with appropriate set of knowledge and skills found in the curricula; it is ensuring attainment of the expected graduates' competence. In practice, the competence of fresh graduates of construction management and expectation of the industry shows a significant gap [25]. This is potentially associated with effectiveness of the adopted learning methods. Contribution of active learning methods (ALMs) is instrumental in ensuring the attainment of the expected competence [12, 22]. Case-based learning method is one of the ALMs expressed in different terminologies; e.g. project based and problem-based learning [6, 8, 16]. However, in this specific study the term case-based learning (CBL) is adopted as it is commonly used in construction education [18], and it integrates project based and problem-based learning.

CM profession is technical in nature and the practice is project-based. Hence, the teaching method has to consider this nature of the professional practice [25]. Real world focus learning is significant in improving students' performance [8]; in addition it helps to enhance employability of the graduates. The purpose of internship program in construction education is exposing the students to the real world [17]. To strengthen this, integrating the practice to the different course related activities would help to achieve the intended quality of education. There is positive perception towards ALMs by faculty in general, however, they seldomly practice it. This is attributed to various institutional and individual instructor level challenges.

There is a dearth of studies focusing on engineering education in Ethiopia [17]. Beyond engineering education, different aspects of higher education have been addressed by previous studies; i.e., higher education quality [2], and teaching and learning methods [23]. The existing construction related studies focus on knowledge and skill gap of engineering graduates [20]. Yizengaw [27] has identified the causes of skill gaps and mismatches of competence of engineering graduates. Similarly, Mengistu and Mahesh [17] have indicated the shortfalls of internship program coordination in construction education. However, research has not been conducted on teaching method with a focus on CM courses. Focusing on the details of the teaching learning methods would help to improve the graduates' competence. Recently in the year 2022/2023 a national level university exit examination for graduating students was started in Ethiopian universities. It is a new initiative by the Ministry of Education (MoE) to assess the competence of graduates. However, as per the report of the MoE the pass rate of the two consecutive examinations is poor which is below 50%. This result can be associated with the effectiveness of learning methods in ensuring attainment of the expected graduates' competence.

The teaching methods commonly adopted by the CM course instructors does not consider nature of the industry and the professional practice. The learning method is lecture dominated; which has less utility in ensuring visualization of the real-world practice. CBL is an ideal method to address this gap, hence, identifying the approaches to enhance case-based learning is crucial. Ensuring effectiveness of the learning method needs considering different factors [14]. Appropriate design of the different dimensions is important such as developing good lesson plans [1], allocate adequate durations for the classroom activities, framing appropriate group size, integrate different approaches of CBL to suit with the specific outcomes and adopting suitable assessment methods [6]. Hence, this study focuses on effective application of CBL through mainly addressing implementation level of the three dimensions: the lesson plan, assessment method and implementation approach, and the associated challenges.

2. LITERATURE REVIEW

ALMs are instrumental in ensuring students active engagement and graduates' success in attaining the required competences. CBL is one of the effective ALMs to focus on the real world practice [16] and it enhances student engagement. CBL is a method in which different cases are brought to the classroom to simulate the real-world practice [21]; i.e., uncertainty and involvement of various stakeholders in a project are among the major characteristics of the construction industry which can better be delivered through CBL. For example, this can be addressed in a class through role playing; the activities can be undertaken against client perspective, consultant perspective and contractor perspective [3]. These would help the students to develop critical thinking and problem-solving skills which are among the major soft skills expected from CM graduates [17, 27]. Effective implementation of CBL needs appropriate design of the major dimensions: lesson plan, assessment methods and implementation approaches. Implementation of these dimensions is affected by different challenges. Ensuring maturity of the CBL practice and the continual improvement need regular evaluation of the practice in bringing the intended impacts on the students' engagement and performance. The subsequent paragraphs focus on addressing the CBL dimensions; the challenges and the key performance indicators (KPIs) to evaluate students' engagement and performance.

Developing a good lesson plan needs giving the due attention to the contents focusing on the outcomes, and it has to be flexible to accommodate the circumstances [28]. Achieving the intended purpose of CBL needs developing lesson plans that encourages active engagement of the students [1]. Lesson planning is a challenging task; it needs addressing the outcomes, allocating adequate duration for students – teacher activities and ensure integration with the adopted CBL implementation approaches.

Assessment is an integral part of teaching and learning process [24]. Depending on the adopted CBL implementation approaches; appropriately integrating the relevant assessment method is important. In CBL both the traditional and innovative assessment methods are applied [6]. The common innovative assessment methods are peer review and self-assessment, however the traditional assessment are final exam and attendance [15]. The common assessment methods used are: case analysis and presentation, assignment/written case report, applied projects, examination/ test, and in-class response and feedback from instructor. In addition to the grading, assessment has to help to improve teaching and learning, and student success [24]. It is used to check and enhance learning [28]. Hence, it is important to adopt assessment types that enhances learning. Construction management practice needs teamwork effectiveness in addition to individual competence. Hence, in the CBL it is important to use continuous assessment with the purpose of assessing individuals' as well as teams' concerted effort.

Considering coverage level as a category of approach, CBL can be implemented at course level, cross course level, and curriculum level [6]. The other parameters of the approach are associated with the duration which is number of lessons, and group size which is number of students. The duration could be session-based and/or cross session-based which could be semester long project. The group size could be individual-based and/or group-based students' activity. The third dimension is integrating the CBL with other experiential and practical learning such as site visit and invited industry guest speaker. Practicability of the approach is affected by the nature of the objectives and outcomes. However, it is expected to create the opportunity to enhance individual engagement and team building to address nature of the CM profession.

There are different potential challenges that affect effective implementation of experiential and practical learning methods [4, 22]. The major challenges can be classified as individual level – teacher and students, institutional level and cultural level [6]. The teacher related challenges include: lack of instructional skill training, challenges of assessments, and interpersonal skill challenges. The student related challenges are lack of teamwork skills, lack of self-learning skills, and low interest to devote. Institutional level challenges are associated with lack of institutional support and inadequacy of the provisions in the curriculum in addressing enforceable requirements. Cultural level challenges are associated with language barriers, multiculture and cultural sensitivity. The challenges are caused by social, technological, economic, environmental, and political situations.

The previous studies can be categorized into two: the first category are generic topics focusing on higher education quality [2], the second category focuses on certain dimension(s) of the teaching and learning method [4, 22, 23]. This particular study addresses this gap; focusing on a single learning method and addressing all the relevant dimensions: the lesson plan, assessment method and implementation approach. This would help to understand details of the learning method and broadens the specific knowledge domain.

3. RESEARCH METHODOLOGY

In this study a mixed-method approach comprising of structured questionnaire, semi structured interview and case studies were used in collecting the quantitative and qualitative data. The detailed procedure adopted had two steps. The first step was evaluating the practice level of the CBL dimensions and the associated challenges. Questionnaire one was developed to achieve these two specific objectives. The questionnaire consisted of four main parameters: lesson plan, assessment methods, approach to implement CBL and the challenges. The respondents were requested to rate implementation level of the variables indicated under the lesson plan, assessment methods, and approach to implement CBL. In the rating, a 5-point Likert scale was used where; [1] = Very Low, [2] = Low, [3] = Moderate, [4] = High, [5] = Very High. Similarly, the respondents were requested to rate impact level of the potential challenges on effectiveness of the practice. Randomly selected CM course instructors from Construction Technology and Management (CoTM) department of seven universities participated in the questionnaire survey: Dilla University, Hawassa University, Mizan Tepi university, Wachemo University, Woldia University, Wolita Sodo University and Wolkite University. The CoTM curriculum has three category of courses without considering the basic science courses offered at the first year: construction engineering, construction technology and construction management [13]. The curricula across the universities are substantially equivalent with few differences, hence, considering maximum twelve major CM courses per curriculum, the questionnaire was distributed to eighty-four instructors under the seven universities. Totally seventy instructors duly completed the questionnaire which is resulted in 83.33% response rate. The interview question focuses on the same issues; however, it targets the instructors teaching the selected courses to apply the intervention.

The second step had two activities: applying the CBL intervention on the selected courses and evaluating the impact on the students' leaning performance. The first activity of the second step of the study was applying the CBL intervention. For implementation of the intervention, three CM courses were selected purposively based on willingness of the course instructors from Hawassa University. The selected courses were: Performance Management and Resource Optimization (CoTM5074), Construction law (CoTM4102), and Specification and Quantity Surveying (CoTM3081). The courses are offered for 5th, 4th and 3rd year students respectively. The intervention focuses on all the three dimensions of CBL: lesson plan, assessment methods and implementation approaches. Effective implementation of CBL needs appropriate design of the different dimensions of the CBL. Hence, the major expected contents and activities were listed in a form of template to be employed by the selected course instructors. Concerning specific experience of the course instructors, the course instructors have taught the selected course for two, six and eight consecutive semesters. Totally 9 cases were applied among which 6 are single session-based and 3 are cross session-based. Implementation of the intervention took 5 consecutive weeks.

The second activity of the second step was evaluating the impact of the CBL intervention on the students' learning performance through questionnaire and collecting feedback form the instructors who applied the intervention. Similar approaches was used by previous study conducted on CBL [18]. Impact of the CBL intervention was evaluated using a questionnaire in which 81 students were participated; all the students who took the selected courses duly completed the questionnaire. The questionnaire mainly focused on six parameters: (i) confidence to ask questions, (ii) confidence to participate and contribute, (iii) self-initiative, (iv) independence, (v) regular attendance and (vi) attitude to learning. The interview questions are designed to gather instructors' feedback on the use of CBL method in teaching the CM courses. Evaluation of the students' performance improvement across the six parameters is conducted based on the students' and instructors' perception relative to the students' previous performance before application of CBL.

The respondents' profile indicates that 77.15% of the instructors have more than 5 years' experience, 62.86% have taken HDP training and 98.57% have good perception towards HDP training. Higher Diploma Program (HDP) is a yearlong professional instructional training offered for academic staffs in Ethiopian universities. The proportion of instructors who are HDP trained and have industry experience are 41.75 %. The students' distribution indicates that 3rd, 4th and 5th year students are 31, 34 and 16 respectively.

Cronbach alpha test was conducted to test internal consistency of the scale and the results are: 0.876 (lesson plan), 0.752 (assessment methods), 0.804 (implementation approaches), 0.782 (challenges of CBL), and 0.903 (effectiveness of the CBL). All the results were found to be higher than 0.7, which is in the acceptable range and, hence, the instrument found to be reliable [11]. Frequency and descriptive statistics were used in evaluating the quantitative data. The practice level was interpreted using frequency of response as it indicates individual instructors' level of practicing; specifically, the High and Very High were added together for interpretation. The mean and standard deviation were used to rank impact level of the challenges and effectiveness of the intervention in achieving the intended impact. The qualitative data was analyzed thematically focusing on specific objectives of the study. Results from questionnaire survey and interview were triangulated to validate the results.

4. ANALYSIS AND DISCUSSION

4.1 Practice Level of Case-Based Learning

The CBL practice was evaluated focusing on the three dimensions: (i) the lesson plan, (ii) the assessment methods and (iii) approaches in implementing case-based learning. Results of the three CBL dimensions are summarized in Table 3.1.

Table 3.1 Effectiveness of the three CBL dimensions

The three dimensions of case-based learning	Frequency (%)	
	[1+2+3]	[4+5]
Lesson plan parameters		
Developing lesson plans that encourage active involvement of students	54.29	45.71
Identifying the specific lesson outcome(s)	40.00	60.00
Matching the problem/case types with the lesson outcome(s)	42.86	57.14
Allocating adequate duration to ensure appropriate balance between the teachers and students' activity	51.43	48.57
Matching the group size to the nature of intended activities of the students	48.57	51.43
Assessment methods		
Case analysis and presentation	51.43	48.57
Assignment/written case report	22.86	77.14
Applied Projects	24.29	75.71
Written examination/ test	15.71	84.29
In-class response and feedback from instructor	45.71	54.29
Approaches		
Single session based	61.43	38.57
Cross-sessions based	50.00	50.00
Integrating the CBL method with site visit	40.00	60.00
Integrating the CBL method with industry guest speaker	48.57	51.43
Individual based approach	58.57	41.43
Group based approach	27.14	72.86

Source: Compiled by authors

Lesson plan: Summary of the results in Table 3.1 indicate that developing lesson plan that encourage active involvement of students, allocating adequate duration to ensure appropriate balance between the teachers and students' activity and matching the group size to the nature of intended activities of the students are identified as areas that need major improvement with the frequency of 45.71%, 48.57% and 51.43% respectively. These show that more than half of the staffs are not implementing these lesson plan parameters effectively.

The interview findings indicate that overall, the practice is poor. The nationally harmonized curriculum that is used by all universities contains the major parameters to be included in the lesson plan: course objective and outcome, course description, course requirements such as site visit and semester project depending on the nature of the course [13]. Hence, considering these and matching the case type with these parameters in the lesson plan development is expected to be significantly high, however, implementation of these parameters is found to be 60.00% and 57.14% respectively. The respondent profile indicates that HDP trained instructors are 62.86% which is not consistent with evaluation of the practice; this shows that the HDP trained instructors are not exercising what they have gained through the HDP training.

Assessment methods: The analysis results in Table 3.1 show that the dominant assessment method is written examination/ test with the frequency of 84.29%. The least practiced method is case analysis and presentation with the frequency of 48.57%. These analysis results show that the practice has missed two important aspects: the first is one dimension of purposes of assessment and the second is nature of CM profession. The purpose of assessment is to check and enhance learning [26]. However, dominantly using written examination/ test as an assessment method would not help to ensure continual enhancement of learning. Using case analysis and presentation would help the teacher to evaluate individuals understanding and the team's concerted effort in analyzing the case to give alternative solutions to the practical problem.

The practice is also against provision of the curriculum which suggest continuous assessment with a mix of modalities/methods [13]. The mix of different assessment methods can be applied in CBL to ensure attainment of the lesson objectives; e.g., while cases analysis and presentation, and applied projects would help to measure both individuals' and team's effort, assignment and test can majorly evaluate individuals' performance. Hence, including the range of assessment in the lesson plan is important to capture the different characteristics of the industry and ensure effectiveness of CBL. The major reasons raised by the instructors, for not using case analysis and presentation, and in-class response and feedback as an assessment method, were large number of students and students' low appetite to work. However, this is associated with poor lesson planning practice; the challenges can be mitigated through good lesson planning [26].

Approaches in implementing case-based learning: The results in Table 3.1 show that the dominant CBL implementation approach is group-based approach and integrating CBL with site visit with frequency of 72.86% and 60% respectively. Single session based and individual based approach is found to be the least applied approach with the frequency of 38.57% and 41.43% respectively. The instructors also reflected as the dominant approaches in implementing CBL is group based and it is commonly semester long project. However, it was found that the assessment lacks progress evaluation and final oral presentation. Providing semester-based project creates the opportunity to build team spirit, associating it with progress assessment would give the opportunity to assess individuals' engagement and teams' concerted effort. Team building and adoptability to changing business environment is important soft skills for CM professionals [17]. Hence, the shortfall lies with the assessment methods, not the approach.

Due to the nature of CM profession, integrating more cases in teaching the courses is important. Hence, single session based and individual based approaches also need improvement. CBL practice at the course level can be effectively implemented from one class to one semester, through combing different approaches [6]. Group based approach can better be delivered through role playing [28]. Different groups can represent different stakeholders: client, contractor or consultant. Similarly, within a group individuals can represent different professional positions. Group size need to be optimum to ensure active engagement of all the group members; a groups of four or five is considered optimal size for productive discussion and effective collaboration [7].

4.2 Challenges of Case-Based Learning

As indicated in Table 3.2, the major challenge is lack of instructors' exposure to real world practice/ lack of professional practical experience with mean score of 3.67. The respondents' profile indicates that 67.14% of the respondents have industry experience. Construction education is a discipline in which industry experience should be an important requirement for the course instructors [5]. Hence, the remaining 32.86% is significant number, thus, it is important to devise strategies to create the exposure to the course instructors. Mean score of the other challenges are within the range of 2.8–3.36 which is not insignificant. In addition, the respondents' profile indicates that only 41.75% of the instructors have both industry experience and HDP trained; this indicates that significant number of the instructors lack pedagogical skill and the exposure to real world practice. The challenges are associated with these profiles of the instructors; e.g., allocating role of the stakeholders to the groups/within group to the members and considering dynamic nature of the construction industry needs understanding the practical world [3]. One of the advantages of having industry exposure is creating a professional network, hence, lack of professional network would result in difficulty in getting relevant and up-to-date practical cases. Lack of instructional skills would result in difficulty in allocating sufficient time for in-depth case analysis and discussion, difficulty in integrating CBL with other teaching methods, and difficulty in assessing and evaluating student learning in case-based activities [9, 19]. Hence, these challenges affect all the three dimensions of CBL: the lesson plan, assessment methods and the implementation approach; thus, need to be mitigated to ensure effective implementation of CBL.

Table 3.2   Challenges of case-based learning in teaching construction management courses

Potential challenges	Mean	SD	Rank
Difficulty in getting relevant and up-to-date case studies	3.36	0.99	2
Difficulty in adapting case studies to match course objectives and student level	2.99	1.05	3
Difficulty in facilitating effective group discussions and collaboration	2.79	1.03	7
Difficulty in assessing and evaluating student learning in case-based activities	2.80	0.97	6
Difficulty in integrating case-based learning with other teaching methods	2.96	1.06	5
Difficulty in allocating sufficient time for in-depth case analysis and discussion	2.99	1.24	4
Lack of instructors' exposure to real world practice/ lack of professional practical experience	3.67	1.40	1

Source: Compiled by authors

The challenges fall under the category of individual instructor and institutional level. Some of the challenges need initiative of both instructor and the institution, e.g., as indicated above the institutions have provide HDP training, however, there is a gap in the practice by the individual instructors. Results of the quantitative data summarized in Table 3.2 majorly focus on individual instructors' level. The challenges raised by the instructors during the interviewees that falls under the category of institutional level are: curriculum inadequacy in addressing the enforceable requirements, lack of the required infrastructure, budget related challenges to arrange logistics for site visit and weak university industry linkage to create industry exposure to the instructors. Evaluating the practice of HDP trained instructors and creating an externship opportunity for the instructors to get practical exposure is important. Hence, the improvement strategies should consider these aspects.

4.3 Case Study

This is the second step of the study that focuses on applying the CBL intervention on the selected courses and evaluating the impact on the students' learning performance.

Action strategies employed/ the intervention: In implementing the intervention, the course instructors were advised to consider the five parameters indicated in Table 3.1 for effectiveness of their lesson plan with the major emphasis on the parameters which were found to be the major area of improvement; e.g., developing lesson plans that encourage active involvement of students, allocating adequate duration to ensure appropriate balance between the teachers and students' activity and matching the problem/case types with the lesson outcome. Similarly, they were advised to consider the five assessment methods indicated in Table 1 with an emphasis on using a combination of the assessment methods depending on the situation and specific nature of their course. There is no single best approach that suits all circumstances; hence, a combination of approaches was suggested.

The cases used by the instructors had various nature; during the discussion the instructors explained that most of the cases had a series of relevant follow up questions. The collected cases were evaluated against the course objectives and the appropriateness in addressing characteristics of the industry. As an example, from the course CoTM4102; the students were provided with a case in which they had to analyze the role and responsibilities of the contracting parties and understand significance of ensuring completeness of the contract document during procurement planning. In another case, from the course CoTM5074; the students were instructed to visit a nearby construction site in a group of three members and write a report focusing on selected performance dimensions: time, cost, quality, health and safety and resource wastage. In addition, alternative site layout design to enhance the productivity was part of the assignment. It was a two weeks assignment; it had site observation and interviewing project teams about the project performance. Similarly, in the course CoTM3081 the students were provided with complete working drawings of a building project, different project for different group, and instructed to prepare complete takeoff sheet and bill of quantities as per the national building specifications and methods of measurement.

Nature of the course determines approach of the cases; e.g., the case for course CoTM5074 had a presentation in which series of follow up questions were raised that created sharing of experiences among the different groups. The case for the course CoTM 3081 took half of the semester where there were a serious of consultations and presentation in the consecutive weeks. The key difference introduced in the intervention was the deliberate planning of lessons, assessment methods, and the approaches. The instructors reflected on their past practices, noting that they had been applying CBL without conscious planning of the CBL dimensions, hence, their past experience was not effective.

Evaluating impact of case-based learning: The students were requested to rate their performance improvement against the six parameters to evaluate impact of the intervention relative to their past performance. The result indicating impact of the intervention on the students' aspects of engagement and performance is summarized in Table 3.3. Improvement is observed in all the parameters, however, the regular attendance, independence and attitude to learning showed the higher improvement with the mean score of greater than 3.5. The same was reflected by the course instructors noting that the students were more engaged and participated actively in the discussions and group activities. Similarly, Gunawan et al. [10] have reported effectiveness of CBL in teaching CM courses. A combination of assessment methods was used; case analysis and presentation, applied projects and written reports coupled with the series of relevant follow up questions had helped the students to actively engage in the learning process. As reflected by the course instructors, not only were students more engaged, but they also demonstrated a deeper understanding of the course content.

Table 3.3 Effectiveness of CBL

Students' engagement and performance	Mean	SD	Rank
Confidence to ask questions [Hint: students feel empowered to ask questions, seek clarification, and actively engage with the instructor and their peers during case-based discussions]	3.33	0.94	5
Confidence to participate and contribute [Hint: increase in the frequency and depth of their participation]	3.46	0.88	4
Self-initiative [Hint: seeking out additional resources, proposing solutions, or taking on leadership roles within their teams]	3.28	1.07	6
Independence [Hint: ability to work autonomously on case-based assignments and projects]	3.57	1.02	3
Regular attendance	3.93	0.89	1
Attitude to learning	3.64	1.08	2

Source: Compiled by authors

5. CONCLUSION

Investigating the CBL practice level and visioning to realize the practice to be a culture through institutional as well as individual instructor level at academic institutes is important to equip the graduates with the required competence. Findings of this study indicates that there is a room for improvement in all dimensions of the CBL practice. From the lesson plan perspective allocating adequate duration to ensure appropriate balance between the teacher and students' activity and matching the group size to the nature of intended activities were identified as areas that need major improvement. Dominantly using written examination would not give the opportunity to assess individuals' and teams' concerted effort. Hence, the findings indicate that the shortfall lies with the assessment methods not the CBL implementation approach. The major challenge identified was lack of instructors' exposure to real world practice.

Integration of CBL in teaching allow students to engage with real-world scenarios, connect theoretical concepts to practical applications, and develop critical thinking and problem-solving skills which are the major soft skills required to be competent professional in the construction industry. The findings have implications in indicating the different instructional roles for CM course instructors to effectively implement CBL. In addition, it is important to effectively exercise institutional role. The academic institutions have significant role in creating enabling environment through instructors' capacity building, effective university-industry linkage, infrastructure provision, allocation of adequate budget for site visits and ensuring continual curriculum relevance in including enforceable requirements. The instructor needs to focus on the three dimensions of CBL and expected to be self-reflective practitioner to ensure continual improvement of the practice. In addition to monitoring and controlling of the dimensions of CBL, both the institution and individual instructor have to measure the overall students' engagement and performance regularly. Similarly, government need to support the system through regulatory framework. This study majorly focusses on the course level implementation of CBL. Further, focusing on major institutional issues, including more case studies, and application of other statistical analysis techniques that can indicate the casual relationship between the enablers and the practice would consolidate the concept and broaden the knowledge domain.

REFERENCES

[1] Adeyeye, K. (2009). Teaching construction contracts: mutual learning experience. *Journal of Legal Affairs and Dispute Resolution in Engineering and Construction*, 1(2), 97–104. https://doi.org/10.1061/(ASCE)1943-4162.

[2] Akalu, G. A. (2014). Higher education in ethiopia: expansion, quality assurance and institutional autonomy. *Higher Education Quarterly*, 68(4), 394–415. https://doi.org/10.1111/hequ.12036.

[3] Bhattacharjee, S. (2014). Effectiveness of role-playing as a pedagogical approach in construction education. In 50th ASC Annual International Conference Proceedings.

[4] Børte, K., Nesje, K., & Lillejord, S. (2023). Barriers to student active learning in higher education. *Teaching in Higher Education*, 28(3), 597–615. https://doi.org/10.1080/13562517.2020.1839746.

[5] Boyer, D. M., & Walker, E. B. (2020). The impact of industry expert adjuncts on students' course experiences. *International Journal of Innovative Teaching and Learning in Higher Education*, 1(2), 16–28. https://doi.org/10.4018/ijitlhe.2020040102.

[6] Chen, J., Kolmos, A., & Du, X. (2021). Forms of implementation and challenges of PBL in engineering education: a review of literature. *European Journal of Engineering Education*, 46(1), 90–115. https://doi.org/10.1080/03043797.2020.1718615.

[7] Cohen, E. G., & Lotan, R. A. (2014). Designing groupwork: strategies for the heterogeneous classroom. In Communities (Thrid, Vol. 2nd). Teachers College Press, Columbia University.

[8] Gallagher, S. E., & Savage, T. (2023). Challenge-based learning in higher education: an exploratory literature review. *Teaching in Higher Education*, 28(6), 1135–1157. https://doi.org/10.1080/13562517.2020.1863354.

[9] Geletu, G. M. (2022). The effects of teachers' professional and pedagogical competencies on implementing cooperative learning and enhancing students' learning engagement and outcomes in science: practices and changes. *Cogent Education*, 9(1), 2153434. https://doi.org/10.1080/2331186X.2022.2153434.

[10] Gunawan, T. J., Wang, J., & Liao, P. C. (2022). Factors of project-based teaching that enhance learning interest: evidence from construction contract management course. *Sustainability (Switzerland)*, 14(22), 15314. https://doi.org/10.3390/su142215314

[11] Hair, J. F., Black, W. C., Babin, B. J., & Anderson, R. E. (2019). Multivariate Data Analysis, (8th edn.). Cengage Learning. EMEA. https://doi.org/10.1002/9781119409137.ch4.

[12] Hartikainen, S., Rintala, H., Pylväs, L., & Nokelainen, P. (2019). The concept of active learning and the measurement of learning outcomes: A review of research in engineering higher education. *Education Sciences*, 9(4), 9–12. https://doi.org/10.3390/educsci9040276.

[13] HU-IoT (2022). Hawassa University – Institute of Technology, Curriculum for Bachelor of Science in Construction Technology and Management, Department of Construction Technology and Management.

[14] Leung, M. Y., Ng, S. T., & Li, Y. K. (2004). Evaluating learning approaches of construction students in Hong Kong through a matrix framework. *Journal of Professional Issues in Engineering Education and Practice*, 130(3), 189–196. https://doi.org/10.1061/(ASCE)1052-3928(2004)130:3(189).

[15] Lutsenko, G. (2018). Case study of a problem-based learning course of project management for senior engineering students. *European Journal of Engineering Education*, 43(6), 895–910. https://doi.org/10.1080/03043797.2018.1454892.

[16] Maros, M., Korenkova, M., Fila, M., Levicky, M., & Schoberova, M. (2023). Project-based learning and its effectiveness: evidence from Slovakia. *Interactive Learning Environments*, 31(7), 4147–4155. https://doi.org/10.1080/10494820.2021.1954036.

[17] Mengistu, D. G., & Mahesh, G. (2019). Construction education in Ethiopia: knowledge and skills level attained and effectiveness of internship program. *Higher Education, Skills and Work-Based Learning*, 9(3), 510–524. https://doi.org/10.1108/HESWBL-06-2018-0062.

[18] Mesthrige, J. W., Lam, P. T. I., Chiang, Y.-H., & Samarasinghalage, T. I. (2021). Effectiveness of case-based learning: views of construction and real estate students. *International Journal of Construction Education and Research*, 17(4), 318–332. https://doi.org/10.1080/15578771.2020.1758254.

[19] Molla, A., Yayeh, M., & Bisaw, A. (2023). The current status of faculty members' pedagogical competence in developing 21st century skills at selected universities in Ethiopia. *Cogent Education*, 10(2), 2228995. https://doi.org/10.1080/2331186X.2023.2228995.

[20] Negeri, A. T., & Oumar, J. (2023). Engineering graduates' skill acquisition and employers skill need as applied to science education in ethiopia. *International Journal of Research in STEM Education*, 5(2), 46–60. https://doi.org/10.33830/ijrse.v5i2.1607.

[21] Puolitaival, T. (2013). Challenges of project-based learning in construction management education. In 38th Australasian Universities Building Education Association (AUBEA) Conference.

[22] Qureshi, M. A., Khaskheli, A., Qureshi, J. A., Raza, S. A., & Yousufi, S. Q. (2023). Factors affecting students' learning performance through collaborative learning and engagement. *Interactive Learning Environments*, 31(4), 2371–2391. https://doi.org/10.1080/10494820.2021.1884886.

[23] Sewagegn, A. A., & Diale, B. M. (2021). Modular/Block teaching: practices and challenges at higher education institutions of Ethiopia. *Teaching in Higher Education*, 26(6), 776–789. https://doi.org/10.1080/13562517.2019.1681391.

[24] Suskie, L. (2018). Assessing Student Learning: A Common Sense Guide, (3rd edn.). Jossey-Bass.

[25] Tezel, E., & Çakmak, P. I. (2024). Skills, competencies and knowledge for construction management graduates. *İstanbul Teknik Üniversitesi Journal of the Faculty of Architecture*, 21(1), 1–14.

[26] Thompson, C., Spenceley, L., Tinney, M., Battams, E., & Solomon, A. (2024). The Ultimate Guide to Lesson Planning: Practical Planning for Everyday Teaching. London and New York: Routledge, Taylor & Francis Group.

[27] Yizengaw, J. Y. (2018). Skills gaps and mismatches: Private sector expectations of engineering graduates in Ethiopia. *IDS Bulletin*, 49(5), 55–70. https://doi.org/10.19088/1968-2018.174.

[28] Zhang, J., Xie, H., & Li, H. (2019). Improvement of students problem-solving skills through project execution planning in civil engineering and construction management education. *Engineering, Construction and Architectural Management*, 26(7), 1437–1454. https://doi.org/10.1108/ECAM-08-2018-0321.

Advances in Construction, Real Estate, Infrastructure and Project Management – Anil Kashyap et al. (eds)
© 2026 Taylor & Francis Group, London, ISBN 978-1-041-13433-6

4

Urban History Courses as a Foundational Education for Future Construction and Design Professionals

Ella Howard[1]
Professor of History,
School of Sciences and Humanities Wentworth Institute of Technology,
Boston, USA

Allison K. Lange[2]
Associate Professor of History,
School of Sciences and Humanities Wentworth Institute of Technology,
Boston, USA

■ **ABSTRACT:** Construction Management and Design students benefit from understanding the dynamics that have shaped the historical development of cities. Once they know the political, social, and economic forces that have created the modern environment, they are better able to make sound decisions as students and as professionals. As history professors at a technology-focused university, we teach general education courses to create strong foundations that equip students for the workplace. Our urban history classes, "Boston History" and "American Urban History," provide examples of this teaching.

In this case study, we provide a brief overview of the goals of the courses and discuss examples of our pedagogy that could be adapted for similar courses elsewhere. We employ active learning, a cornerstone of our university's mission, to engage students in the study of the built environment beyond textbooks and historical documents. To accomplish this, we lead walking tours of Boston in both classes to familiarize students with the histories that shaped the growth of their own city. We discuss models for creating walking tours for any city. We also share a miniature case study that exemplifies how we connect the city's past to its present.

■ **KEYWORDS:** Case study, Pedagogy, Urban history, Walking tour

1. INTRODUCTION

To design and build in Boston—or any city—professionals must know its history. Students in HIST 4300 American Urban History and HIST 4223 Boston History, study the history of cities in the

[1]howarde@wit.edu, [2]langea@wit.edu

DOI: 10.1201/9781003669814-4

United States. In this case study, we will discuss the aims of each course and one of its major assignments. This case study demonstrates the significance of understanding a city's history to help students become better professionals in design and construction management.

HIST 4300 and HIST 4223 are general education courses designed for students in all majors at Wentworth Institute of Technology. These courses are taken at various points in the program of study, and thus must meet all levels of students. Students are not expected to have any prior historical knowledge about this specific topic. Wentworth prides itself on engaging students in hands-on learning through projects and activities. Educational research has clearly demonstrated that such pedagogical approaches ensure that students understand and retain a high level of information [8].

The courses are designed to appeal to Wentworth students with a range of professional interests. Majors in construction management and design, including architecture, industrial design, and civil engineering, often take this class. This style of teaching is grounded in constructivist learning theory, which shows that students learn by integrating new knowledge into their existing experiences and frame of reference [16]. Students are already taking relevant classes within their majors, and this history course provides a new angle and enriches their existing knowledge.

The students learn about the variety of historical factors that influenced the development of urban areas into the ones we recognize today. Numerous themes affected the development of cities, from geographical and governmental influences to racial, gendered and economic factors. To learn about these topics, students perform close readings of primary source historical documents, watch documentaries, listen to podcasts, make posters, and visit historical sites. By reviewing this array of content, analyzing it, and producing their own insights, students perform a range of skills necessary for their future careers [4].

In addition to learning skills, students gain important subject knowledge. For example, understanding the history of Boston's geography is essential for those working in design and construction in this city. Students learn about Boston's emergence as a port city during the colonial era. American colonists and the British government depended on ships for communication, transportation, and trade. As a result, Boston's significance during the American Revolution was due in part to its geographic location and its centrality for all of these vital aspects of colonial governance [3, 17]. After the American Revolution, as overland travel such as trains became more widespread, Boston became less central to American politics and culture. The city never regained the prominence it had during the colonial era. The city's historical footprint as a port city still influences its aesthetic style and its layout.

To support shipping, the city of Boston originally was on a small piece of land that was almost an island. But, over the 19th century, Bostonians filled in parts of the marshy waters surrounding it to add more land. Consequently, many of the city's wealthiest commercial and residential areas, notably the Back Bay, are built on landfill. Landfill requires significant maintenance, especially since many historical structures are built on wooden pylons that must remain fully submerged in water to remain effective. And, while the pylons require specific water levels, rising sea levels also threaten the entire landfill area [13].

2. DESCRIPTION OF THE COURSES

2.1 Case Study of Walking Tours: American Urban History Assignment

To learn more about the history of cities, students must read and produce their own research. In HIST 4300, students complete research using historical documents and scholarly secondary sources to create a history of a specific place in Boston. Rather than submit a traditional essay, students convey their research in the form of a short TikTok-style video. They focus on a specific theme and tell the history of four specific locations in the form of a historical walking tour. In addition to discussing historical facts, the videos must convey an argument about their topic. This assignment builds on the work of scholars who have documented the effectiveness of walking tours in establishing a sense of place for learners [1, 10, 11].

Students pursue research in a range of local and national archives through online databases. They select historical documents such as photographs, newspapers, magazines, and newsreels. First, they use these sources to produce a script. The script includes quotes from their historical texts and context information from their secondary scholarly sources. Next, students edit their script to improve them. Then, develop content for their short videos. They include images of their historical documents and videos of the area as it looks today.

The assignment is divided into six phases and lasts throughout much of the semester. For the first phase, students propose the themes of their walking tours. They must list a creative and descriptive title of their tour that helps convey their main ideas. Then they describe the theme. They must include at least two main ideas that they want to convey. These two main points drive their research, since they need evidence to support them.

For phase two, students continue to edit their title and main ideas, but they also generate a walking map of their tour. This is an opportunity for students to better understand the geography of the location that they are analyzing. They can also have to think through the logistics of navigating the city. In this phase, students describe each of the four locations they are focusing on. They discuss what they plan to talk about and the types of historical documents they need to tell their story. Finally, they must include a preliminary bibliography. Their list of sources must include at least three historical documents, at least two relevant scholarly sources, and at least three additional sources of any kind. By the time they finish this phase, they should have a clear idea of the historical and scholarly research they must do to support their main ideas.

Students must draft their script for phase three. They are required to write at least 250 words for an introduction. Their introduction must deliver their argument in a compelling and clear manner. Students discuss each of the four locations they selected. The script must cite their historical documents as well as their scholarly documents to support their points. They can integrate quotes from historical documents and refer to historical images as evidence as well. At the end of this assignment, they also include a works cited section to emphasize the importance of using evidence to support their ideas.

During phase four, students edit their script. They work with their peers and the professor to ensure that they have the evidence needed to tell their story. Students hone their arguments and ensure that their main ideas are clear. Students continue to update their bibliography and perform research to develop their narrative.

Students submit their short videos for phase five. They are required to convey clear main ideas in the introduction and throughout the video. The videos must include details from historical documents and secondary scholarly sources to tell their story. The videos feature their main ideas clearly and with evidence to support them. Students are required to have a compelling and entertaining presentation style. The videos are presented at the School of Sciences and Humanities Undergraduate Research Showcase at the end of each semester. Students need to convey their ideas to their classmates familiar with the topic but also to colleagues in other classes who are not unfamiliar with this subject.

For the final phase of their walking tour, students submit a finalized bibliography for their project. This bibliography includes any additional resources that they added as they revised their materials. Then students write a reflection on their work. The reflection requires the class to consider what they learned while researching and completing the assignment that they did not know before. The reflection also asks students to think more about how what they learned will influence their experience of the city and how it might influence their future careers and other engagement with urban life.

2.2 Case Study of the Persistent Dilemma of Eminent Domain: Boston History Assignment

Students in Boston History engage in a variety of learning activities, including participating in guided walking tours, researching the historical precedents of a local current event of their choice, and presenting their work as a poster at our School of Sciences and Humanities Research Showcase. Each week they also engage in active learning through in-class activities. Below is a miniature case study used in the course that focuses on eminent domain and the built environment.

Transportation is a critical and controversial issue in Boston, Massachusetts. In this dramatized case study, the Director of the Boston Planning & Development Agency (BPDA) faces a challenging decision concerning a fictionalized proposal to expand the subway system run by the Massachusetts Bay Transportation Authority (MBTA) subway system, popularly referred to as the "T."

Many Boston residents do not currently have equitable access to public transit. A proposal has been put forward to enhance access to transit in Roxbury by expanding the Orange Line. This project would add new stations to the T. This would most likely result in economic growth for small businesses. It could also lead to elevated housing values in the area, due to the already tight housing market in the city [2].

For this expansion of the Orange Line to take place, several existing buildings need to be claimed through the power of eminent domain. Eminent domain is a long-standing legal power in the United States that allows governmental bodies and their representatives to claim formerly private properties for necessary public use. When traditional eminent domain is used, properties are surveyed, and valuations are issued. Property owners are compensated for the transaction. However, they are not allowed to refuse the transfer of property [12].

The use of eminent domain is highly controversial. In Boston in particular, it has an infamously complex and unjust history because it has been used in ways that led to negative impacts on low-income and minoritized communities. Because of this painful legacy, the Director cannot make a careless decision about the proposed expansion project for the T.

The Director will analyze a historic example drawn from Boston's history of mid-20th century urban renewal. The redevelopment that was carried out in the West End is widely regarded as an example of poor urban planning and execution that led to unjust economic, social, and political outcomes. The project was part of a federal program of urban renewal that emerged from the New Deal. As part of the government's plan to revitalize urban centers, money was provided for slum clearance. Property values were then marked down as the sites were sold to private developers to create new, mixed-use and often mixed-income developments. The area's former residents, now displaced, were sometimes relocated to the public housing developments that formed the third major aspect of this revitalization agenda [10].

Boston's West end neighborhood was a densely populated area when it was selected for redevelopment. Many families in the area had developed tight bonds and described a strong sense of community. It was a working-class neighborhood, home to primarily white residents, many of whom were immigrants. Many area residents disputed the claims that their neighborhood was a slum. Nonetheless, the project proceeded [7].

The urban renewal carried out in the West End through eminent domain demolished approximately 40 acres of the area. Over 7, 000 residents were displaced, as were many businesses. In place of the old neighborhood rose the Charles River Plaza, a residential and shopping complex designed by architect Victor Gruen. The Government Center complex, Massachusetts General Hospital, and North Station were also built. [14].

The urban renewal of the West End is frequently cited as an example of the wrong way to carry out urban planning and development because it lacked both transparency of communications and consideration of issues of equity and social justice. The case study also raises critical questions about the relative weight that should be given to various parties when such projects are implemented.

Ethical Frameworks: Utilitarianism and Property Rights

Approaching a case study this complex requires the use of multiple ethical frameworks. Utilitarianism is central to all discussions of eminent domain. Utilitarianism is a moral theory that argues the best action in each situation will maximize the greatest good for the greatest number of people.

A utilitarian approach to situations involving eminent domain requires the assessment of the potential benefits and risks that may result from the project. Will the expansion of the T lead to economic development, better traffic flows, and a higher quality of life for city residents? What is the weight to be given to the shattered community bonds and displacement of residents and businesses? What decisions can be made to maximize the benefits for the greatest number of people [9]?

Property rights also need to be considered in this type of situation. In the United States, property rights are central to historic notions of individual citizenship and liberty. Thomas Jefferson famously framed property ownership as a key part of the American experience and this ethos has continued through centuries of work and rhetoric around "the American dream." Many Americans resist eminent domain proposals because they see them as a fundamental denial of an inherent right [15].

Roxbury's History

The fictionalized proposal to expand the Orange Line is further complicated by its route through Roxbury, a neighborhood whose residents have faced historical discrimination. Boston's Black community moved from Beacon Hill to the South End, and then to Roxbury, Dorchester, and Mattapan, many neighborhoods of which remain primarily Black due to Boston's history of economic and racial segregation [6].

In the mid-20th century, the planned construction of Interstate 93 began. This highway was to run through Roxbury, displacing many residents and businesses. An unlikely coalition of neighborhood residents, activists, and political leaders came together to block this project. The physical scars from the demolition that were carried out remain, as do the bitter feelings of some area residents regarding the way the situation was handled [5].

Decision Time

After reflecting on the way that urban renewal was carried out in the West End, the ethical frameworks of utilitarianism and property rights, and the history of Roxbury, the Director is ready to make a recommendation on the planned expansion of the Orange Line.

They will write a memo to their staff communicating their decision and outlining the reasons for it. The memo will address the following questions:

- What are the potential benefits and drawbacks of using eminent domain in this case?
- How will the decision impact the Roxbury community?
- What steps can be taken to mitigate any negative impacts?
- How does the West End Redevelopment Project inform the decision?
- How does a utilitarian perspective influence the decision?
- How do different perspectives on property rights and justice inform the decision on whether to use eminent domain for the T expansion project?

The Director's memo will be well-reasoned, clearly articulated, and supported by evidence. It will also demonstrate their understanding of the complex ethical and social issues surrounding the use of eminent domain.

3. CONCLUSION

Students in construction management and design become more informed professionals by understanding the history of the places where they work. While students cannot learn all of this in a single course, a class can teach them the importance of this knowledge and how to engage with it during their careers. Students in American Urban History and Boston History learn to find the information they need using available resources, process and analyze the data, develop a compelling narrative, and deliver it to colleagues in a clear, engaging manner. These skills are invaluable in the 21st-century workplace.

REFERENCES

[1] Borucka, J. (2019). City walk: a didactic innovative experiment in architectural education. *World Transactions on Engineering and Technology Education*, 17, 158–163. https://mostwiedzy.pl/pl/publication/city-walk-a-didactic-innovative-experiment-in-architectural-education, 152531-1.

[2] Brinker, A. (2024). A sobering statistic': the typical house here now costs $950, 000. *Boston Globe.* https://www.bostonglobe.com/2024/05/21/business/greater-boston-home-prices.

[3] Carp, B. L. (2010). Defiance of the Patriots: The Boston Tea Party and the Making of America. Yale University Press.

[4] Corey, S. H. (2010). Pedagogy and place: merging urban and environmental history with active learning. *Journal of Urban History*, 36(1), 28–41. https://doi-org.ezproxy.neu.edu/10.1177/0096144209349882.

[5] Crockett, K. (2018). People Before Highways: Boston Activists, Urban Planners, and a New Movement for City Making. University of Massachusetts Press.

[6] Elton, C. (2020). How has Boston gotten away with being segregated for so long? *Boston Magazine*, 8. https://www.bostonmagazine.com/news/2020/12/08/boston-segregation/.

[7] Fisher, S. M., & Hughes, C. (1992). The Last Tenement: Confronting Community and Urban Renewal and Boston's West End. Bostonian Society.

[8] Freeman, S., Eddy, S. L., McDonough, M., & Wenderoth, M. P. (2014). Active learning increases student performance in science, engineering, and mathematics. *Proceedings of the National Academy of Sciences*, 111(23), 8410–8415. https://doi.org/10.1073/pnas.1319030111.

[9] Herzog, D. J. (1985). Against utilitarianism. In Without Foundations: Justification in Political Theory, (pp. 110–160). Cornell University Press.

[10] Hirsch, A. R. (1983). Making the Second Ghetto: Race and Housing in Chicago, 1940–1960. Cambridge University Press.

[10] Holton, W. (1990). Walking tours for teaching urban history in boston and other cities. *OAH Magazine of History*, 5(2), 14–19. https://doi-org.ezproxy.neu.edu/10.1093/maghis/5.2.14.

[11] Klaniczay, J. (2024). The urban walking tour as an experience-based methodology for built environment education in Budapest. *Periodica Polytechnica Architecture*, 55(1), 60–71. https://doi.org/10.3311/PPar.23235.

[12] Lehavi, A., & Licht, A. N. (2007). Eminent domain, Inc. *Columbia Law Review*, 107(7), 1704–1748. https://papers.ssrn.com/sol3/papers.cfm?abstract_id=967970#.

[13] Newman, W. A., & Holton, W. E. (2006). Boston's Back Bay: The Story of America's Greatest Nineteenth-Century Landfill Project. Northeastern University Press.

[14] O'Connor, T. H. (1993). Building a New Boston: Politics and Urban Renewal. Northeastern University Press.

[15] Post, D. M. (1986). Jeffersonian revisions of locke: education, property-rights, and liberty. *Journal of the History of Ideas*, 47(1), 147–157. https://doi.org/10.2307/2709601.

[16] Vong, S. (2016). A constructivist approach for introducing undergraduate students to special collections and archival research. *Rare Books and Manuscripts Librarianship*, 31(2), 118–130. https://rbm.acrl.org/index.php/rbm/article/viewFile/9666/11112.

[17] Zabin, S. R. (2020). The Boston Massacre: A Family History. Houghton Mifflin Harcourt.

Advances in Construction, Real Estate, Infrastructure and Project Management – Anil Kashyap et al. (eds)
© 2026 Taylor & Francis Group, London, ISBN 978-1-041-13433-6

5 A Review on Recent Advances in Construction Safety Management in India

Viraj Kale
P.G. Student,
Department of Civil Engineering,
Dr. Vishwanath Karad MIT World Peace University,
Pune

Pravin Minde*
Assistant Professor,
Department of Civil Engineering,
Dr. Vishwanath Karad MIT World Peace University,
Pune

■ **ABSTRACT:** The construction sector is often characterized by high risks related to employee safety, with a significant prevalence of severe injuries, work-related ailments, and fatalities. According to a 2019 report by the Indian Express, 38 workers die daily in the Indian construction industry due to workplace incidents. However, the last few years have perceived substantial advancements in the field of construction safety management in India, driven by an increased emphasis on worker safety and regulatory compliance. This paper provides an overview of these advancements, focusing on key initiatives, technological innovations, and regulatory changes aimed at enhancing safety standards. Major advancements include accepting new technologies like Building Information Modelling (BIM), real-time monitoring systems, and mobile applications for safety inspections and reporting. AR and VR have revolutionized construction management by enabling immersive applications to visualise complex scenarios, enhance risk-prevention knowledge, and provide safe training environments. These technologies help construction companies identify and mitigate safety hazards more effectively. Stricter safety regulations and enforcement measures by government agencies also play a crucial role. These regulations mandate the usage of PPE, safety training programs, and safety audits. Establishing safety committees and implementing safety management systems have further improved safety practices on construction sites. Industry-wide initiatives such as safety awareness campaigns, skill development programs, and collaboration between industry stakeholders and government bodies have promoted a safety philosophy in the construction industry. Despite these advancements, challenges such as lack of awareness, inadequate enforcement, and resistance to change remain obstacles that need to be addressed to further improve construction safety management practices

*Corresponding author: pravin.minde@mitwpu.edu.in

DOI: 10.1201/9781003669814-5

in India. Overall, the paper offers a comprehensive review of recent advances in construction safety management in India, focusing on virtual and augmented reality applications.

■ **KEYWORDS:** Construction safety, Safety management, Advances, India, Building information modeling (BIM), Regulations

1. INTRODUCTION

1.1 Background

The construction business is a monument to human innovation and ambition in the records of human history. From the majestic pyramids of ancient Egypt to the soaring skyscrapers of contemporary cities, builders have never stopped pushing beyond the achievable limits. However, danger accompanied advancement, and safety risk management evolved to become a crucial part of the construction story. The construction sector is one of the riskiest in the world due to its consistently high accident and fatality rates [17]. Over centuries, the construction business has seen significant changes on the global stage. Though they had few safeguards in place, ancient civilizations set the foundation for later architectural wonders. Construction methods improved along with civilizations, but safety frequently lags. Safety concerns had not become widely known until the Industrial Revolution when worker rights movements and terrible accidents that occurred during this time gained traction [9].

1.2 Need

Numerous articles covering a wide range of subjects related to construction safety have been released. This study's findings can be used as the basis for developing construction safety management [25]. Across the globe and in India, safety risk management has turned out to be a crucial aspect of construction. Effective safety standards became increasingly necessary as project scale and complexity rose. Safety is now a strategic necessity for success in the construction sector, as well as a moral or regulatory requirement. It is more difficult to identify hidden safety concerns and provide the appropriate information at the right moment to the relevant workforces during construction work because the safety information used on sites does not match the aspects involved in real construction work settings [15]. Safety plays a variety of roles in the construction industry. It defends investments, increases productivity, promotes sustainable development, and protects the lives and well-being of workers. Safety risk management has become essential to every construction project, ensuring that developments are made without sacrificing human lives due to strict rules and technological advancements [5]. The history of the construction industry serves as a reminder of how far people went and how far people may go with an ongoing commitment to safety in this constantly evolving world [26]. Considering all the sectors worldwide, the construction industry has extensively remained known to have the greatest rate of accidents. For example, in 2014, the Hong Kong Census and Statistics Department reported that 31, 00 construction workers were injured and 24 people died annually during construction—that is, two workers died every month. In addition, the Department of Occupational Safety and Health (DOSH) reports that 7984 accidents occurred in Malaysia's construction industry in 2019, which is a record high when compared to previous years [8, 12]. AR and VR are needed in construction safety management to provide immersive training environments, allowing workers to visualize and practice handling complex scenarios safely, and to enhance risk prevention through advanced simulation and monitoring techniques.

1.3 Aim and Scope

The paper provides a comprehensive review of safety risk management in construction, exploring historical developments, current challenges, and innovative approaches. It emphasizes proactive risk mitigation, systematic strategies, and advancements like AR and VR for immersive training and improved risk prevention. Additionally, it introduces a Modified Framework for Risk Management.

2. LITERATURE REVIEW

The study uses a systematic literature review to integrate current knowledge about developments in construction safety management, particularly in residential projects. The data is then analyzed to reach conclusions and provide recommendations. A flow chart depicts this procedure, highlighting processes from literature search to final synthesis and framework building, ensuring a systematic approach to examining safety risk management improvements. An analysis of the literature review was conducted on selected criteria of safety management in building construction, covering 38 articles. This review aimed to identify key trends, methodologies, and best practices in the field. Figure 5.1 illustrates the detailed process of conducting the literature review, encompassing all required steps.

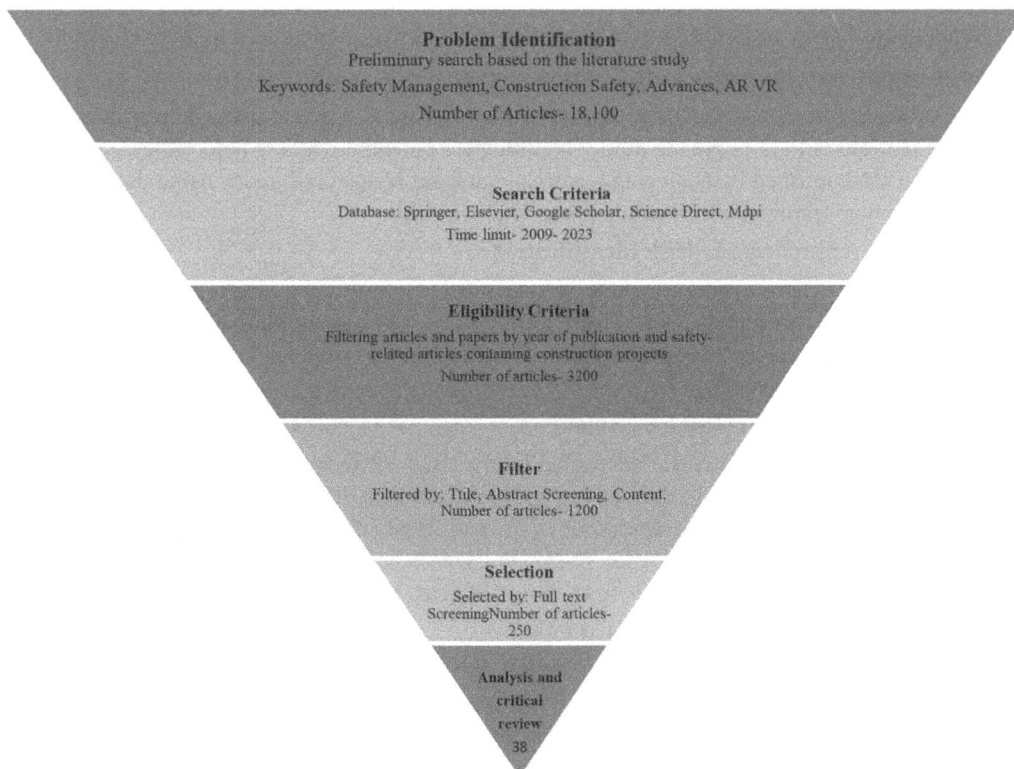

Problem Identification
Preliminary search based on the literature study
Keywords: Safety Management, Construction Safety, Advances, AR VR
Number of Articles- 18,100

Search Criteria
Database: Springer, Elsevier, Google Scholar, Science Direct, Mdpi
Time limit- 2009- 2023

Eligibility Criteria
Filtering articles and papers by year of publication and safety-related articles containing construction projects
Number of articles- 3200

Filter
Filtered by: Title, Abstract Screening, Content.
Number of articles- 1200

Selection
Selected by: Full text ScreeningNumber of articles- 250

Analysis and critical review
38

Figure 5.1 Literature review systematic process

Source: Authors

3. FRAMEWORK FOR RISK MANAGEMENT

3.1 General Framework for Risk Management

A methodical framework for recognizing, evaluating, controlling, and overseeing risks is offered by a risk management framework (RMF) [14]. It helps businesses methodically address uncertainties that can affect their goals [6]. An outline of the essential elements of a thorough risk management framework is provided below [4]:

The General Framework for Risk Management involves a structured approach to handling risks effectively. It begins with goal setting, which considers internal and external factors to establish risk criteria based on likelihood, impact, and tolerance levels. Next is risk identification, employing techniques like SWOT analysis and risk registers to document potential risks from diverse sources. Risk assessment evaluates identified risks through qualitative and quantitative analyses to prioritize their significance. Risk treatment follows, exploring strategies like mitigation, transfer, acceptance, or prevention, with action plans for implementation. Continuous review and monitoring ensure the approach remains relevant by adapting to new data and evolving contexts. Lastly, communication and consultation engage stakeholders, fostering transparency and collaboration.

3.2 Proposed Modified Framework for Risk Management

The proposed framework integrates AI and ML to enhance predictive capabilities, minimize human error, and enable proactive risk management. It focuses on innovation in goal-setting, identification, assessment, reduction, and monitoring for greater efficiency and adaptability. Figure 5.2 shows the Conventional and Proposed Modified Framework.

1. **Strategic Goals for Efficient Risk Management**

   A strategic goal is to enhance predictive skills to achieve efficient management of the risk. The purpose is to improve the ability to anticipate potential dangers before they occur. This can be accomplished by analyzing historical data and identifying trends using data analytics and machine learning, which can indicate future hazards.

2. **Alternate Approaches to Risk identification**

   AI and ML enable proactive risk management by providing more precise and rapid identification of potential risks. These technologies can also minimize the drawbacks and human errors typically associated with the risk detection process, leading to more reliable and effective safety measures.

3. **Different Methods of Risk Assessment**

   Predictive analysis involves forecasting future risks by examining historical and current data through statistical algorithms and machine learning techniques. This method allows organizations to anticipate potential risks and their impacts, enabling proactive risk management. Additionally, it helps prioritize hazards depending on their projected severity and likelihood, facilitating more effective and targeted mitigation strategies.

4. **Innovative Approaches to Risk Reduction**

   AI-powered risk evaluation swiftly identifies and assesses hazards with greater accuracy by analyzing large data sets, improving decision-making, and reducing human error. Collaborative risk management platforms enhance coordination and data sharing among stakeholders, leading to comprehensive risk mitigation. Crowdsourced risk intelligence gathers diverse

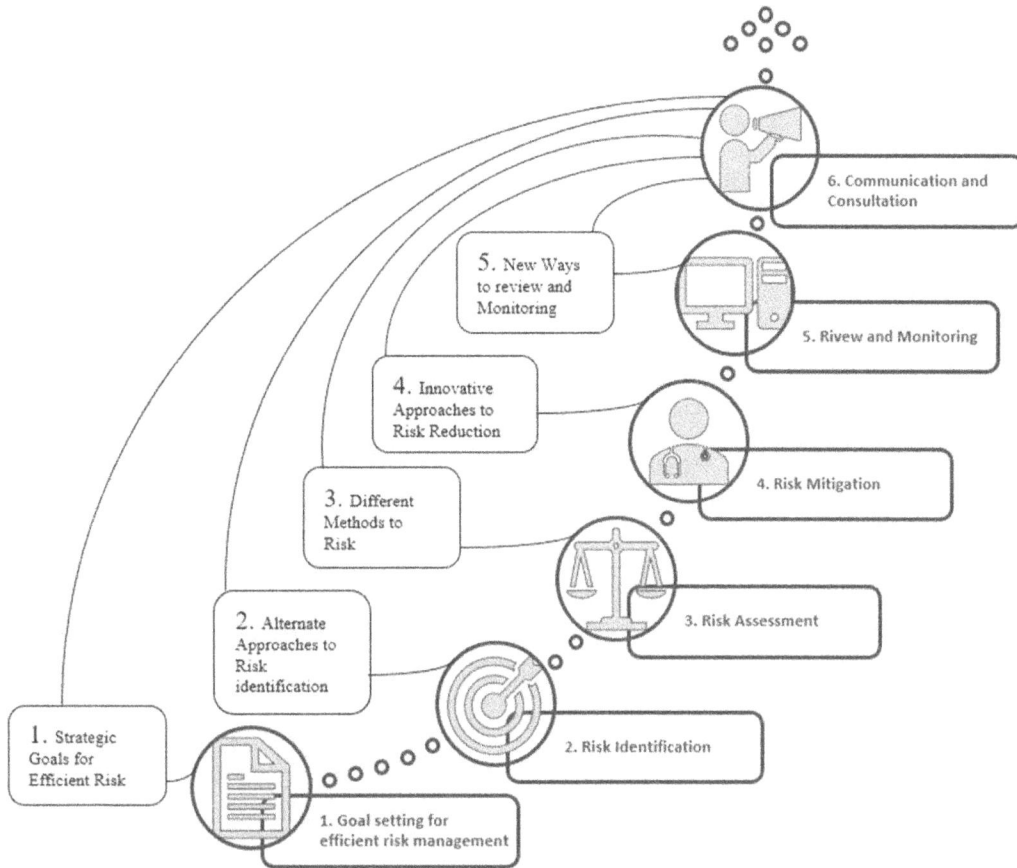

Figure 5.2   Conventional and developed or modified framework

Source: Authors

insights from employees, customers, and experts, uncovering risks that traditional methods might miss and enriching the risk data available for analysis.

5. **New Ways to review and Monitoring**

Through immersive and interactive risk assessment and mitigation approaches, augmented reality (AR) and virtual reality (VR) are revolutionizing safety risk management. Real-time overlay of safety information into physical settings is made possible by AR, which facilitates on-site decision-making and danger detection. Virtual reality (VR) provides secure venues for emergency response training and rehearsal by enabling realistic simulations of high-risk scenarios. Combined, these technologies raise the effectiveness of safety training, increase hazard visualization, and promote proactive risk management by enabling teams to anticipate and address possible safety concerns before they arise.

4. INTRODUCTION TO ADVANCED RISK ASSESSMENT TECHNIQUES

Advanced techniques like PRA, FTA, Delphi, SWOT, and Benchmarking enhance risk evaluation by providing strategic insights, prioritizing critical risks, and integrating best practices for improved safety and decision-making.

4.1 Probabilistic Risk Assessment (PRA)

It is a technique for evaluating hazards in complex systems by predicting the likelihood and severity of unfavourable outcomes. It entails identifying potential failure scenarios, quantifying their probabilities with tools like FTA and ETA, and evaluating the implications. PRA provides a quantitative foundation for decision-making, assisting in prioritizing risk management activities and improving system safety by focusing on the most severe threats [20].

4.2 Fault Tree Analysis (FTA)

It is an organized, rational approach to determining the underlying reasons for probable system breakdowns. It visually maps out the pathways leading to a specific undesirable event, called as the "top event." The analysis commences with the upper event and traces back through a series of logical gates (AND, OR) to determine the underlying causes, which can be basic events or further intermediate events. FTA helps in understanding the complex interactions between different components and processes, prioritizing risk mitigation efforts, and improving system reliability and safety by identifying and addressing the most critical failure points [2].

4.3 Delphi Technique

A panel of experts uses the Delphi technique, a structured communication strategy, to reach an understanding of risk management. Multiple rounds of questioning are used in this continuous process, where specialists offer their viewpoints. The input is then combined and shared with the group, and the process is repeated until an arrangement is established [3].

4.4 SWOT Analysis in Risk Management

A strategic planning tool called a SWOT analysis is used to determine and evaluate a project's or company's Strengths, Weaknesses, Opportunities, and Threats. SWOT analysis aids in risk management by evaluating internal and external variables that may have an impact on goal achievement [19].

4.5 Benchmarking

The systematic process of comparing an organization's procedures, workflows, and performance measures with industry or cross-industry best practices is known as benchmarking. When it comes to risk management, benchmarking is comparing one's risk management procedures to those of top companies to pinpoint problem areas and put best practices into action [18].

5. TECHNOLOGY AND INNOVATION IN SAFETY RISK MANAGEMENT

In recent years, technology and innovation have significantly transformed safety risk management across various industries. Recent technologies like as AI, ML, and the IoT are now essential for more effective and efficient risk identification, assessment, and mitigation. The management of safety risks has been greatly improved by technological improvements, especially in sectors like construction where dangers are common. Drones and sensors are examples of key technologies that are essential for enhancing safety protocols and lowering hazards.

5.1 Drones in Safety Risk Management

Drones are transforming safety risk management in construction through advanced site inspections and data analysis. With real-time monitoring capabilities, drones provide continuous aerial views of sites, quickly identifying hazards like structural issues and compliance violations without endangering workers. They also enable safe inspections in hard-to-reach or dangerous areas, for example, high-rise structures or confined spaces. Equipped with high-definition cameras and sensors, drones create accurate maps and 3D models that support improved planning and risk assessment. Additionally, drones can capture regular images and videos to monitor construction progress, detect safety deviations, and facilitate timely corrective actions [21].

5.2 Sensors in Safety Risk Management

Advanced sensor technology plays a crucial role in monitoring environmental conditions and ensuring occupational safety. Sensors can detect hazardous factors such as humidity, temperature, and gas concentrations, triggering alarms for quick evacuation and preventive action when danger is detected. They also monitor noise and vibration levels to prevent structural issues and hearing damage, allowing for immediate corrective measures [10, 22, 23]. Wearable technology further supports safety by tracking workers' movements and vital signs, alerting supervisors to potential overwork or exposure to hazardous conditions. Additionally, attaching sensors to equipment ensures continuous monitoring of machinery, predicting maintenance needs and avoiding malfunctions to maintain a safe work environment [7].

5.3 Automated Equipment

Automated equipment and robotics are enhancing safety in construction by taking on high-risk tasks such as material handling and demolition, reducing the need for human workers to perform dangerous activities. Wearable exoskeletons are also improving safety and productivity by assisting workers in lifting heavy objects, thereby reducing the risk of muscle injuries and boosting overall efficiency [13].

5.4 Safety Management Software

Software solutions simplify the recording, monitoring, and evaluation of safety incidents, enabling a data-driven approach to detect patterns and implement effective preventive measures [10]. Prominent risk management software includes Oracle, Project Risk Manager, Qualys, Vendor360, and MSP Software.

5.5 Construction Safety Management Using Augmented Reality (AR) and Virtual Reality (VR)

General and Applications

AR and VR is a very recent trend and it has very high potential and concern in construction safety, its applicability is explained further. A developing trend in construction management is the application of virtual reality (VR) and augmented reality (AR), which hold promise for increasing project effectiveness and efficiency. These cutting-edge technologies provide immersive and interactive ways to improve decision-making and streamline workflows. AR and VR can provide realistic and practical safety experiences for construction education, improving knowledge dissemination, reflection, and assessment.

- Training in construction management is changing as a result of virtual reality (VR) and augmented reality (AR). VR offers immersive environments for practising safety procedures and identifying risks without real-world hazards, significantly improving learning outcomes and retention. AR enhances on-site training by overlaying digital instructions onto the physical environment, reducing human error and improving adherence to safety protocols. Together, these technologies make training more engaging, effective, and contextually relevant [11].

- Virtual reality (VR) and augmented reality (AR) enhance hazard analysis and safety planning in construction. VR allows project managers to conduct virtual site walkthroughs, proactively identifying risks and planning safety measures to reduce accidents. It also enables detailed risk assessments by simulating construction stages and highlighting potential dangers. AR complements these efforts by providing live hazard alerts and overlaying critical safety information in real-time, improving on-site awareness. Additionally, AR facilitates seamless communication between on-site workers and off-site managers, ensuring efficient and timely safety management [1].

Figure 5.3 VR glass used for learning and training

Source: Authors

- Virtual reality (VR) is revolutionizing accident analysis in construction by enabling detailed reconstruction of incident scenes to uncover root causes and develop preventative strategies. Additionally, it facilitates training by enabling staff members to virtually experience previous mishaps, offering insightful guidance on how to prevent such occurrences in the future [16, 24]. Figure 5.3 depicts a VR glass device utilized for immersive learning and training, enabling users to experience realistic simulations that enhance skill development and safety awareness in a controlled environment as shown in Figure 5.4.

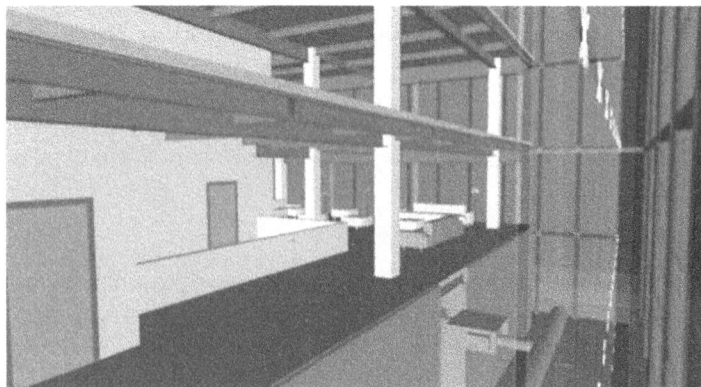

Figure 5.4 Inside view of the building after wearing a VR headset

Source: Authors

- Augmented reality (AR) is enhancing on-site safety by integrating with wearable technology for real-time monitoring of workers' vital signs, such as heart rate and temperature, and alerting them when signs of fatigue, overheating, or risky behaviour are detected. Additionally, AR provides interactive safety protocols and checklists to ensure employees consistently follow safety regulations and guidelines [7, 29]. Figure 5.5 illustrates the fundamental perspective of augmented reality (AR).

- **Advantages of VR and AR for Safety in Construction**
  - Interactive, hands-on training using VR and AR captivates workers, enhancing their learning experience and making training more effective than traditional methods.
  - Implementing VR and AR technology helps cut costs by minimizing the need for physical training spaces and equipment.
  - By offering realistic and hands-on safety training, VR and AR make sure that managers and employees comprehend and follow safety protocols.
  - Proactive safety training with VR and AR helps reduce on-site accidents by preparing workers to respond effectively to potential hazards.
- **Challenges and Considerations**
  - The initial cost of implementing VR and AR technology can be high, making it challenging for smaller companies to adopt.
  - Workers may require time and training to become familiar with and effectively use AR and VR technology.
  - The hardware and software used in AR and VR must be robust and reliable to meet the demands of a construction environment.

Figure 5.5 AR model

Source: Authors

6. CONCLUSION

The paper comprehensively reviews recent advances in construction safety management in India, focusing on virtual and augmented reality applications, while also highlighting a philosophical shift. Construction project risk management can be significantly enhanced by incorporating state-of-the-art technologies and collaboration platforms. Predictive analytics made possible by these technologies can proactively identify and reduce possible hazards, lowering accident rates and improving safety performance. Furthermore, incorporating virtual reality (VR) and augmented reality (AR) into safety training offers engaging, practical experiences that enhance worker readiness and risk mitigation. Automated equipment and exoskeletons also help create a safer, more creative building site. This study emphasizes the value of an integrated strategy for risk management that combines cutting-edge technologies with conventional techniques. By implementing these technologies, the construction sector may better predict and control hazards and promote a continuous improvement and safety culture. This study also aims to uncover how AR and VR technologies can effectively advance as time- and cost-saving tools in the construction industry. The approach could benefit researchers and practitioners by highlighting new VR/AR applications and understanding their research and practical frontiers. Future research should focus on expanding the use of these technologies and developing new tools to address emerging hazards in the evolving construction sector.

REFERENCES

[1] Azhar, S. (2017). Role of visualization technologies in safety planning and management at construction jobsites. *Procedia Engineering*, 171, 215–226. https://doi.org/10.1016/j.proeng.2017.01.329

[2] Bakeli, T., & Hafidi, A. A. (2020). A fault tree analysis (FTA) based approach for construction projects safety risk management. In *Proceedings of the 5th NA International Conference on Industrial Engineering and Operations Management Detroit*, Michigan, USA.

[3] Balkis, A., Janani, M. S., & Gandhi, A. G. M. (2021). A study on critical risk assessment and safety management for a high rise building. *International Research Journal of Engineering and Technology*. www.irjet.net

[4] Charehzehi, A., & Ahankoob, A. (2012). Enhancement Of Safety Performance at Construction Site. In *International Journal of Advances in Engineering & Technology*, 303.

[5] Dejus, T., & Antuchevičiene, J. (2013). Assessment of health and safety solutions at a construction site. *Journal of Civil Engineering and Management*, 19(5), 728–737. https://doi.org/10.3846/13923730.2013.812578

[6] Gillen, M. (2010). The NIOSH Construction Program: Research to practice, impact, and developing a national construction agenda. *Journal of Safety Research*, 41(3), 289–299. https://doi.org/10.1016/j.jsr.2010.04.002

[7] Hammad, A., Garrett, J. H., & Karimi, H. (2004). Location-based computing for infrastructure field tasks. *Telegeoinformatics: Location-based computing and services*, 287–314.

[8] Jaafar, M. H., Arifin, K., Aiyub, K., Razman, M. R., Ishak, M. I. S., & Samsurijan, M. S. (2018). Occupational safety and health management in the construction industry: a review. *International Journal of Occupational Safety and Ergonomics*, 24(4), 493–506. https://doi.org/10.1080/10803548.2017.1366129

[9] Keng, T. C., & Razak, A. (2014). Case studies on the safety management at construction site. *Journal of Sustainability Science and Management*, 9, 90–108.

[10] Laufer, A. (1987). Construction Accident Cost and Management Safety Motivation. In *Journal of Occupational Accidents*, 8.

[11] Lee, U. K., Kim, J. H., Cho, H., & Kang, K. I. (2009). Development of a mobile safety monitoring system for construction sites. *Automation in Construction*, 18(3), 258–264. https://doi.org/10.1016/j.autcon.2008.08.002

[12] Li, X., Yi, W., Chi, H. L., Wang, X., & Chan, A. P. C. (2018). A critical review of virtual and augmented reality (VR/AR) applications in construction safety. *Automation in Construction*, 86, 150–162. https://doi.org/10.1016/j.autcon.2017.11.003

[13] Manzoor, B., Othman, I., & Waheed, A. (2022). Accidental safety factors and prevention techniques for high-rise building projects – a review. *Ain Shams Engineering Journal*, 13(5). https://doi.org/10.1016/j.asej.2022.101723

[14] McCann, M. (2006). Heavy equipment and truck-related deaths on excavation work sites. *Journal of Safety Research*, 37(5), 511–517. https://doi.org/10.1016/j.jsr.2006.08.005

[15] Md Sofwan, N., Zaini, A. A., & Mahayuddin, S. A. (2016). Preliminary study on the identification of safety risks factors in the high-rise building construction. *Jurnal Teknologi*, 78(5–3), 13–18. https://doi.org/10.11113/jt.v78.8505

[16] Park, C. S., & Kim, H. J. (2013). A framework for construction safety management and visualization system. *Automation in Construction*, 33, 95–103. https://doi.org/10.1016/j.autcon.2012.09.012

[17] Sacks, R., Perlman, A., & Barak, R. (2013). Construction safety training using immersive virtual reality. *Construction Management and Economics*, 31(9), 1005–1017. https://doi.org/10.1080/01446193.2013.828844

[18] Sanni-Anibire, M. O., Mahmoud, A. S., Hassanain, M. A., & Salami, B. A. (2020). A risk assessment approach for enhancing construction safety performance. *Safety Science*, 121, 15–29. https://doi.org/10.1016/j.ssci.2019.08.044

[19] Serpella, A. F., Ferrada, X., Howard, R., & Rubio, L. (2014). Risk Management in Construction Projects: A Knowledge-based Approach. *Procedia – Social and Behavioral Sciences*, 119, 653–662. https://doi.org/10.1016/j.sbspro.2014.03.073

[20] Tamošaitiene, J., Gaudutis, E., & Kračka, M. (2013). Integrated model for assessment of high-rise building locations. *Procedia Engineering*, 57, 1151–1155. https://doi.org/10.1016/j.proeng.2013.04.145

[21] Tan, S., & Moinuddin, K. (2019). Systematic review of human and organizational risks for probabilistic risk analysis in high-rise buildings. In *Reliability Engineering and System Safety*, 188, 233–250. https://doi.org/10.1016/j.ress.2019.03.012

[22] Vithana, N. D. I. & Dissanayake, D. M. N. C. T., & (2021). Study on Development and Implementation of Safety Inspection Drones with Machine Learning Algorithms to Improve Construction Safety in Sri Lanka

[23] Velev, D., Zlateva, P., Steshina, L., & Petukhov, I. (2019). Challenges of using drones and virtual/augmented reality for disaster risk management. *International Archives of the Photogrammetry, Remote Sensing and Spatial Information Sciences – ISPRS Archives*, 42(3/W8), 437–440. https://doi.org/10.5194/isprs-archives-XLII-3-W8-437-2019

[24] Xu, J., Yan, C., Su, Y., & Liu, Y. (2020). Analysis of high-rise building safety detection methods based on big data and artificial intelligence. *International Journal of Distributed Sensor Networks*, 16(6). https://doi.org/10.1177/1550147720935307

[25] Zhang, M., Cao, T., & Zhao, X. (2017). Applying sensor-based technology to improve construction safety management. *Sensors,* 17(8). https://doi.org/10.3390/s17081841

[26] Zhao, D., Lucas, J., & Thabet, W. (2009). Using virtual environments to support electrical safety awareness in construction. *Proceedings of the 2009 Winter Simulation Conference (WSC)*, 2679–2690. https://doi.org/10.1109/WSC.2009.5429258

[27] Zhou, Z., Goh, Y. M., & Li, Q. (2015). Overview and analysis of safety management studies in the construction industry. *Safety Science,* 72, 337–350. https://doi.org/10.1016/j.ssci.2014.10.006

[28] Zhou, Z., Irizarry, J., & Li, Q. (2013). Applying advanced technology to improve safety management in the construction industry: a literature review. *Construction Management and Economics*, 31(6), 606–622. https://doi.org/10.1080/01446193.2013.798423

[29] Wang, X., Truijens, M., Hou, L., Wang, Y., & Zhou, Y. (2014). Integrating Augmented Reality with Building Information Modelling: On-site construction process controlling for liquefied natural gas industry. *Automation in Construction*, 40, 96–105. https://doi.org/10.1016/j.autcon.2013.12.003

Advances in Construction, Real Estate, Infrastructure and Project Management – Anil Kashyap et al. (eds)
© 2026 Taylor & Francis Group, London, ISBN 978-1-041-13433-6

Sustainable Development in the Construction Industry: A Review of Biochar's Impact on Concrete Performance

6

Afan S. Tamboli*, Sourabh S. Patil
UG Student,
Department of Civil Engineering,
Kasegaon Education Society's Rajarambapu Institute of Technology,
Affiliated to Shivaji University,
Sakhrale, India

Mayur M. Maske and Savita N. Patil
Assistence Professor,
Department of Civil Engineering,
Kasegaon Education Society's Rajarambapu Institute of Technology,
Affiliated to Shivaji University,
Sakhrale, India

■ **ABSTRACT:** The building sector is a significant contributor to global carbon emissions, leading to an increase in global temperature. Embracing sustainable practices and minimizing environmental impacts in this industry is crucial. Biochar, a carbon-rich material derived from biomass carbonization, can decrease the carbon footprint of concrete through carbon sequestration. This study examines how different types of biochar, when substituted for cement, affect the fresh and hardened properties of concrete. This study explored potential advantages, including enhanced strength and improved sound and thermal insulation. It also addresses possible drawbacks such as reduced strength and workability at high replacement levels, increased water penetration, and shrinkage. Analysis of the collected data suggests that an optimal biochar dosage of approximately 5% is necessary to enhance mechanical properties. The characteristics of concrete containing biochar are influenced by factors such as the biochar content, fineness, chemical composition, production method, and temperature. This study recommends further investigation into life cycle assessment and cost-effective biochar production techniques to fully harness the potential of biochar for a more environmentally friendly construction industry.

■ **KEYWORDS:** Biochar, Biomass pyrolysis, Carbon sequestration, Carbon footprint, Cement replacement

*Corresponding author: afantamboliat5@gmail.com

DOI: 10.1201/9781003669814-6

1. INTRODUCTION

The Industrial Revolution has brought great ease to human life, but has also given rise to many environmental problems. Industrial activities, such as increased production and burning of fossil fuels, have resulted in a surge of GHG emissions, leading to global climate change. It is projected that more than 40 billion tons of greenhouse gases are released annually and continue to grow at an alarming rate. Specifically, CO_2 is the largest source accounting for about 74% of all greenhouse gas emissions (Greenhouse Gas Emissions, Climate Watch). A significant source of global carbon dioxide (CO_2) emissions is the building sector, with cement production alone responsible for 0.5–0.6 tons of CO_2 per ton of cement manufacturing, which is approximately 7–8% of the world's anthropogenic CO_2 emissions [24]. If current CO_2 emissions levels continue then between 2030–2052, it is likely to reach 1.5°C global temperature [9].

Extensive research has been conducted worldwide to create methods that could lower CO_2 emissions from cement manufacturing industries while maintaining the performance of cement [52]. Biochar is a carbon-rich substance created through the thermochemical combustion of organic matter, including biomass, waste from animals, and municipal refuse, under conditions of limited oxygen. [3]. Researchers are interested in this material because of its special qualities, including stability, cation exchange capacity, high surface area, porosity, and functional groups. These qualities make it suitable for use in carbon sequestration applications. Because of its cost efficiency, eco-friendliness, reutilization, and ease of manufacture, biochar is another advantageous sustainable resource [38]. Current studies indicate that biochar can serve as a substitute for cement or aggregate in the production of sustainable concrete, without compromising the concrete's performance within certain thresholds [18–19, 26–27, 33–34, 44].

This review aims to critically analyse the current state of knowledge on biochar concrete. We examined the production and properties of different biochar varieties. The fresh, hardened, and durability properties of biochar compared with those of conventional concrete. In addition, we review the economic and environmental effects over time. The different applications of cementitious materials and their challenges are discussed. This analysis seeks to offer new perspectives on the potential and feasibility of biochar-based concrete as a sustainable alternative in the building industry by synthesizing the current literature and identifying areas that require further study.

2. BIOCHAR PRODUCTION, PROPERTIES AND MORPHOLOGY

Pyrolysis operates in the absence of oxygen and rapidly decomposes biomass material, leading to the creation of a carbon-rich material known as biochar. To ensure the integrity of this material and its use in various applications, some parameters must be optimized, such as the heat flow, temperature, and time [48]. Biomass feedstock suitable for pyrolysis includes materials such as wood waste, paddy straw, sawdust, bamboo, coconut shells, and food waste. However, because there are different types of pyrolysis techniques, such as flash pyrolysis, gasification, slow and high-temperature pyrolysis, temperature, and the feedstock used significantly impact the yield. In particular, slow and flash pyrolysis tend to yield more biochar, liquid biofuels are produced by fast pyrolysis and gaseous biofuels are produced most by gasification [25, 32].

The bamboo biochar morphology is illustrated in Figure 6.2, whose microstructure and pore structure have microcracks as interconnected voids that are developed within it during the pyrolysis

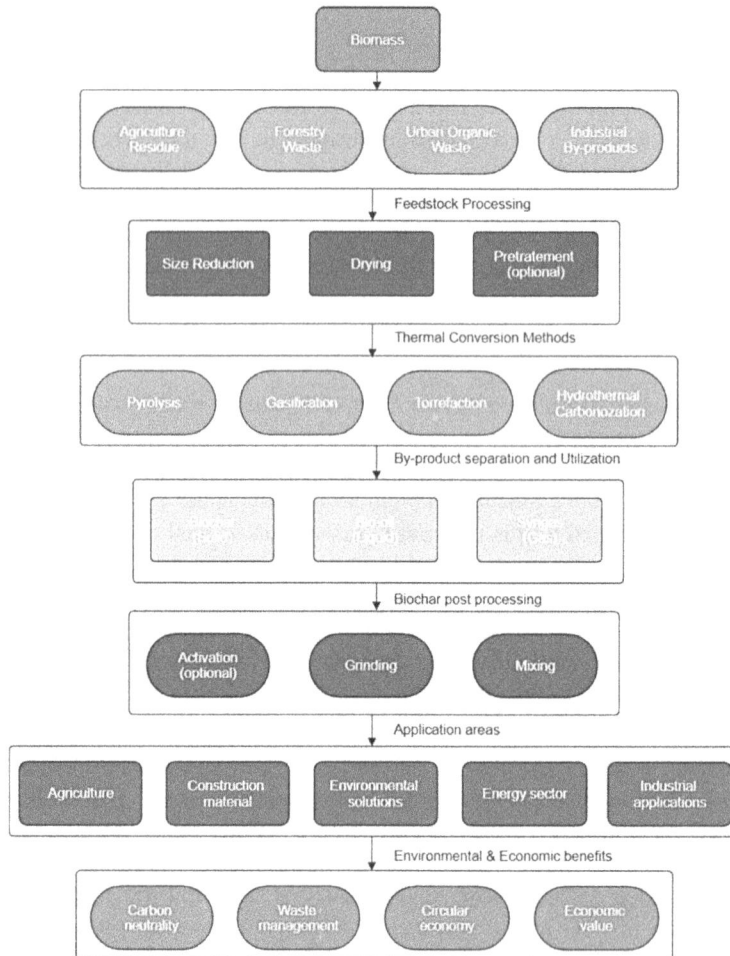

Figure 6.1 Biochar production, application, and benefits

Source: Compiled by authors

Figure 6.2 SEM images of bamboo biochar

Source: Authors

process. This structure enhances the water absorption capacity of biochar, improves its interaction with cement paste, and refines the microstructure, ultimately increasing the workability of concrete composites [10]. However, despite these advantages, the natural porosity and high carbon content of biochar present challenges, potentially compromising the strength of the material and limiting its effectiveness in specific applications.

Table 6.1   Properties of various biochar

Source	Pyrolysis Temp. (°C)	Surface area (m²/g)	Particle size	Pore volume (cm³/g)	pH	C (%)	H (%)	N (%)	Reference
Rise husk	-	500	-	-	7–10	50–80	2–6	0.4–5	[4]
Peanut hull	500	2.15	23 µm	0.0065	8.17	56	2.60	5.20	[12]
Wood dust	300–500	196.92	500µ-5µm	0.071	-	62.25	-	-	[15]
Saw dust	850	8.76	-	0.015	5.16	82.30	3.20	0.40	[20]
Food waste digest	550–750	14.1	0.1–1 mm	0.047	-	32.9	3.0	4.0	[5]
Rise husk	500	-	6.4 µm	-	10.7	61.05	0.53	2.47	[33]
Olive stone	350–750	9.44	0.04–40 µm	-	-	76.60	2.56	0.28	[21]
Mix wood	500	196.15	2 to 300 µm	0.07122	-	87.13	-	-	[14]
Food waste	500	9.70	2 to 300 µm	0.00288	-	70.90	-	-	[14]
Sewage sludge	400	33.4	-	0.783	7.76	4.64	-	-	[23]
Bamboo	700	365.6	0.86 nm	0.310	11	-	-	-	[11]

Source: Compiled by authors

Note: C-Carbon, H-Hydrogen, N-Nitrogen

3. BIOCHAR IN CEMENTITIOUS MATERIALS

3.1 Workability

Biochar incorporation appears to negatively influence the workability of cementitious materials. Sirico et al. [35], Zeidabadi et al. [42], and Sacdalan et al. [44] have worked on this topic and consistently reported reduced workability with increased biochar content. There are two reasons for this: the increased water demand and high surface area of the porous particles in biochar. More hydration water is absorbed by the porous nature of biochar; therefore, more water must be added so that the mix remains uniform. This means that water was retained at the surfaces for up to 15 min. Therefore, less free water is available for ultrafine-grained concrete, as observed from the results of Pauzi et al. [45], where the workability declined from medium to low as the biochar content increased (0–15%). Moreover, even when they added 2.5 wt. %, there was still enough flowability for about 50 min in mortars with biochar, despite gradual hindrance to their workability over time, according to Sirico et al. [34]. A possible solution could be to increase the superplasticizer dosage or water content, according to Sirico et al. [35]. However, adding too much water may lead to deteriorating strength. Finding an optimal biochar content and potentially using workability-enhancing admixtures are crucial for the successful implementation of biochar concrete.

3.2 Compressive Strength

Research has indicated that the incorporation of biochar into cement-based materials can substantially affect the compressive strength of concrete. Sinha et al. [33] demonstrated that incorporating biochar as a cement substitute up to 5% by volume improved the development of C-S-H gel and increased the overall strength. This improvement was attributed to the compaction of the cement paste owing to finer biochar particles. Similarly, Gupta et al. [14] and Puazi et al. [45] noted the positive impacts of biochar on strength and highlighted its role in carbon sequestration. Nair et al. [26] suggested that pretreatment methods, such as acid treatment, could help maintain strength levels. However, Praneeth et al. [28] reported a decrease in strength when sand was replaced with biochar, which was attributed to the increased porosity and weak interfacial zones. Ling et al. [18] stressed the importance of selecting the right type and quantity of biochar, as these factors can influence hydration product formation and optimize compressive strength. Overall, further analysis is required to determine the ideal biochar content, pretreatment methods, and material combinations to strike a balance between strength and sustainability in biochar-incorporated concrete. Unlike compressive strength, the incorporation of biochar into cementitious materials seems to have a positive effect on their flexural strength. Khushnood et al. [17] examined that toughness was notably enhanced while the study also revealed that flexural strength significantly improved by 66%. Praneeth et al. [28] observed a significant increase (26%) in flexural strength for mortars with 20% biochar replacement (B-20) relative to the control mix (B-0). This was because of the crack deflection ability of biochar particles, where an improvement in flexural performance led to an increase in its ability to resist cracking, hence increasing from 5.02 MPa to 6.3 MPa. Additionally, Gupta and Kua [51] and Gupta and Kua [13] supported this by asserting that the enhancement of flexural behaviour is attributed to the high fracture energy provided by the tough structure of biochar's, resulting in the ability to resist the propagation of cracks during loading. Moreover, Sirico et al. [34] observed a negligible influence on flexural properties owing to the use of just 1% biochar. The flexural strength also appeared to be influenced by the curing method. Gupta et al. [13] mentioned that specimens cured in moist conditions possessed higher flexural strength as opposed to those that were air-cured, which could have resulted from pre-soaked biochar particles providing internal curing and reducing the negative effects of air curing on flexural strength.

3.3 Split Tensile Strength

However, it is difficult to evaluate the effects of biochar on the tensile strength of cementitious materials. For example, Zeidabadi et al. [42] found that the splitting tensile strength of concrete increased with rice husk ash up to 15% but weakened at 20% owing to excessive porous microaggregates. It should also be noted that at 28 days, the compressive strength remained almost unchanged, but a reduction in split tensile strength was noticed by Gupta et al. [15] in concrete mixes containing biochar compared to control samples. However, Gupta et al. [12] discovered an increase of 24% in split tensile strength for biochar-incorporated mortar after seven days, which they ascribed to the filler effect and matrix densification by biochar. This is in addition to the fact that curing conditions also matter. In addition, Gupta et al. [13] observed that while dry biochar caused a decrease in tensile strength, these particles did not result in any drop when incorporated in mortars followed by the air-curing process, as opposed to pre-soaked ones.

3.4 Durability

The durability of cementitious materials is influenced by biochar and relies on its content and microstructure. Sinha et al. [33] found that the incorporation of biochar into concrete lowered the water absorption by up to 5% because the microstructure became more compact, thereby reducing the porosity. Conversely, Praneeth et al. [28] found a straight-line increase in water absorption and voids with increased biochar levels, suggesting high levels of porosity at these levels. This underscores the necessity for an optimum biochar level. Moreover, Gupta et al. [12] observed that keeping unwashed biochar salt leads to greater shrinkage owing to a higher dosage of the same. However, in the case of air curing, Gupta et al. [13] found a substantial decrease in water absorption and penetration depth in pre-soaked biochar mortars, suggesting that there could be fewer linked-up pore channels and, hence, better resistance to chemicals. Ling et al. [18] backed up this claim by stating that a 1–3% incorporation of biochar caused an appreciable reduction in carbonation depth as well as the chloride diffusion coefficient, which indicated a significant improvement in concrete's resistance to degradation processes. Nevertheless, the increased shrinkage potential, careful placement, and choice of biochar content are necessary considerations regarding the use of biochar to optimize the water absorption and durability attributes of cementitious materials. Although biochar provides benefits to the concrete environment, additional research is needed to determine its effects on the corrosion of steel reinforcements. Zanotto et al. [41] observed that the unique properties of biochar could affect the immediate vicinity of the rebar. The retention ability of the large-sized pores in biochar can create a moist environment surrounding the steel, possibly leading to early local corrosion. This can also impede oxygen access from the atmosphere and chlorides toward the rebar surface during natural passivation film repair processes. In addition, strong interfacial bonding between biochar particles and cement can impact the overall resistance of reinforcing bars to corrosion. Therefore, careful dosages are required when biochar is added to concrete structures for long-term durability in the long run. Table 6.2 outlines data from different studies, such as various types of biochar, their usage in cementitious materials, the percentage added, curing times, the tests conducted, and the resulting impacts, along with references.

3.5 Microstructure Analysis

Biochar's effect on the production of cementitious hydration products has been studied in many investigations using microstructure analysis. Sinha et al. [33] utilized multiple techniques, such as FESEM, XRD, and TGA, to prove that the introduction of biochar, even at a 5% replacement, increased the development of hydration products. SEM images showed that the 5% biochar sample had more compact and denser hydration products (C-S-H, $Ca(OH)_2$, ettringite), which is supported by the porous nature of biochar. These findings are in agreement with those obtained from XRD and TGA, which showed an increase in CH peak intensity, C-S-H peak intensity, and an elevated peak at 105–540°C for 5% biochar owing to the increased formation of hydration products. Bamboo biochar concrete with different cement replacement levels (0%, 5%, 10%, 15%) is represented in Figure 6.3 using SEM images (scanning electron microscope). This shows that a higher biochar content results in an increased porosity and a more irregular microstructure. Additionally, the study

Table 6.2 Comparison of different studies of biochar cement composite

Biochar used	Biochar Usage	Percentage added	Curing time	Test performed on Cementitious material	Impact	Source
Coconut shell	Cement replacement	1, 2, **5**	7, 28	Compressive strength, Carbon dioxide adsorption	Pozzolanic activity enhanced, Strength not compromised, Carbon dioxide sorption increased	[26]
Bamboo	Cement replacement	1, **2**, 3	3, 7, 28	Compressive strength, Flexural strength	Strength improved by water curing with 0.45 w/c ratio	[31]
Rice husk	Cement replacement	**5**, 10, 15	7, 14, 28	Workability, Compressive strength, Water absorption	Strength increased up to 25%	[45]
Rice husk	Cement replacement	3, **5**, 10	28	Setting time, Porosity, Density, Water absorption, Compressive strength, XRD, FESEM, TGA, Thermal conductivity	Setting time decreased, Porosity reduction up to 8.3%, Water absorption reduction up to 14.2%, Thermal conductivity reduced by 0.86 W/m.K.	[33]
Poultry litter	Sand replacement	**10**, 20, 40	28	Water absorption, Density, Compressive strength, Flexure strength, SEM, XRD, Micro-CT EDS, Thermal conductivity	Density reduced, Flexure strength increased up to 26%, Thermal conductivity reduction up to 26.2%	[28]
wood waste	Cement replacement	1, **3**, 5, 10	7, 28	Compressive strength, Flexural strength, Carbonation, Chloride diffusion, SEM, XRD	Reductions of up to 17.9% and 32% in carbonation depth and chloride diffusion, respectively	[18]
Bamboo	Cement replacement	2, **5**, 8	7, 28	Workability, Compressive strength, CO_2 adsorption, Carbonation	Workability not compromised, CO_2 adsorption increased up to 83.92 %	[44]
Mix woodchip	Additive	**1**, 2.5	7, 28	Workability, Density, Compressive strength, Flexure strength	Fracture energy increased	[34]

Biochar used	Biochar Usage	Percentage added	Curing time	Test performed on Cementitious material	Impact	Source
Poultry litter, Rice husk, Pulp and paper sludge	Cement replacement	**0.1**, 0.25, 0.5, 0.75, 1	7, 14, 28	Water absorption, Compressive strength, Flexure strength, Split tensile strength, ATR-FTIR, SEM, XRD	Water absorption improved	[1]
Rice husk, Sugarcane bagasse	Cement replacement	**5**, 10	28	Compressive strength, Split tensile strength, SEM, EDS	Tensile strength increased up to 78%	[42]
Dry Distillers Grains	Replacement for sand or aggregate	1, 2, **3**, 5, 10, 12, 15	28	Density, Sound absorption Compressive strength, Thermal conductivity	Density decreased, Thermal insulation enhanced, Sound adsorption coefficient ranged from 200 to 2000 Hz	[8]
Grey Borgotaro	Filler, cement replacement	0.8, 1, 1.5, **2**, 2.5	7. 28	Flexure strength, fracture energy	Increased flexure strength and fracture energy	[30]
Waste wood sawdust	Filler	0.5, **1**, 2.5	7, 28	Compressive strength, SEM, XRD, TGA	Strength increased by 8.9%	[37]
Peanut	Admixture	1, **3**	7, 28	Drying shrinkage, Density, Compressive strength	Density and drying shrinkage increased	[16]
Softwood	Cement replacement	0.8, **1**	7, 28	Compressive strength, Flexural strength, FESEM, Fracture energy	Flexure strength and fracture energy reduction	[7]
Wood chips, Rice husk, Olive stone waste	Cement replacement	0.5, 1, 2, **4**	7, 28	Water absorption, Compressive strength, Flexure strength, SEM, XRD, TGA	Workability reduction, no significant effect on water absorption, Hydration products promoted	[21]
Wood waste	Filler	2.5, **5**, 7.5, 10	7, 28, 100, 365	Workability, Density, Shrinkage, Compressive strength, Flexure strength, Split tensile, SEM	30% Compressive strength increased after 365 days, No significant effect on tensile strength	[35]
Corn stover	Filler	2, **4**, 6, 8	3 (CO2 curing), 28 (Water curing)	Compressive test, XRD, FTIR, SEM	Increased CO2 uptake and carbonation depth, Early strength improved	[27]

Biochar used	Biochar Usage	Percentage added	Curing time	Test performed on Cementitious material	Impact	Source
Rice husk	Filler	**10**, 20, 30, 40	1, 7, 28	Workability, Density, Capillary absorption, Porosity, Compressive strength, Hydration kinetics	Lower permeability by 9–25%, 15 to 20% higher retention of strength	[15]
Waste peanut shell	Cement replacement	1, **3**	7, 28	Degree of hydration, Compressive strength, Flexure strength, Carbonation, Water absorption, Sorptivity, shrinkage, SEM	Carbon sequestration improved by 7–13 % (by weight of cement), High drying shrinkage	[12]
Hardwood	Cement replacement	**5**, 10, 15, 20	7, 28	Workability, Water absorption, Compressive strength	Workability reduction, better water retention capacity	[6]
Rice husk, coconut shell, bamboo	Cement replacement	2.5, 5, 7.5, **10**	28	Compressive strength, Thermal conductivity, SEM, Infrared heat transfer	Thermal conductivity reduction, Thermal performance improved	[39]
Coffee powder, hazelnut shells	Cement replacement	**0.5, 0.8**, 1	7, 28	Compressive strength, Flexural strength	Improved overall mechanical properties	[29]
Raw peanut shell and hazelnut shell	Additive	0.025, 0.05, **0.08, 0.2**, 0.5, 1	28	Fracture toughness, Flexural strength, Shielding effectiveness, Raman analyses, FESEM	Increased toughness and flexural strength, 353% higher shielding capability	[17]
Alum sludge	Cement replacement	**5**, 10, 15, 20	7, 28, 90	Slump cone, Compressive strength, Split tensile strength, flexure strength, Permeability, Abrasion resistance, SEM, XRD, FTIR analysis	Improved durability, Reduction of Global warming potential by 32.2 %, Scored 0.95 in Multi-criteria decision-making framework	[22]

Source: Compiled by authors

Note: bold figures in column "Percentage added" *signify optimum blending % based on 28th day strength* noted that higher levels of biochar resulted in a reduced bond strength at the interfacial transition zone and higher voids, which may result in lower strength. Nevertheless, it has been proposed that monophasic biochar inclusion can improve the properties of lightweight concrete, provided there is proper dispersion and hydration.

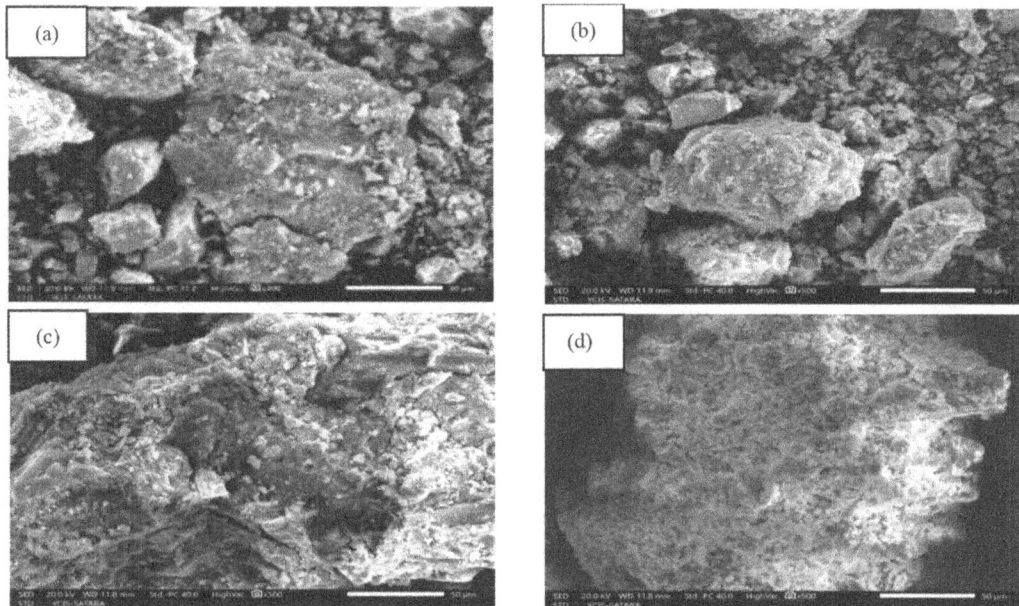

Figure 6.3 SEM images of bamboo biochar concrete (a) 0% (b) 5% (c) 10% (d) 15

Source: Authors

Using XRD and TGA techniques, Maljaee et al. [21] confirmed these results by reporting the synthesis of additional calcium hydroxide and carbonate and the role played by biochar as a site for nucleation in hydrated pastes. Nevertheless, Yang and Wang [40] observed different outcomes through their XRD peaks, where it was not clear if more hydration products were formed as a result of adding biochar compared to the control paste. However, they realized that the voids created by the incorporated biochar provided space for the formation of hydrate-containing products. Homogeneity in the dispersion of biochar particles, with superior interactions between the surface of the biochar and the cement matrix, was found by Sirico et al. [35]. In addition, Ling et al. [18] noted that at 3% biochar, there is a microfiller effect, which leads to denser and more uniform microstructures with increased nucleation sites for cement hydration.

3.6 Carbon Dioxide Sequestration

Studies have indicated how biochar can enhance CO_2 capture. Adding just 1% biochar to concrete, for example, will raise its CO_2 uptake by 42% [49], while the study carried by Javed et al. [47] showed that biochar itself exhibits an adsorption capacity of approximately 1.25–1.72 mmol/g of CO_2. According to Nair et al. [26], the combustion of a candle resulted in up to 75.30% adsorption in mortar cubes containing biochar, and Sacdalan et al. [44] employed the same technique, which gave similar results of 83.92% and 77.88% for single and double candles, respectively, with 8% biochar content in concrete. Dissanayake et al. [46] also provided evidence demonstrating that a biochar's micropore surface area determines its CO_2 adsorption ability, with engineered biochar displaying even greater potential capture as a result of controlled modification of their surfaces; therefore, these findings have great importance in determining carbon sequestration mechanisms in soils and the atmosphere.

4. ENVIRONMENTAL AND ECONOMIC ANALYSIS

Biochar is a sustainable construction solution that minimizes CO_2 emissions and enhances the attributes of building materials. According to Praneeth et al. [28], the carbon emission factor of biochar production is negative, whereas partially replacing sand with biochar in cement mortars significantly lowers the overall CO_2 footprint compared to conventional materials. In practical terms, this amounts to Gupta et al.'s [14] suggestion that using 40% biochar instead of cement reduces the CO_2 emissions of plastered areas by 20%. Furthermore, it can be seen that even though more studies are required on the combined economics of biochar and silica fume as replacements, they offer a potentially cheaper way of reducing cement consumption. Apart from these environmental gains, biochar has something extra to offer from a functional perspective. Zhang et al. [43] showed that the inclusion of biochar into pervious concrete increases its water uptake capacity, thereby mitigating the urban heat island effect. Biochar's unique structure also helps in reducing thermal conductivity in cement composites, hence enhancing insulation properties by 25%. This is consistent with Tan et al.'s [36] review, which revealed this aspect as one where biochar could mitigate global warming and store carbon in construction materials. Biochar provides an alternative source of cement through utilization of waste resources.

5. APPLICATION OF BIOCHAR-CEMENT COMPOSITE

Biochar is a promising, sustainable, and functional additive for cement applications. It can act as a filler to improve the flow and overall performance of the cement composites. Biochar can boost the strength, durability, and fracture resistance of concrete, potentially leading to the development of carbon-negative building materials. Furthermore, biochar can be used as an alternative aggregate in concrete production, reducing its environmental impact while maintaining desirable properties. Another benefit is the ability of biochar cement to immobilize hazardous waste materials. The inherent properties of biochar also contribute to moisture regulation and carbon sequestration within concrete. Moreover, biochar can be used in special types of concrete, such as self-sensing concrete, 3D-printable concrete, and self-healing concrete [43]. Biochar is not only versatile but also environmentally safe for construction. For instance, Aman et al. [2] indicated that it can be used in agriculture, wastewater treatment, and chemical recovery, whereas its particular characteristics make it a valuable substance for the industry. For example, biochar is utilized in different composites such as geo-polymers, bitumen materials, red clay binders, and inorganic clay composites.

6. CHALLENGES AND FUTURE SCOPE

Although biochar concrete offers a promising approach to sustainable construction, there are still challenges to address and exciting areas for future exploration.

- Optimizing dosage and type of biochar: It must be realized that finding the right balance between maximizing CO_2 capture and retaining the workability and strength of concrete is important. Different types of biochar may require dosage adjustments to obtain optimal results.
- Durability: More work should be done to identify how well biochar concrete performs over a long period under different environmental conditions, such as freeze-thaw cycles, moisture exposure, and chemical attack.
- Production techniques: The ability to modify various aspects (surface area and porosity) of biochar by employing activation or modification could enhance CO_2 capture and potentially improve the performance of concrete. Maintain the reliability and performance of biochar in concrete applications through consistent production methods and quality control measures.
- Combined cementitious materials: Combining biochar with other supplementary cementitious materials, such as fly ash or silica fumes, could offer synergistic effects that optimize both environmental and performance benefits.
- Sustainability: Extensive evaluation of the environmental impact of biochar production, transportation, and disposal is required throughout the life of concrete. Research on large-scale, cost-efficient biochar production methods is crucial for their widespread adoption in the construction industry.

7. CONCLUSION

The incorporation of biochar as a filler or supplementary cementitious material offers a promising solution for both the construction sector and environment. In this study, the chemical and physical characteristics of biochar are discussed, along with their effects on the properties of concrete. The characteristics of biochar are significantly affected by the type of feedstock, temperature, and method of production. For high biochar yields, slow pyrolysis conducted between 300–500°C is preferred. Biochar's presence within cement paste enhances hydration product formation by providing a large surface area. Moreover, its water absorption characteristics render it suitable for internal curing purposes.

Concrete performance is significantly influenced by the type, amount, and particle size of biochar. Although an increase in biochar content reduces workability and increases drying shrinkage, it has been reported that an optimal dosage of 5% biochar should be used. In addition, carbon sequestration may be improved while maintaining strength by utilizing activation or modification methods. Moreover, the flexural strength, tensile strength, fracture energy, thermal insulation, and sound insulation were enhanced by biochar incorporation. Initially, water absorption decreases with increasing content, but ultimately increases because it is hydrophilic. The potential of biochar to create carbon-negative concrete, combined with its status as a waste material, makes it an attractive and economical alternative. In summary, this review highlights biochar as an eco-friendly material that enhances the longevity of structural concrete, promotes waste recycling, reduces carbon emissions, and helps combat climate change.

REFERENCES

[1] Akhtar, A., & Sarmah, A. K. (2018). Novel biochar-concrete composites: Manufacturing, characterization and evaluation of the mechanical properties. *Science of the Total Environment*, 616–617, 408–416.
[2] Aman, A. M. N., Selvarajoo, A., Lau, T. L., & Chen, W. (2022). Biochar as cement replacement to Enhance concrete Composite properties: a review. *Energies*, 15(20), 7662.

[3] Bhatia, S. K., Palai, A. K., Kumar, A., Bhatia, R. K., Patel, A. K., Thakur, V. K., et al. (2021). Trends in renewable energy production employing biomass-based biochar. *Bioresource Technology*, 340, 125644.

[4] Campos, J., Fajilan, S., Lualhati, J., Mandap, N., & Clemente, S. (2020). Life cycle assessment of biochar as a partial replacement to Portland cement. *Earth and Environmental Science*, 479(1), 012025.

[5] Chen, L., Zhu, X., Zheng, Y., Wang, L., Poon, C. S., & Tsang, D. C. (2024). Development of high-strength lightweight concrete by utilizing food waste digestate based biochar aggregate. *Construction & Building Materials*, 411, 134142.

[6] Choi, W. C., Yun, H. D., & Lee, J. Y. (2012). Mechanical properties of mortar containing Bio-Char from pyrolysis. *Han-guk Gujomul Jindan Yuji Gwalli Gonghakoe Nonmunjip/Journal of the Korea Institute for Structural Maintenance and Inspection*, 16(3), 67–74.

[7] Cosentino, I., Restuccia, L., Ferro, G. A., & Tulliani, J. (2019). Type of materials, pyrolysis conditions, carbon content and size dimensions: The parameters that influence the mechanical properties of biochar cement-based composites. *Theoretical and Applied Fracture Mechanics*, 103, 102261.

[8] Cuthbertson, D., Berardi, U., Briens, C., & Berruti, F. (2019). Biochar from residual biomass as a concrete filler for improved thermal and acoustic properties. *Biomass & Bioenergy*, 120, 77–83.

[9] Fawzy, S., Osman, A. I., Doran, J., & Rooney, D. W. (2020). Strategies for mitigation of climate change: a review. *Environmental Chemistry Letters*, 18(6), 2069–2094.

[10] G, N. P. K., Tee, K. F., Gimbun, J., & Chin, S. C. (2023). Biochar in cementitious material—A review on physical, chemical, mechanical, and durability properties. *AIMS Materials Science*, 10(3), 405–425.

[11] Ge, Q., Li, P., Liu, M., Xiao, G., Xiao, Z., Mao, J., & Gai, X. (2023). Removal of methylene blue by porous biochar obtained by KOH activation from bamboo biochar. *Bioresources and Bioprocessing*, 10(1).

[12] Gupta, S., & Kashani, A. (2021). Utilization of biochar from unwashed peanut shell in cementitious building materials – Effect on early age properties and environmental benefits. *Fuel Processing Technology*, 218, 106841.

[13] Gupta, S., & Kua, H. W. (2018). Effect of water entrainment by pre-soaked biochar particles on strength and permeability of cement mortar. *Construction & Building Materials*, 159, 107–125.

[14] Gupta, S., & Kua, H. W. (2019). Combination of biochar and silica fume as partial cement replacement in mortar: Performance evaluation under normal and elevated temperature. *Waste and Biomass Valorization*, 11(6), 2807–2824.

[15] Gupta, S., & Kua, H. W. (2020). Application of rice husk biochar as filler in cenosphere modified mortar: Preparation, characterization and performance under elevated temperature. *Construction & Building Materials*, 253, 119083.

[16] Han, T. (2020). Application of peanut biochar as admixture in cement mortar. IOP Conference Series. *Earth and Environmental Science*, 531(1), 012061.

[17] Khushnood, R. A., Ahmad, S., Restuccia, L., Spoto, C., Jagdale, P., Tulliani, J., & Ferro, G. A. (2016). Carbonized nano/microparticles for enhanced mechanical properties and electromagnetic interference shielding of cementitious materials. *Frontiers of Structural and Civil Engineering*, 10(2), 209–213.

[18] Ling, Y., Wu, X., Tan, K., & Zou, Z. (2023). Effect of biochar dosage and fineness on the mechanical properties and durability of concrete. *Materials*, 16(7), 2809.

[19] Liu, W., Li, K., & Xu, S. (2022). Utilizing bamboo biochar in cement mortar as a bio-modifier to improve the compressive strength and crack-resistance fracture ability. *Construction & Building Materials*, 327, 126917.

[20] Madzaki, H., KarimGhani, W. A. W. A., NurZalikhaRebitanim, N., & AzilBahariAlias, N. (2016). Carbon dioxide adsorption on Sawdust Biochar. *Procedia Engineering*, 148, 718–725.

[21] Maljaee, H., Paiva, H., Madadi, R., Tarelho, L. A., Morais, M., & Ferreira, V. M. (2021). Effect of cement partial substitution by waste-based biochar in mortars properties. *Construction & Building Materials*, 301, 124074.

[22] Mekky, K. M., Ibrahim, M. G., Sharobim, K., Fujii, M., & Nasr, M. (2024). Evaluating environmental and Economic benefits of using biochar in Concrete: a life cycle assessment and multi-criteria decision-making framework. *Case Studies in Construction Materials*, 21, e03712.

[23] Méndez, A., Terradillos, M., & Gascó, G. (2013). Physicochemical and agronomic properties of biochar from sewage sludge pyrolysed at different temperatures. *Journal of Analytical and Applied Pyrolysis*, 102, 124–130.

[24] Mikulčić, H., Klemeš, J. J., Vujanović, M., Urbaniec, K., & Duić, N. (2016). Reducing greenhouse gasses emissions by fostering the deployment of alternative raw materials and energy sources in the cleaner cement manufacturing process. *Journal of Cleaner Production*, 136, 119–132.

[25] Murtaza, G., Ditta, A., Ullah, N., Usman, M., & Ahmed, Z. (2021). Biochar for the management of nutrient impoverished and metal contaminated soils: preparation, applications, and prospects. *Journal of Soil Science and Plant Nutrition*, 21(3), 2191–2213.

[26] Nair, J. J., Shika, S., & Sreedharan, V. (2020). Biochar amended concrete for carbon sequestration. IOP Conference Series. *Materials Science and Engineering*, 936(1), 012007.

[27] Praneeth, S., Guo, R., Wang, T., Dubey, B. K., & Sarmah, A. K. (2020). Accelerated carbonation of biochar reinforced cement-fly ash composites: enhancing and sequestering CO2 in building materials. *Construction & Building Materials*, 244, 118363.

[28] Praneeth, S., Saavedra, L., Zeng, M., Dubey, B. K., & Sarmah, A. K. (2021). Biochar admixtured lightweight, porous and tougher cement mortars: Mechanical, durability and micro computed tomography analysis. *Science of the Total Environment*, 750, 142327.

[29] Restuccia, L., & Ferro, G. A. (2016). Promising low cost carbon-based materials to improve strength and toughness in cement composites. *Construction & Building Materials*, 126, 1034–1043.

[30] Restuccia, L., Ferro, G. A., Suarez-Riera, D., Sirico, A., Bernardi, P., Belletti, B., et al. (2020). Mechanical characterization of different biochar-based cement composites. *Procedia Structural Integrity*, 25, 226–233.

[31] Selvanathan, N. P., Gunasekaran, N. P. K., Lai, N. Y. F., Anand, N. N., & Chin, N. S. C. (2024). Curing Effect on the Strength of Cement Mortar with Bamboo Biochar. International *Journal of Nanoelectronics and Materials*, 17(3), 465–471.

[32] Senadheera, S. S., Gupta, S., Kua, H. W., Hou, D., Kim, S., Tsang, D. C., & Ok, Y. S. (2023). Application of biochar in concrete: a review. *Cement & Concrete Composites*, 143, 105204.

[33] Sinha, S., Pandey, A., B, S. N., & Prasad, B. (2023). Preliminary Study of Agricultural Waste as Biochar Incorporated into Cementitious Materials. *Journal of Architectural Environment & Structural Engineering Research*, 6(2), 59–79.

[34] Sirico, A., Bernardi, P., Belletti, B., Malcevschi, A., Dalcanale, E., Domenichelli, I., et al. (2020). Mechanical characterization of cement-based materials containing biochar from gasification. *Construction & Building Materials*, 246, 118490.

[35] Sirico, A., Bernardi, P., Sciancalepore, C., Vecchi, F., Malcevschi, A., Belletti, B., et al. (2021). Biochar from wood waste as additive for structural concrete. *Construction & Building Materials*, 303, 124500.

[36] Tan, K., Qin, Y., & Wang, J. (2022). Evaluation of the properties and carbon sequestration potential of biochar-modified pervious concrete. *Construction & Building Materials*, 314, 125648.

[37] Wang, L., Chen, L., Tsang, D. C., Guo, B., Yang, J., Shen, Z., Hou, D., Ok, Y. S., & Poon, C. S. (2020). Biochar as green additives in cement-based composites with carbon dioxide curing. *Journal of Cleaner Production*, 258, 120678.

[38] Yaashikaa, P., Kumar, P. S., Varjani, S., & Saravanan, A. (2020). A critical review on the biochar production techniques, characterization, stability and applications for circular bioeconomy. *Biotechnology Reports*, 28, e00570.

[39] Yang, S., Wi, S., Lee, J., Lee, H., & Kim, S. (2019). Biochar-red clay composites for energy efficiency as eco-friendly building materials: Thermal and mechanical performance. *Journal of Hazardous Materials*, 373, 844–855.

[40] Yang, X., & Wang, X. (2021). Hydration-strength-durability-workability of biochar-cement binary blends. *Journal of Building Engineering*, 42, 103064.

[41] Zanotto, F., Sirico, A., Balbo, A., Bernardi, P., Merchiori, S., Grassi, V., et al. (2024). Study of the corrosion behaviour of reinforcing bars in biochar-added concrete under wet and dry exposure to calcium chloride solutions. *Construction & Building Materials*, 420, 135509.

[42] Zeidabadi, Z. A., Bakhtiari, S., Abbaslou, H., & Ghanizadeh, A. R. (2018). Synthesis, characterization and evaluation of biochar from agricultural waste biomass for use in building materials. *Construction & Building Materials*, 181, 301–308.

[43] Zhang, Y., He, M., Wang, L., Yan, J., Ma, B., Zhu, X., et al. (2022). Biochar as construction materials for achieving carbon neutrality. *Biochar*, 4(1).

[44] Sacdalan J. L. D., Bartolo C. A. D., Bautista C. A. D., Cruz G. G., Muñoz J. C., (2023). Bamboo-derived biochar as partial cement replacement in concrete for carbon dioxide capturing and sequestration, *Chemical Engineering Transactions*, 106, 409–414.

[45] Nur Irfah Mohd Pauzi, Abdul Shakoor Musa and Mohd Shahril Mat Radhi, (2020). Biochar usage for improving concrete mix as cement replacement, *International Journal of Advanced Research in Engineering and Technology*, 11(8), 594–601.

[46] Dissanayake, P. D., Choi, S. W., Igalavithana, A. D., Yang, X., Tsang, D. C., et al. (2020). Sustainable gasification biochar as a high efficiency adsorbent for CO2 capture: A facile method to designer biochar fabrication. *Renewable and Sustainable Energy Reviews*, 124, 109785.

[47] Hasan, I. M. U., Javed, H., Hussain, M. M., Shakoor, M. B., Bibi, I., Shahid, M., et al. (2022). Biochar/nano-zerovalent zinc-based materials for arsenic removal from contaminated water. *International Journal of Phytoremediation*, 25(9), 1155–1164.

[48] Hussain, R., Kumar, H., Bordoloi, S., Jaykumar, S., Salim, S., Garg, A., Ravi, K., Sarmah, A. K., Gogoi, N., & Sreedeep, S. (2024). Effect of biochar type and amendment rates on soil physicochemical properties: Potential application in bioengineered structures. *Advances in Civil Engineering Materials*, 13(1), 1–20.

[49] Mishra, R. K., & Mohanty, K. (2023). A review of the next-generation biochar production from waste biomass for material applications. *The Science of the Total Environment*, 904, 167171.

[50] Zhou, B., Perel, P., Mensah, G. A., & Ezzati, M. (2021). Global epidemiology, health burden and effective interventions for elevated blood pressure and hypertension. *Nature Reviews Cardiology*, 18(11), 785–802.

[51] Gupta, S., & Kua, H. W. (2017). Biochar as a carbon sequestering construction material in cementitious mortar. *Academic Journal of Civil Engineering*, 35(2), 563–568.

[52] Mensah, R., Shanmugam, V., Narayanan, S., Razavi, N., Ulfberg, A., Blanksvärd, T., Sayahi, F., et al. (2021). Biochar-Added Cementitious Materials—A review on mechanical, thermal, and environmental properties. *Sustainability*, 13(16), 9336.

Advances in Construction, Real Estate, Infrastructure and Project Management – Anil Kashyap et al. (eds)
© 2026 Taylor & Francis Group, London, ISBN 978-1-041-13433-6

7

Exploring the Scope of Various Integrated Methods of Lean Construction and BIM in the Construction Industry: A Systematic Literature Review Using PRISMA and TCCM Approaches

Ullas M.[1], Ayush Jha[2],
D. Sai Vaishnav[3], Akash Deep[4]
Student,
NICMAR University Pune

Deepak M. D.[5]
Assistant Professor,
NICMAR University Pune

■ **ABSTRACT:** The construction sector plays a vital role in the economy and adds significantly to the Gross Domestic Product (GDP) of countries around the world. However, the industry faces challenges such as inefficiency, cost and time related delays, which affect its growth. To tackle this problem, a lot of research has been going on to improve efficiency and effectiveness in the building sector. Therefore, this research emphasizes on examining how Lean Construction (LC) and Building Information Modeling (BIM) are combined in the building sector. A systematic literature review was conducted using the PRISMA (Preferred Reporting Items for Systematic Reviews and Meta-Analysis) approach to carefully find, screen, and analyze relevant information from journals, conferences, and reports. A total of sixty-one papers have been identified from the study. Furthermore, the scrutinized papers were analyzed using the TCCM (Theme, Context, Characteristics, Methodology) approach to identify key themes, contexts, characteristics, and methodologies related to the integration of Lean Construction and BIM in the building industry. Through the synthesis of findings from the systematic literature review, this research seeks to provide significant insights into the various methods, approaches, difficulties, and possibilities related to the combination of Lean Construction and BIM in construction.

■ **KEYWORDS:** Lean construction, Building information modelling, PRISMA, TCCM, Construction industry

[1]P2370187@student.nicmar.ac.in, [2]P2370183@student.nicmar.ac.in, [3]P2370180@student.nicmar.ac.in, [4]P2370281@student.nicmar.ac.in, [5]deepakmd@nicmar.ac.in

DOI: 10.1201/9781003669814-7

1. INTRODUCTION

1.1 Background

Construction industry greatly contributes to the economic growth of many countries in the world by improving the Gross Domestic Product (GDP) and Gross National Product (GNP), because of which many developing countries often prioritize the construction sector for their progress. The construction sector directly influences various industries such as steel, cement, stones, paints, and many more which are essential for construction [13]. Thus, construction projects need efficient use scarce items like capital, time, and resources. The chosen delivery system will play a significant role in the success of the project. Traditional delivery techniques are less effective in managing complicated projects, which leads to cost overruns, schedule overruns, quality and environmental issues, and disagreements between various stakeholders. Such methods also face many challenges in maintaining and coordinating throughout the process. The integration of Lean Construction (LC) and Building Information Modelling (BIM) holds considerable potential to revolutionize the construction industry by combining their respective strengths. LC is defined as a way to minimize the waste of materials, and time, and maximize the value of the project [8]. BIM is defined as a revolutionary development that is quickly reshaping the Architect-Engineering-Construction (AEC) industry [1]. Adopting integrated LC and BIM approaches can improve construction projects by boosting efficiency, reducing waste, improving quality, and fostering better collaboration among stakeholders. Despite the potential, the practical implementation of integrated LC and BIM is challenged by various problems such as opposition to change, absence of professional knowledge, and high initial investment.

Practicing traditional approaches, Indian construction sector facing many problems to achieve desired success. Many construction projects are facing the delays in delivery of the project in India. According to this data, construction delays are a big issue in the Indian construction industry, with significant economic implications. Numerous scholars have sought to address issues of overruns in construction projects, resulting in developing a significant body of literature. Various studies have been done to address this topic. It has seen that 9 out of 10 infrastructure projects are deviating from the desired outcome. With reference the study conducted in Australia, average cost overrun for a construction project was 12.22% [6]. It was claimed that infrastructure projects in Columbia have cost overruns of up to 110% and delays of up to 342% [6]. India has invested 4.9 trillion dollars in construction projects, which is a substantial amount of money and cannot be simply neglected. This research paper seeks to systematically find various integrated methods of LC and Building Information Modelling within the construction sector, using the PRISMA framework and TCCM approach. By conducting a systematic literature review, this study seeks to find different integration methods, assess their outcomes, and understand the challenges to the implementation.

2. LITERATURE REVIEW

2.1 Lean Construction (LC)

A project delivery method known as Lean Construction, emphasises on increasing value and reducing waste during the construction process. It aims to improve how projects are delivered by promoting quality, efficiency, and teamwork among everyone involved. Lean principles, tools, and techniques were developed mainly to address two key issues: poor collaboration among people and inefficient planning processes in construction.

The lean process helps in adding value to construction, which brings a high return on investment by reducing waste [9]. One major benefit is that it helps complete projects on time, avoiding major delays. This reduces conflicts between stakeholders, especially clients and contractors [14]. It also boosts productivity, supports better communication and teamwork, and minimizes mistakes and rework. Moreover, the construction industry is growing faster because of these lean principles.

2.2 Building Information Modelling (BIM)

To improve delivery of the project, the building sector uses several information and communication technologies (ICT) tools [3]. Studies show that as ICT usage increases, productivity in construction also improves. The most well-known example of effective ICT use in the building sector is BIM [4].

Building Information Modelling (BIM) is an advanced technology that that helps create, share, and analyze digital models of buildings. This improves teamwork, reduces mistakes, and leads to better project results [11, 12]. BIM is also a tool for visualizing every part of a project, including structures, components, and materials, along with their details and features [7]. BIM improves efficiency, reduces errors, and supports better decision-making throughout the construction phases, making it a valuable tool in the modern building sector.

2.3 Integration of LC and BIM Concepts

For projects to be more efficient, the architectural, engineering, and construction (AEC) departments must be integrated, which is a key challenge in the construction sector. This issue can be resolved by integrating LC techniques with BIM. Such integration offers many benefits, including better collaboration and communication among stakeholders, smoother workflows, reduced waste, and improved decision-making. Together, LC and BIM create a powerful synergy that enhances project performance in cost, schedule, quality, and sustainability outcomes [2].

The construction sector generates significant waste through unsafe practices, poor time management, lack of visual tools, unnecessary movements, and poor coordination, which motivates the integration of LC and BIM. Traditional project delivery methods struggle with complex projects, leading to cost and time related delays, quality compromises, and disagreements between the parties involved. The building sector could change as a result of integration, leading to significant improvements in project results [17, 18].

3. RESEARCH METHODOLOGY

3.1 Systematic Literature Review (SLR) using "PRISMA"

A systematic literature review (SLR) facilitates recognizing of relevant papers according to the quality of the study and helps in transparently synthesizing the findings. The first stage of SLR involves using the PRISMA framework that helps to systematically identify and screen articles based on the focus area. The detailed methodology is presented in Figure 7.1. The first step includes defining the search strategy to identify the articles from the Scopus database. Based on the keyword search from the database, around 574 relevant articles were identified. Next, certain screening criteria were applied to obtain articles based on geographical region, year of publication, and research methods to gain a clearer understanding of the progress made on the topic.

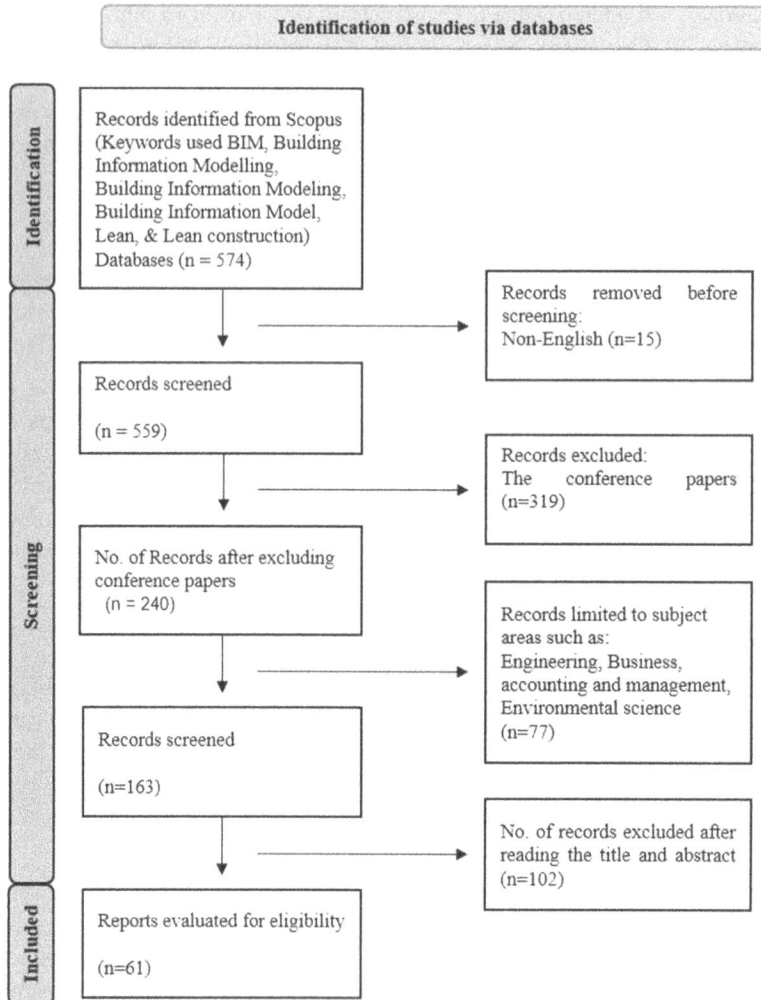

Figure 7.1 PRISMA framework adopted for this study

Source: Authors

Based on the detailed PRISMA framework, the results comprising of geographical region are shown in Figure 7.2. With 11 publications on the integration of LC and BIM, the UK emerged as the country with the largest concentration of articles, followed by Turkey, the USA, and India (4 each). Countries like Brazil, Peru, Syria, and Sri Lanka have also contributed to the research, which reveals that the studies on integration techniques are not only limited to developed countries but have gained traction in developing countries as well.

Analysis of the year of publication revealed that the research on integration techniques has gained traction since the beginning of 2021, as shown in Figure 7.3. Alongside, methodology of publication analyses of the 61 reports has also been done as shown in Figure 7.4. According to the data, 32.39% of research chose to use surveys in their studies. A literature review was applied in 28.16% of the research methods, and case studies were employed in 25.16% of the methodologies used. Only

Figure 7.2 Articles published over regions

Source: Authors

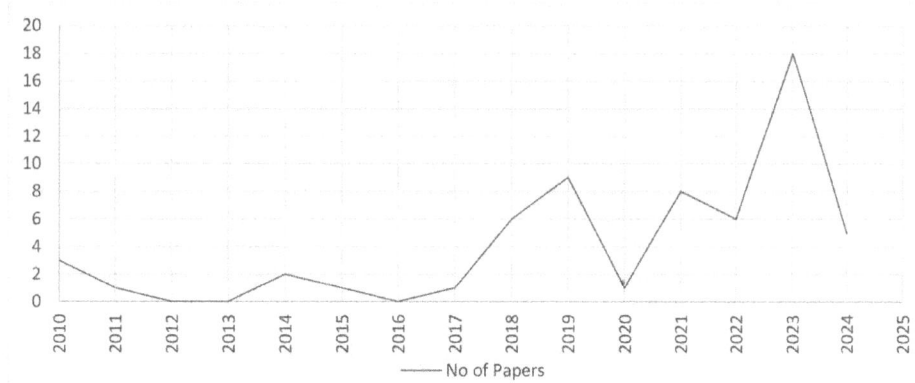

Figure 7.3 Articles published over years

Source: Authors

14.08% of conceptual studies are adopted, which is relatively low. Also, 66.19% of the research methodologies adopted have been since the beginning of 2021, which proves that more detailed research is being done on the topic.

The PRISMA framework was instrumental in ensuring a systematic and unbiased identification of relevant studies for this review. Its structured process of identification, screening, and inclusion allowed the study to comprehensively explore the combination of Lean Construction (LC) with Building Information Modeling (BIM). For instance, through PRISMA, 61 high-quality articles were identified after a rigorous screening process that included filtering out non-English articles, irrelevant subject areas, and studies lacking sufficient relevance based on titles and abstracts.

However, the PRISMA methodology comes with its own set of limitations. It involves extensive literature searches, data extraction, and analysis which can be found difficult. If the search strategy is not

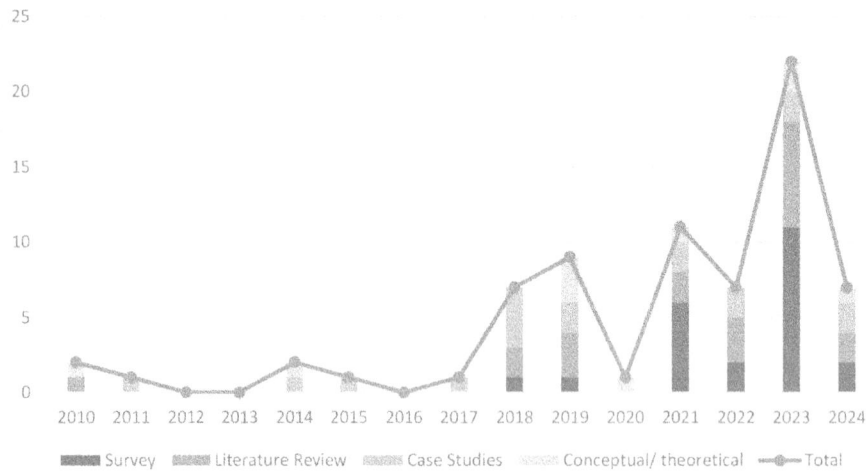

Figure 7.4 Methodologies adopted during data collection

Source: Authors

exhaustive or if relevant databases are not included, important studies might be missed. Additionally, the extensive effort required for screening and analysis can make the process time-intensive.

3.2 Literature Review using "TCCM" Method

The second stage of SLR involves the TCCM (Theory, Context, Characteristics, Methodology) model which helps organize and analyze literature by focusing on theoretical underpinnings, contextual factors, specific characteristics, and methodologies of the studies being reviewed. The motive of this model to identify the gaps.

3.3 Theory

Benefits of integrating BIM with various LC techniques, namely Big Room approach, Integrated Project Delivery (IPD) have been explored in the papers. IPD has been focused more on all the possible methods, that integrate people, systems, and processes to optimize project results by reducing waste and maximizing efficiency through all the project phases. This method can be used to develop sustainable construction projects [7]; and can be used in the design phase [5] of the project, which would help in optimizing the cost of projects.

3.4 Context

The research papers give an idea of how the integration of LC and BIM varies across different countries as the papers have different focus areas and geographical locations. The papers also analyze differences in adoption and implementation across small, medium, and large enterprises. Application of integration in various sectors of the construction industry such as residential, commercial, and infrastructure projects have been explored.

3.5 Characteristics

The exploration of innovative methodologies can enhance project efficiency and performance. It emphasizes the waste reduction and improvement of the project by fostering collaboration among all the stakeholders [9]. The papers examine how IPD fosters collaboration, risk-sharing, and making choices regarding projects [15]. Additionally, they explore the modular construction of projects, including the planning, execution, and evaluation, using the case study approach [10].

3.6 Methodology

The papers include a case study analysis [16], a literature review, a survey, interviews [5], and workshops to gather insights into the implementation and effectiveness of construction projects. Qualitative analysis was used to find challenges, facilitators, methods, and benefits affiliated to the integration of innovative methodologies. Creating and validating framework adoption models would support the implementation of combining LC and BIM techniques.

The TCCM framework's structured approach helps identify research gaps and trends while offering a comprehensive understanding of both conceptual and real-world aspects. However, its application presents some challenges. Categorizing research into the four TCCM components requires subjective interpretation, as different researchers may have varying views on theoretical, contextual, or methodological elements. This subjectivity can lead to inconsistencies in the analysis, especially when dealing with studies that have overlapping or unclear characteristics. As a result, the findings may vary depending on the researcher's viewpoint.

4. RESULTS AND DISCUSSIONS

4.1 Findings

The integration of LC and BIM offers significant benefits for the construction industry. While techniques like the LPS, IPD, and Big Room existed before BIM, their effectiveness is enhanced when combined with BIM. For example, 4D and 5D BIM, which include time and cost factors, allow planners to view the project schedule in real-time, test different budget scenarios, and identify potential issues early. This results in more accurate cost and time estimates, helping to prevent overruns.

The visual and data-rich nature of BIM improves communication and collaboration by making plans and schedules easier to understand for all stakeholders. It also acts as a central source of information, capturing everything from design details to material properties, fostering transparency and teamwork. Additionally, integrating BIM helps finding and addressing potential issues during the design and planning stages, enabling better risk management. Despite the promising outcomes, the implementation of these integrated techniques comes with a set of challenges.

4.2 Challenges for Implementing LC and BIM Concepts

Applying LC and BIM in the building sector offers several advantages but also presents various challenges. Lack of government support, high investments, and resistance to change are some of the few challenges. One critical factor for successful implementation is unwavering backing and dedication from high level management within an organization. This leadership involvement is essential for fostering adoption.

However, the integration of LC and BIM often involves substantial upfront costs, which many organizations hesitate to invest in due to concerns over return on investment. A lack of awareness, insufficient professional expertise, and outdated educational approaches further contribute to the difficulties, creating educational barriers. Additionally, limited government investment in research and development, coupled with resistance from individuals and organizations towards adopting new technologies, further impedes progress.

Such obstacles prevent the full potential of LC and BIM from being realized. Nevertheless, their integration can result in major improvements in project outcomes. These include reduced costs and schedules, minimized waste, enhanced quality and safety standards, improved collaboration among stakeholders, and more sustainable project delivery practices.

5. CONCLUSION AND FUTURE DIRECTIONS

This review explored combining Lean Construction (LC) to Building Information Modelling (BIM) in the construction sector using PRISMA and TCCM approaches. Techniques like VCD, IPD, and LPS benefit from BIM integration, improving scheduling, waste reduction, quality, safety, and collaboration. However, challenges like cultural resistance, lack of training, and high costs hinder adoption. Addressing delays, cost overruns, inefficiencies, and stakeholder issues is crucial for sustainable growth in the sector.

Building upon the insights derived from this literature review, our future research will focus on quantitatively analysing the awareness, implementation, and integration of LC and BIM within the construction sector of India. The research will aim to assess the current levels of awareness regarding LC and BIM, identify factors driving their implementation, and evaluate their influence on the outcomes of the project such as cost-effectiveness, timely completion, and resource optimization. Special emphasis will be placed on understanding the significance of IPD in facilitating effective collaboration and addressing common project inefficiencies.

To achieve a comprehensive understanding, the study will utilize Structural Equation Modeling (SEM) as the primary analytical tool. This method identifies complicated correlations among different parameters, providing useful knowledge into barriers, IPD factors, and their impact on the outcome of the project.

REFERENCES

[1] Azhar, S., Khalfan, M., & Maqsood, T. (2015). Building information modelling (BIM): now and beyond. *Construction Economics and Building,* 12(4), 15–28. http://doi:10.5130/AJCEB.v12i4.3032.

[2] Bayhan, H. G., Demirkesen, S., Zhang, C., & Tezel, A. (2023). A lean construction and BIM interaction model for the construction industry. *Production Planning and Control*, 34(15), 1447–1474. https://doi.org/10.1080/09537287.2021.2019342.

[3] Bui, N., Merschbrock, C., & Munkvold, B. E. (2016). A review of building information modelling for construction in developing countries. *Procedia Engineering*, 164, 487–494. https://doi.org/10.1016/j.proeng.2016.11.649.

[4] Eadie, R., Browne, M., Odeyinka, H., McKeown, C., & McNiff, S. (2013). BIM implementation throughout the UK construction project lifecycle: an analysis. *Automation in Construction*, 36, 145–151. https://doi.org/10.1016/j.autcon.2013.09.001.

[5] El Mounla, K., Beladjine, D., Beddiar, K., & Mazari, B. (2023). Lean-BIM approach for improving the performance of a construction project in the design phase. *Buildings*, 13(3), 654. https://doi.org/10.3390/buildings13030654.

[6] Gómez-Cabrera, A., Salazar, L. A., Ponz-Tienda, J. L., & Alarcón, L. F. (2020). Lean tools proposal to mitigate delays and cost overruns in construction projects. In IGLC 28 – 28th Annual Conference of the International Group for Lean Construction 2020. https://doi.org/10.24928/2020/0049.

[7] Ismail, N. A. A., Hasbullah, I. S., Mohamed, M. A., Marhani, M. A., Rooshdi, R. R. R. M., Sahamir, S. R., et al. (2023). Lean-BIM collaborative approach for sustainable construction projects in Malaysia. *Journal of Advanced Research in Applied Sciences and Engineering Technology*, 33(1), 356–366. https://doi.org/10.37934/araset.33.1.356366.

[8] Koskela, L., Howell, G., Ballard, G., & Tommelein, I. (2002). The foundations of lean construction. In Best, R., & de Valence, G. (Eds.), Design and Construction: Building in Value. Routledge.

[9] Madusha, M. D. Y., Francis, M., & Liyanawatta, T. N. (2023). Applicability of bim technology for enhancing the lean construction process in Sri Lanka. *World Construction Symposium*, 1(July), 174–184. https://doi.org/10.31705/WCS.2023.15.

[10] McHugh, K., Dave, B., & Craig, R. (2019). Integrated lean and BIM processes for modularised construction – a case study. In 27th Annual Conference of the International Group for Lean Construction, IGLC 2019. https://doi.org/10.24928/2019/0252.

[11] Moradi, S., & Sormunen, P. (2023). Integrating lean construction with BIM and sustainability: a comparative study of challenges, enablers, techniques, and benefits. *Construction Innovation*, 24(7), 188–203. https://doi.org/10.1108/CI-02-2023-0023.

[12] Musharavati, F. (2023). Optimized integration of lean construction, building information modeling, and facilities management in developing countries: a case of Qatar. *Buildings*, 13(12), 3051. https://doi.org/10.3390/buildings13123051.

[13] Ofori, G. (2015). Nature of the construction industry, its needs and its development: a review of four decades of research. *Journal of Construction in Developing Countries*.

[14] Pal, A., & Nassarudin, A. (2020). Integrated project delivery adoption framework for construction projects in India. In IGLC 28 – 28th Annual Conference of the International Group for Lean Construction 2020, (pp. 337–348). https://doi.org/10.24928/2020/0018.

[15] Rashidian, S., Drogemuller, R., & Omrani, S. (2023). Building information modelling, integrated project delivery, and lean construction maturity attributes: a delphi study. *Buildings*, 13(2). https://doi.org/10.3390/buildings13020281.

[16] Sampaio, A. Z., Fernandes, V., & Gomes, A. (2023). The use of BIM-based tools to improve collaborative building projects. *Procedia Computer Science*, 219, 2027–2034. https://doi.org/10.1016/j.procs.2023.01.504.

[17] Sari, E. M., Irawan, A. P., Wibowo, M. A., Siregar, J. P., Tamin, R. Z., Praja, A. K. A., et al. (2023). Challenge and awareness for implemented integrated project delivery (IPD) in Indonesian projects. *Buildings*, 13(1), 262. https://doi.org/10.3390/buildings13010262.

[18] Sharma, N., & Laishram, B. (2024). Understanding the relevance and impact of the cost of quality in the construction industry: a systematic literature review using PRISMA. *Construction Innovation*. https://doi.org/10.1108/CI-08-2023-019.

Advances in Construction, Real Estate, Infrastructure and Project Management – Anil Kashyap et al. (eds)
© 2026 Taylor & Francis Group, London, ISBN 978-1-041-13433-6

8 Strategic Approaches for Successful Pre-Cast Construction Implementation in India

Shivam Singh[1]
MBA-CEQS,
RICS School of Built Environment, Amity University,
Noida, India

VPS Nihar Nanyam[2]
Senior Lecturer in Quantity Surveying,
School of the Built Environment, London Metropolitan University,
United Kingdom

Swati Sinha[3]
Research Scholar,
School of Energy and Environment, NICMAR University,
Pune, India

Mohd. Suhail Khan[4]
Assistant Professor,
RICS School of Built Environment, Amity University,
Noida, India

■ **ABSTRACT:** As a result of the increasing need for housing and infrastructure, numerous advanced construction methods have been developed and analyzed. Among these, the "pre-cast construction technology" has surfaced as a notable modern approach. This technique is gaining traction in India, particularly for projects requiring swift construction timelines. However, its implementation in India has encountered several challenges. This research aims to identify the Critical Success Factors (CSFs) relevant to the pre-cast industry in India to address these challenges effectively. A thorough examination of pertinent research publications regarding the identification of CSFs in the pre-cast sector was undertaken. This review facilitated the identification of CSFs specific to the Indian context. Subsequently, these the relative relevance of CSFs was determined by ranking them using a set of questions employing a Likert scale and ultimately expert judgment. Additionally, the effect of the CSFs on the pre-cast industry was analyzed by applying the Partial Least Square Method. The results of this research are anticipated to aid contractors, owners, architects, engineers, vendors, and other stakeholders in

[1]shvmsingh.qs@gmail.com; [2]nihar057a@gmail.com, n.nanyam@londonmet.ac.uk; [3]swatisinha.hi@gmail.com;
[4]suhailkhan@ricssbe.edu.in

DOI: 10.1201/9781003669814-8

the construction sector in identifying and prioritizing key areas for the effective implementation of pre-cast technology. This will enable the development of strategic approaches for the effective adoption of pre-cast construction in India.

■ **KEYWORDS:** Pre-cast construction technology, Critical success factors, CSFs

1. INTRODUCTION

The steady growth of India's population and the rising requirement for essential housing and infrastructure have prompted extensive study on construction techniques that might facilitate rapid and high-quality development. One method of construction that can be employed is pre-cast construction, wherein building elements are fabricated in a controlled factory setting and subsequently brought to the construction site for assembly in accordance with precise specifications [21]. There is a growing popularity of pre-cast technology in India, with many contractors and customers attempting to include this approach [22]. Nevertheless, several issues must be resolved in order to effectively adopt this technology. Therefore, research on the Critical Success Factors (CSFs) of pre-cast technology is necessary [26].

Multiple research investigations have been carried out in order to determine the CSFs related to construction technology [27, 32, 33]. However, it is necessary to study the CSFs specific to pre-cast technology in order to ensure its appropriate and successful implementation. Currently, the utilization of pre-cast technology in the Indian construction sector is barely 2%, mostly confined to major projects like roads and bridges [20].

The challenges in the implementation of precast construction technology are multifaceted and include difficulties in understanding the concept [9], resistance to change [1], affordability [18], setting up cost of pre-cast unit [25], viability of the pre-cast industry [10], and transportation cost [22].

1.1 Critical Success Factors

These key areas should be prioritized since they have the potential to essure the successful completion of the project and align with the project manager's objectives. Identifying the CSFs has become critical for the effective delivery of the project [33].

The following steps illustrate how to identify and construct CSFs.

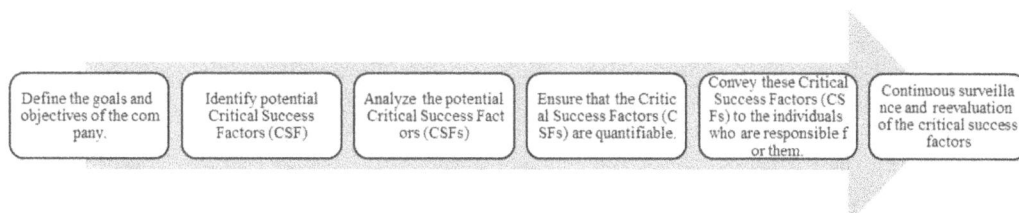

| Define the goals and objectives of the company. | Identify potential Critical Success Factors (CSF) | Analyze the potential Critical Success Factors (CSFs) | Ensure that the Critical Success Factors (CSFs) are quantifiable. | Convey these Critical Success Factors (CSFs) to the individuals who are responsible for them. | Continuous surveillance and reevaluation of the critical success factors |

Figure 8.1 Steps to identify and develop the CSFs

Source: Authors

1.2 CSFs Pertaining to the Pre-Cast Construction Technique

The process of identifying and classifying CSFs related to the viability of prefabricated construction was shown using the particular case of Houshenshan Hospital, which was constructed within a just 10 days amidst the COVID-19 epidemic [33]. The CSFs associated with the off-site building approach were established, in addition to the implementation of the Interpretive Structural Modelling technique to determine the link between these CSFs and the pre cast construction technique [19]. The expert team determined the most significant CSFs for industrial modularization. The notion of CSF enablers was created to enhance the attention on certain CSFs, ensuring their effective accomplishment [27].

The establishment of the CSFs is based on the CSFs related to pre-cast technologies that were previously found in past studies. Regarding pre-cast construction technique in India, there are several more crucial success elements that might be considered. Furthermore, the prioritization of these identified important success elements will be determined by the subjective viewpoints of certain specialists, which might vary from individual to individual and over time. This study prioritizes the important success factors beyond the primary aim of the pre-cast construction method. This research is going to be relevant to the precast sector in India as well as other countries. However, it should be noted that the important success elements and their ranking or hierarchy may vary depending on the region and building trends.

2. METHODOLOGY

The intent of the study is to rank the CSFs and assess their influence on the goals of pre-cast construction technology. This research has three primary objectives: to identify the CSFs and objectives of pre-cast construction technology in India, to rank the identified CSFs, and to determine the influence of these CSFs on the objectives of the pre-cast construction technology. The methodology involves using a questionnaire based on the Mean Score Method and Partial Least Squares Structural Equation Modeling (PLS-SEM) to rank the CSFs and understand their influence on the objectives. The connection between the objectives and the selected approach is illustrated below:

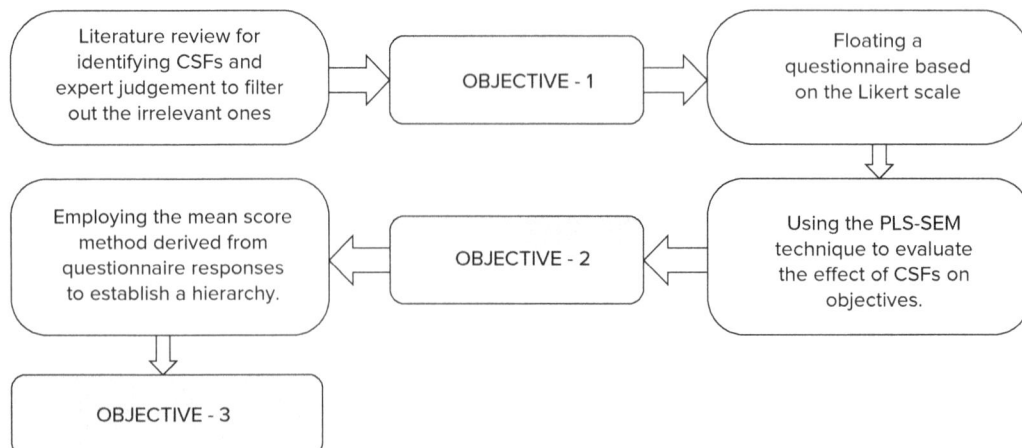

Figure 8.2 Projected diagram linking objectives and approach

Source: Authors

2.1 Questionnaire

By leveraging expert opinion and conducting a literature study, we have identified 20 Critical Success Factors (CSFs) and 7 objectives specific to the pre-cast sector. Based on these findings, we have developed a questionnaire to assess and evaluate the relevance of the identified CSFs in relation to pre-cast construction technology. The mean score was determined using the replies, and the CSFs were sorted according to their scores. Furthermore, an additional part was incorporated into this questionnaire to ascertain the correlation between the stablished CSFs and the goals of the pre-cast industry. The survey results were introduced into the Smart-PLS 3 software to calculate significant coefficients. Additionally, a prototype was created to depict the connections among the CSFs and the objectives.

2.2 Identification of CSFs

The following CSFs were identified:

Table 8.1   Identified CSFs

S.No.	CSF	References
F1	Durable sketching and precise details	[14, 32]
F2	Efficiently structured schedule	[30, 32]
F3	Transportation capabilities	[27, 30, 32]
F4	Mass-produced components being standardized	[11, 30]
F5	Availability of skilled labours	[11, 13, 30, 32]
F6	Utilization of information and communication technologies in a proactive manner.	[3, 14, 24, 29]
F7	Collaboration among participants at each step	[6, 11, 32]
F8	Full participation of three members across all stages	[30, 32]
F9	Assistance from the government	[11, 30]
F10	Ecological surroundings	[14, 30]
F11	Effective cooperation	[16, 23]
F12	Feasibility and economic analysis	[27, 32]
F13	Resilient schemes and incentive	[4, 5, 24]
F14	Standardized designs	[6, 7, 27, 30, 31]
F15	Efficient risk mitigation	[17, 28, 30, 32]
F16	Iterative enhancement and acquisition of knowledge	[21, 30, 32]
F17	Optimal procurement strategy	[8, 30]
F18	Recognition of cost reductions	[27]
F19	Contractor's expertise and track record	[27]
F20	Vendor participation	[27]

Source: Compiled by authors

The table below presents the specified objectives of the pre-cast construction technology.

Table 8.2 Objectives

Objectives		References
1.)	Project planning and design	
2.)	Technological advancements in construction methods	
3.)	Information, communication, along with collaboration	[21, 33]
4.)	Expertise along with knowledge	
5.)	Supply chain	
6.)	External environment	
7.)	Cost savings	

Source: Compiled by authors

2.3 Ranking of the Identified CSFs

To accomplish objective 2, which involved ranking the discovered Critical Success Factors (CSFs), a unified questionnaire was sent. The following data displays the proportion of experience among the respondents.

Ranking of the CSFs was done according to the mean score for the CSFs as shown in Table 8.3.

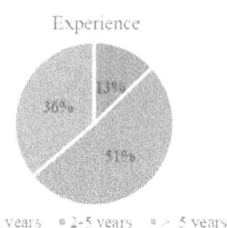

Experience

■ < 2 years ■ 2-5 years ■ > 5 years

Figure 8.3 Experience percentages of respondents

Source: Authors

Table 8.3 Mean score of the CSFs

CSFs	Mean Score	Rank
18. Recognition of cost reductions	4.26	1
15. Efficient risk mitigation	4.24	2
12. Feasibility and economic analysis	4.22	3
17. Optimal procurement strategy	4.22	3
14. Standardized designs	4.15	5
1. Durable sketching and precise details	4.13	6
5. Availability of skilled labours	4.11	7
3. Transportation capabilities	4.07	8
6. Utilization of information and communication technologies in a proactive manner.	4.05	9
11. Effective cooperation	4.03	10
19. Contractor's expertise and track record	4.03	10
4. Mass-produced components being standardized	4.00	12
7. Collaboration among participants at each step	4.00	12
2. Efficiently structured schedule	3.98	14
8. Full participation of three members across all stages	3.96	15
16. Iterative enhancement and acquisition of knowledge	3.96	15
20. Vendor participation	3.92	17
13. Resilient schemes and incentive	3.88	18
10. Ecological surroundings	3.75	19
9. Assistance from the government	3.69	20

Source: Compiled by authors

2.4 SEM Model

The focus of Objective 3 was to determine the correlation between the selected CSFs and the goals of pre-cast building technology in India. The survey pertaining to the third objective entailed examining potential connections between the CSFs and the goals. Subsequently, participants were requested to evaluate the intensity of these connections. The hypothesized association is displayed in Table 8.4 below:

Table 8.4 CSFs and objectives correlation

Objectives	CSFs
Project planning and design (PPD)	F1, F2, F8, F14, F15, F17
Technological advancements in construction methods (TMC)	F4, F5, F6, F14
Information, communication, along with collaboration (ICC)	F8, F9, F11, F13, F16
Expertise along with knowledge (EK)	F5, F15, F16, F19
Supply chain (SC)	F3, F8, F17, F20
External environment (EE)	F9, F10, F13
Cost savings (CS)	F2, F12, F15, F18,

Source: Compiled by authors

The CSFs related to a target were identified based on that objective. For instance, F1 represents a robust drawing and specifications referred to as PPD1, whereas F2 denotes a well-organized timetable known as PPD2, and so on. Furthermore, F4 refers to the standardization of factory-made components, categorized as TMC1. Similarly, F5 pertains to the availability of trained labor, classified as TMC2. The purpose of this was to prepare the data for import into SmartPLS 3 software, which is necessary for doing the analysis needed for objective 3.

2.5 SEM Model using Smart PLS

The data obtained from the questionnaire replies was imported into SmartPLS to create a hypothetical model. All further analyses were conducted using this model. The diagram illustrating the model is presented Figure 8.4.

After constructing the model (Figure 8.5), the Partial Least Squares (PLS) method was executed, yielding the following outcomes (with a maximum iteration limit of 500).

2.6 Measurement Model

Discriminant validity is determined to evaluate if the conceptions are separate and distinguishable from one another.

2.7 Construct Reliability

Table 8.5 Construct reliability and validity

	Cronbach's Alpha	rho_A	Composite Reliability	Average Variance Extracted (AVE)
CS	0.940	0.956	0.957	0.848
EE	0.843	0.853	0.905	0.761
EK	0.905	0.915	0.934	0.780
ICC	0.892	0.905	0.921	0.700
PPD	0.856	0.884	0.887	0.569
SC	0.896	0.906	0.928	0.764
TMC	0.941	0.952	0.957	0.849

Source: Compiled by authors

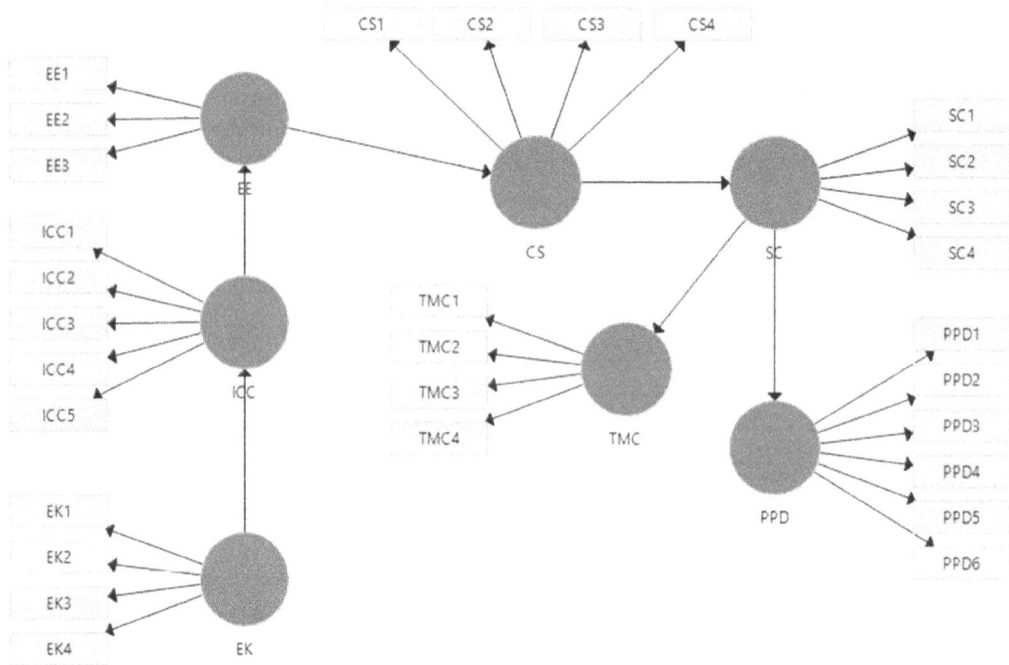

Figure 8.4 Hypothetical model to be tested

Source: Authors

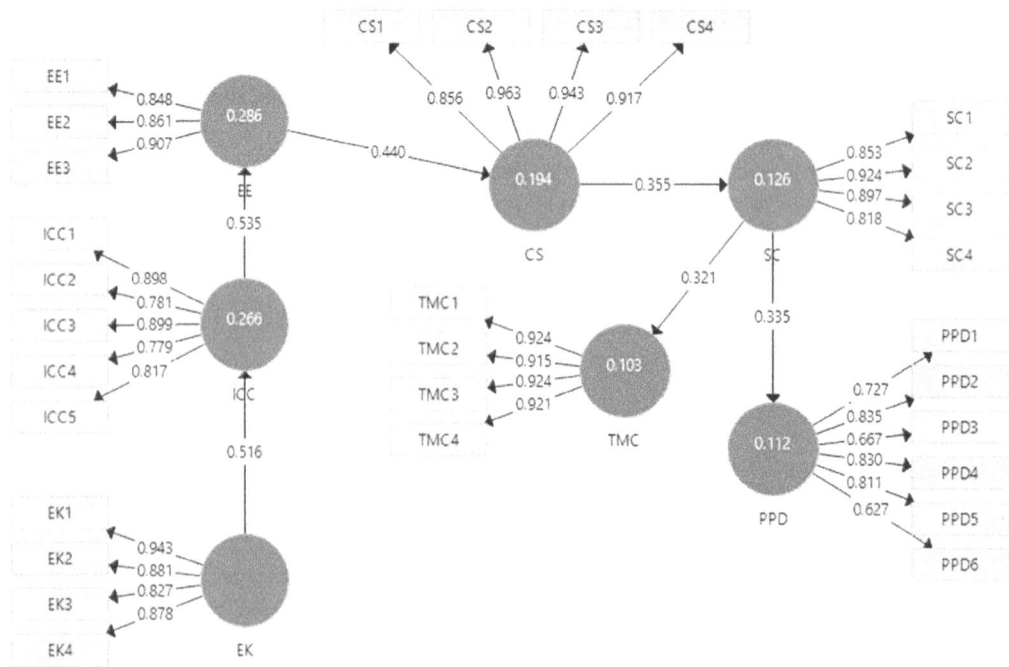

Figure 8.5 MODEL after running PLS-algorithm

Source: Authors

2.8 Cronbach's Alpha

All of the constructions have surpassed the threshold value of 0.70. Therefore, our reliability is excellent.

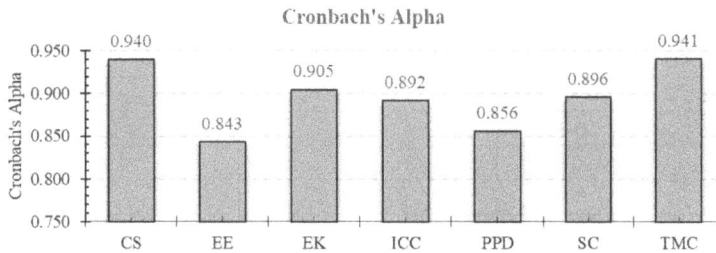

Figure 8.6 Cronbach's alpha

Source: Authors

The model's composite reliability exceeds 0.7 [2]. **It means consistency.**

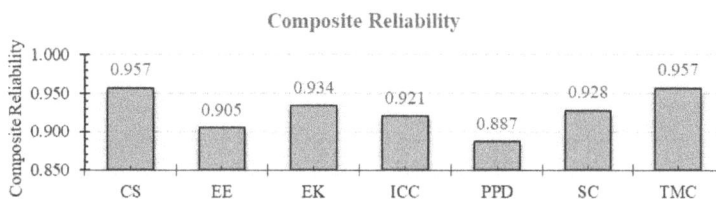

Figure 8.7 Composite reliability

Source: Authors

The factor loadings values must exceed 0.70. [12] (Table 8.6).

2.9 Convergent Validity

The Average Variance Extracted metric indicates the convergent validity. The AVE values for the model exceed the threshold value of 0.50, indicating that convergent validity has been demonstrated [2].

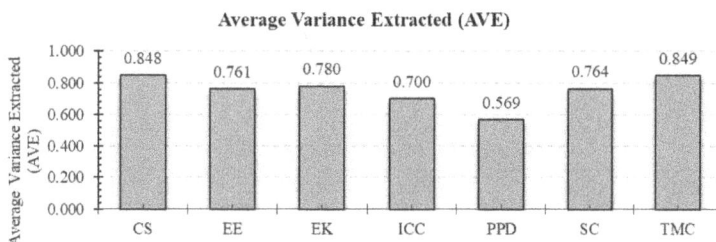

Figure 8.8 Average variance extracted

Source: Authors

2.10 Discriminant Validity

The Fornell-Larcker criteria values were determined to meet the requirement, which states that the square root of the average variance extracted (AVE) of a construct should be higher than its association with other constructs. This indicates that the discriminant validity has been demonstrated (Table 8.7).

Table 8.6 Figure 8.7. Table 8.6: Factor loadings

	CS	EE	EK	ICC	PPD	SC	TMC
CS1	0.856						
CS2	0.963						
CS3	0.943						
CS4	0.917						
EE1		0.848					
EE2		0.861					
EE3		0.907					
EK1			0.943				
EK2			0.881				
EK3			0.827				
EK4			0.878				
ICC1				0.898			
ICC2				0.781			
ICC3				0.899			
ICC4				0.779			
ICC5				0.817			
PPD1					0.727		
PPD2					0.835		
PPD3					0.667		
PPD4					0.830		
PPD5					0.811		
PPD6					0.627		
SC1						0.853	
SC2						0.924	
SC3						0.897	
SC4						0.818	
TMC1							0.924
TMC2							0.915
TMC3							0.924
TMC4							0.921

Source: Compiled by authors

Table 8.7 Fornell – larcker criterion

	CS	EE	EK	ICC	PPD	SC	TMC
CS	0.921						
EE	0.440	0.872					
EK	0.258	0.409	0.883				
ICC	0.307	0.535	0.516	0.837			
PPD	0.181	0.206	0.230	0.215	0.754		
SC	0.355	0.352	0.437	0.555	0.335	0.874	
TMC	0.187	0.191	0.133	0.120	0.129	0.321	0.921

Source: Compiled by authors

The Cross Loadings values were observed to exceed those of any other concept, indicating that discriminant validity has been proven.

Table 8.8 Cross – loadings

	CS	EE	EK	ICC	PPD	SC	TMC
CS1	0.856	0.411	0.125	0.166	0.135	0.177	0.205
CS2	0.963	0.400	0.199	0.274	0.203	0.333	0.224
CS3	0.943	0.408	0.209	0.243	0.174	0.275	0.039
CS4	0.917	0.405	0.373	0.406	0.152	0.472	0.169
EE1	0.267	0.848	0.411	0.503	0.082	0.257	0.268
EE2	0.410	0.861	0.328	0.412	0.279	0.418	0.116
EE3	0.461	0.907	0.338	0.486	0.177	0.254	0.126
EK1	0.310	0.383	0.943	0.520	0.171	0.400	0.192
EK2	0.202	0.331	0.881	0.427	0.273	0.405	0.014
EK3	0.083	0.301	0.827	0.446	0.136	0.442	0.117
EK4	0.308	0.432	0.878	0.418	0.243	0.291	0.133
ICC1	0.356	0.440	0.402	0.898	0.243	0.435	0.146
ICC2	0.205	0.432	0.380	0.781	0.168	0.364	0.032
ICC3	0.492	0.532	0.523	0.899	0.240	0.591	0.118
ICC4	0.048	0.337	0.385	0.779	0.090	0.484	0.133
ICC5	0.109	0.468	0.443	0.817	0.138	0.427	0.077
PPD1	0.217	0.324	0.287	0.208	0.727	0.164	−0.133
PPD2	0.255	0.236	0.181	0.088	0.835	0.239	0.019
PPD3	0.028	0.128	0.013	0.113	0.667	0.196	0.260
PPD4	0.095	0.074	0.163	0.046	0.830	0.322	0.121
PPD5	0.133	0.135	0.243	0.369	0.811	0.327	0.156
PPD6	0.172	0.124	0.190	0.064	0.627	0.023	0.188
SC1	0.354	0.419	0.381	0.510	0.180	0.853	0.287
SC2	0.308	0.365	0.436	0.530	0.465	0.924	0.233
SC3	0.262	0.206	0.341	0.438	0.238	0.897	0.380
SC4	0.325	0.236	0.363	0.460	0.256	0.818	0.231
TMC1	0.166	0.108	0.086	0.076	0.073	0.275	0.924
TMC2	0.174	0.167	0.080	0.091	0.031	0.252	0.915
TMC3	0.232	0.205	0.059	0.040	0.167	0.304	0.924
TMC4	0.123	0.211	0.242	0.218	0.177	0.339	0.921

Source: Compiled by authors

Heterotrait – Monotrait Ratio (HTMT Ratio) was found out to be less than 0.90 establishing the discriminant validity [15].

Table 8.9 HTMT ratio

	CS	EE	EK	ICC	PPD	SC	TMC
CS							
EE	0.490						
EK	0.275	0.472					
ICC	0.312	0.610	0.565				
PPD	0.226	0.266	0.281	0.244			
SC	0.374	0.408	0.482	0.616	0.320		
TMC	0.202	0.216	0.153	0.137	0.217	0.348	

Source: Compiled by authors

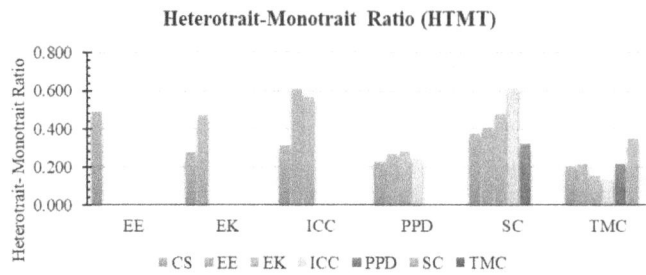

Figure 8.9 HTMT ratio

Source: Authors

The **Structure Model** defines the interconnections between the constructs in the proposed model. To analyze the structural model, we conducted bootstrapping with 5000 sub-samples. The output includes the mean, standard deviation, T-values, and P-values.

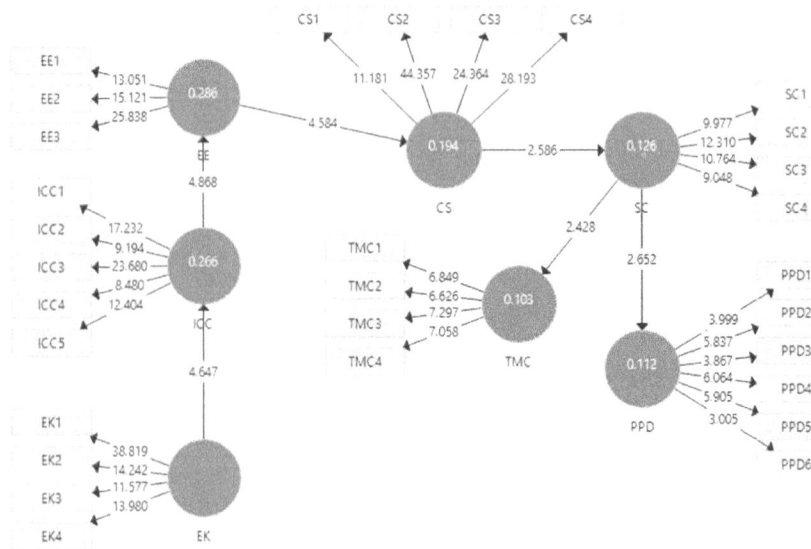

Figure 8.10 MODEL after preforming BOOTSTRAPPING

Source: Authors

The **T-Statistics** result surpassed the threshold of 1.96, suggesting that the expected associations in the model are statistically significant.

The **P values** are less than 0.05, suggesting that the predicted connections are statistically significant.

Table 8.10 T-values and P-values

	Original Sample	Sample Mean	Standard Deviation	T Statistics	P Values
CS → SC	0.355	0.369	0.137	2.586	0.010
EE → CS	0.440	0.448	0.096	4.584	0.000
EK → ICC	0.516	0.526	0.111	4.647	0.000
ICC → EE	0.535	0.562	0.110	4.868	0.000
SC → PPD	0.335	0.396	0.126	2.652	0.008
SC → TMC	0.321	0.336	0.132	2.428	0.015

Source: Compiled by authors

3. RESULT AND DISCUSSION

The values of the relevant coefficients, including Cronbach's Alpha, Rho, AVE, Factor loadings, cross loadings, T value, and P value, were all verified to fall within the necessary range and meet the corresponding qualifying criteria. This suggests that the model holds statistical significance. The calculated coefficients and associated metrics, such as p-values and t-values, fall within acceptable ranges, indicating that the relationships and interactions captured by the model are not due to random chance. Consequently, the model provides a reliable and robust representation of the impact of CSFs within the pre-cast construction industry.

4. CONCLUSION

This study identified the CSFs of the pre-cast construction industry in India through the analysis of questionnaire data and expert assessments. The questionnaire responses revealed a total of 20 CSFs relevant to the pre-cast construction sector. Subsequently, the Mean Score technique was employed to rank these 20 CSFs based on the responses. The factors of cost savings recognition and effective risk management were ranked first and second, respectively, with mean scores of 4.264 and 4.245, indicating their paramount significance. Although the natural environment and government assistance received the lowest rankings, this does not imply that these CSFs should be disregarded, as the variations in average values across all 20 CSFs were minimal. A model was designed to illustrate the connections between the CSFs and the objectives of the pre-cast construction method, mapping out how each factor influences the various goals within the industry. Additionally, the interconnections among various aims of pre-cast construction technology were delineated. All coefficients and ratios were found to be within acceptable limits, indicating that the model was accurate and all indicated correlations were statistically significant.

The findings of this research may catalyze further research on the CSFs of the pre-cast industry in India. There is potential for exploring additional CSFs and their impacts on the pre-cast industry. Moreover, further models can be constructed to represent other types of interactions, thereby enhancing the understanding of pre-cast technology management.

REFERENCES

[1] Abdelmoneim, A. M., Alb6attah, M., & Almuhairi, A. (2022). Single users' perception towards acceptance of precast technology in the construction industry: case of Abu Dhabi, UAE. In ZEMCH International Conference, (pp. 831–839).

[2] Ali, F., Rasoolimanesh, S. M., Sarstedt, M., Ringle, C. M., & Ryu, K. (2018). An assessment of the use of partial least squares structural equation modeling (PLS-SEM) in hospitality research. *International Journal of Contemporary Hospitality Management*, 30(1), 514–538. https://doi.org/10.1108/IJCHM-10-2016-0568/FULL/XML.

[3] Kamar, K. A. M., & Hamid, Z. A. (2011). Supply chain strategy for contractor in adopting industrialised building system (IBS). *Australian Journal of Basic and Applied Sciences*, 5(12), 2552–2557.

[4] Arif, M., & Egbu, C. (2010). Making a case for offsite construction in China. *Engineering, Construction and Architectural Management*, 17(6), 536–548. https://doi.org/10.1108/09699981011090170.

[5] Azam Haron, N., Abdul-Rahman, H., Wang, C., & Wood, L. C. (2015). Quality function deployment modelling to enhance industrialised building system adoption in housing projects. *Total Quality Management and Business Excellence*, 26(7–8), 703–718. https://doi.org/10.1080/14783363.2014.880626.

[6] Azhar, S., Lukkad, M. Y., & Ahmad, I. (2013). An investigation of critical factors and constraints for selecting modular construction over conventional stick-built technique. *International Journal of Construction Education and Research*, 9(3), 203–225. https://doi.org/10.1080/15578771.2012.723115.

[7] Barlow, J., Childerhouse, P., Gann, D., Hong-Minh, S., Naim, M., & Ozaki, R. (2003). Choice and delivery in housebuilding: lessons from Japan for UK housebuilders. *Building Research and Information*, 31(2), 134–145. https://doi.org/10.1080/09613210302003.

[8] Blismas, N., & Wakefield, R. (2009). Drivers, constraints and the future of offsite manufacture in Australia. *Construction Innovation*, 9(1), 72–83. https://doi.org/10.1108/14714170910931552/FULL/XML.

[9] Chippagiri, R., Bras, A., Sharma, D., & Ralegaonkar, R. V. (2022). Technological and sustainable perception on the advancements of prefabrication in construction industry. *Energies*, 15(20), 7548. https://doi.org/10.3390/en15207548.

[10] Dang, P., Niu, Z., Gao, S., Hou, L., & Zhang, G. (2020). Critical factors influencing the sustainable construction capability in prefabrication of Chinese construction enterprises. *Sustainability (Switzerland)*, 12(21), 1–22. https://doi.org/10.3390/su12218996.

[11] Gan, X., Chang, R., Zuo, J., Wen, T., & Zillante, G. (2018). Barriers to the transition towards off-site construction in China: An Interpretive structural modeling approach. *Journal of Cleaner Production*, 197, 8–18. https://doi.org/10.1016/J.JCLEPRO.2018.06.184.

[12] Gefen, D., & Straub, D. (2005). A practical guide to factorial validity using PLS-graph: tutorial and annotated example. *Communications of the Association for Information Systems*, 16(1), 5. https://doi.org/10.17705/1CAIS.01605.

[13] Gibb, A. G. F. (2001). Standardization and pre-assembly – distinguishing myth from reality using case study research. *Construction Management and Economics*, 19(3), 307–315. https://doi.org/10.1080/01446190010020435.

[14] Hasan, A., Baroudi, B., Elmualim, A., & Rameezdeen, R. (2018). Factors affecting construction productivity: a 30 year systematic review. *Engineering, Construction and Architectural Management*, 25(7), 916–937. https://doi.org/10.1108/ECAM-02-2017-0035.

[15] Henseler, J., Ringle, C. M., & Sarstedt, M. (2015). A new criterion for assessing discriminant validity in variance-based structural equation modeling. *Journal of the Academy of Marketing Science*, 43(1), 115–135. https://doi.org/10.1007/S11747-014-0403-8.

[16] Hwang, B. G., Shan, M., & Looi, K. Y. (2018). Knowledge-based decision support system for prefabricated prefinished volumetric construction. *Automation in Construction*, 94, 168–178. https://doi.org/10.1016/J.AUTCON.2018.06.016.

[17] Ismail, F., Yusuwan, N. M., & Baharuddin, H. E. A. (2012). Management factors for successful IBS projects implementation. *Procedia – Social and Behavioral Sciences*, 68, 99–107. https://doi.org/10.1016/J.SBSPRO.2012.12.210.

[18] Ji, Y., Chang, S., Qi, Y., Li, Y., Li, H. X., Qi, K., et al. (2019). A BIM-based study on the comprehensive benefit analysis for prefabricated building projects in china. *Advances in Civil Engineering*, 2019, 3720191. https://doi.org/10.1155/2019/3720191.

[19] Jung, S., Lee, S., & Yu, J. (2021). Identification and prioritization of critical success factors for off-site construction using ISM and MICMAC analysis. *Sustainability*, 13(16), 8911. https://doi.org/10.3390/SU13168911.

[20] Kaja, N., Jauswal, A., & Professor, A. (2021). Review of precast concrete technology in India. *International Journal of Engineering Research and Technology*, 10(6), 867–872. https://doi.org/10.17577/IJERTV10IS060400.

[21] Li, L., Li, Z., Wu, G., & Li, X. (2018). Critical success factors for project planning and control in prefabrication housing production: a china study. *Sustainability*, 10(3), 836. https://doi.org/10.3390/SU10030836.

[22] Liu, C., Zhang, F., & Zhang, H. (2020). Comparative analysis of off-site precast concrete and cast-in-place concrete in low-carbon built environment. *Fresenius Environmental Bulletin*, 29(3), 1804–1812.

[23] Luo, L., Qiping Shen, G., Xu, G., Liu, Y., & Wang, Y. (2019). Stakeholder-associated supply chain risks and their interactions in a prefabricated building project in Hong Kong. *Journal of Management in Engineering*, 35(2), 05018015. https://doi.org/10.1061/(ASCE)ME.1943-5479.0000675.

[24] Mao, C., Shen, Q., Asce, M., Pan, W., & Ye, K. (2013). Major barriers to off-site construction: the developer's perspective in china. *Journal of Management in Engineering*, 31(3), 04014043. https://doi.org/10.1061/(ASCE)ME.1943-5479.0000246.

[25] Mekawy, M., & Petzold, F. (2017). Exhaustive exploration of modular design options to inform decision making. In Proceedings of the International Conference on Education and Research in Computer Aided Architectural Design in Europe, (Vol. 2, pp. 107–114). https://doi.org/10.52842/conf.ecaade.2017.2.107.

[26] Nanyam, V. P. S. N., Basu, R., Sawhney, A., Vikram, H., & Lodha, G. (2017). Implementation of precast technology in india–opportunities and challenges. *Procedia Engineering*, 196, 144–151. https://doi.org/10.1016/J.PROENG.2017.07.184.

[27] O'Connor, J. T., O'Brien, W. J., & Choi, J. O. (2014). Critical success factors and enablers for optimum and maximum industrial modularization. *Journal of Construction Engineering and Management*, 140(6), 04014012. https://doi.org/10.1061/(ASCE)CO.1943-7862.0000842/ASSET/5F6EDB48-C964-40E2-8F9B-834D505C9D32/ASSETS/IMAGES/LARGE/FIGURE4.JPG.

[28] Pan, W., Gibb, A. F., & Dainty, A. R. J. (2007). Perspectives of UK housebuilders on the use of offsite modern methods of construction. *Construction Management and Economics*, 25(2), 183–194. https://doi.org/10.1080/01446190600827058.

[29] Walker, D. H. T. (1995). An investigation into construction time performance. *Construction Management and Economics*, 13(3), 263–274. https://doi.org/10.1080/01446199500000030.

[30] Wuni, I. Y., & Shen, G. Q. (2020). Critical success factors for modular integrated construction projects: a review. *Building Research and Information*, 48(7), 763–784. https://doi.org/10.1080/09613218.2019.1669009.

[31] Wuni, I. Y., Shen, G. Q., & Hwang, B.-G. (2019). Risks of modular integrated construction: A review and future research directions. *Frontiers of Engineering Management*, 7(1), 63–80. https://doi.org/10.1007/S42524-019-0059-7.

[32] Wuni, I. Y., Shen, G. Q., & Osei-Kyei, R. (2022). Quantitative evaluation and ranking of the critical success factors for modular integrated construction projects. *International Journal of Construction Management*, 22(11), 2108–2120. https://doi.org/10.1080/15623599.2020.1766190.

[33] Zhao, M. (2021). Critical success factors associated with time performance in prefabricated construction. In E3S Web of Conferences, (Vol. 253, p. 02044). https://doi.org/10.1051/E3SCONF/202125302044.

Advances in Construction, Real Estate, Infrastructure and Project Management – Anil Kashyap et al. (eds)
© 2026 Taylor & Francis Group, London, ISBN 978-1-041-13433-6

9 Development of an IGBC Compliant Software Model for Assessing Buildings

Shridhar Kumbhar[1]

Assistant Professor,
School of Construction, NICMAR University,
Pune, India

Smita Patil[2]

Post Graduate Student,
MBA in Environmental Sustainability,
School of Energy and Environment, NICMAR University,
Pune, India

■ **ABSTRACT:** The construction industry's rapid expansion and the need for sustainable development have made the creation of sophisticated assessment techniques to gauge a building's sustainability performance necessary. The objective of this study was to develop a software model that complied with the Indian Green Building Council's (IGBC) norms and regulations, enabling comprehensive building assessments. The software model was created to be a practical tool that stakeholders, architects, and construction specialists could use to assess and enhance the sustainability elements of their projects. By providing a systematic and consistent methodology, the software model expedited the assessment process and made it possible to quickly evaluate sustainability performance. Extensive research and engagement with industry experts and stakeholders were required during the creation of the software model to ensure the implementation of appropriate IGBC regulations and guidelines. Users managed to identify areas for improvement and make well-informed decisions to improve the overall sustainability of their facilities by using the model, which made it easier to gather and analyze data on a variety of sustainability criteria. The software model underwent rigorous testing and validation to ensure its precision and reliability in assessing building sustainability. This study resulted in a thorough and intuitive software model that offers a productive way to evaluate a building's sustainability compliance with IGBC criteria. By giving stakeholders vital information and empowering them to make informed decisions in the development of ecologically conscious built environments, this idea promotes sustainable building practices.

■ **KEYWORDS:** IGBC, Green buildings, Sustainability, Software model, Assessment

[1]shridhar.kumbhar1984@gmail.com, shridhar.kumbhar@pune.nicmar.ac.in; [2]smitapatil9580@gmail.com, P2377014@student.nicmar.ac.in

DOI: 10.1201/9781003669814-9

1. INTRODUCTION

In terms of energy consumption, the construction sector has experienced the highest rate of increase in the last 10–15 years, which has resulted in significant adverse effects on the natural resources available for buildings and the environment. "Green Buildings" must be used to tackle these challenges. The practice of planning, constructing, and managing buildings with an emphasis on minimizing environmental effects, saving resources, and enhancing occupant health and well-being is known as green construction. Because green building grading systems are crucial for encouraging sustainable building practices, they have become more and more popular in the construction sector. These methods offer frameworks for assessing and accrediting buildings based on their energy efficiency, environmental performance, and general sustainability. This study's objective is to present a comparative analysis of different green building grading schemes, emphasizing their salient features, advantages, and areas of concentration.

One of the most widely used and respected rating systems in the world is the Leadership in Energy and Environmental Design, or LEED, system developed by the United States Green Building Council (USGBC). Sustainable site planning, water efficiency, energy and atmosphere, materials and resources, indoor environmental quality, and innovation are just a few of the factors that go into determining a building's LEED grade. There are four levels of LEED certification: Certified, Silver, Gold, and Platinum. A national grading system for green buildings used in the planning and assessment of new construction is called the Green grading for Integrated Habitat Assessment (GRIHA). It assigns structures a star rating ranging from one to five. It has a five-year validity period.

In India, sustainable construction methods are encouraged by the Indian Green construction Council's (IGBC) rating system. It evaluates buildings based on factors like planning and site selection, materials, indoor environmental quality, water and energy efficiency, and innovation. The four IGBC certification levels are Platinum, Silver, Gold, and Certified. IGBC follows criteria under LEED which are worldwide accepted but in GRIHA only Indian climate is considered hence, it is applicable for certain region only. IGBC has certified double the number of projects rated by GRIHA [5]. IGBC has more number of certified professionals than GRIHA.

Based on the building's overall sustainability performance, IGBC provides four different rating levels:

- Certified: Represents the fundamental performance in sustainability.
- Silver: Represents strong performance in terms of sustainability.
- Gold: Indicates a high level of sustainability.
- Platinum: Denotes the best possible performance in terms of sustainability.

Numerous factors are assessed by the IGBC rating system, such as energy conservation, water management, indoor air quality, and the use of eco-friendly products. It encourages the use of renewable energy sources, environmentally friendly waste management practices, and other modes of transportation.

2. LITERATURE REVIEW

The existing body of research extensively covers various aspects of sustainable building practices. Researchers have discussed the importance of green buildings and addressed challenges within rating systems like LEED, GRIHA, and IGBC. Adjustments are recommended to improve these systems, emphasizing the need for a more unified approach. Discrepancies exist in how each system rates buildings, and there's a call for more specific and comprehensive assessments. Tang et al. [3] highlight missing criteria, particularly in water and energy sectors, which are critical in IGBC but often overlooked elsewhere. Jadhao et al. [6] analyze sustainability across systems, emphasizing mandatory requirements and minimum points for certification. Sharma [4] compares GRIHA and IGBC, noting IGBC's international recognition and higher number of certified professionals. Overall, research points to ongoing improvements and the importance of holistic evaluation in systems for grading green buildings.

The literature on systems for grading green buildings is vast, covering various aspects of sustainable building practices. Mishra and Gour [5], Joseph et al. [2], Jadhao et al. [6] and Sharma [4] discussed the need of green buildings, addressing major challenges in different systems for grading green buildings with LEED, GRIHA, and IGBC. They suggest adjustments to improve these systems, emphasizing the need for a more unified approach. Joseph et al. [2] provided a general checklist for building projects, covering essential areas for evaluation and certification. Their study reveals discrepancies in how each system rates buildings and calls for improvements to make assessments more specific and comprehensive. Tang et al. [3] review ten different green building rating systems across seventeen categories, identifying missing criteria in many systems. They emphasize the importance of water and energy sectors, which are critical in IGBC but often overlooked in other systems. Future research on the use and effectiveness of various rating systems is necessary, as this study indicates. Jadhao et al. [6] use the Excellence in Design for Greater Efficiencies (EDGE) program to study building sustainability for LEED, GRIHA, and IGBC. Their research illustrates the differing advantages of each grading system by concentrating on water, energy, and resources. They highlight the need for mandatory requirements and minimum points for building certification. Sharma [4] provides an overview of GRIHA and IGBC rating systems, comparing them based on criteria, projects completed, and years of validation. She notes that IGBC, being a subset of LEED, has international recognition and a higher number of certified professionals than GRIHA. This gives IGBC an edge in promoting sustainable building practices globally.

3. METHODOLOGY

The methodology for this research involved a comprehensive approach, including below steps:

1. Case Studies Selection: The study involved both existing and new buildings at the Rajarambapu Institute of Technology (RIT), Rajaramnagar, Sangli campus.
2. Data Collection: Several sustainability parameters, such as indoor air quality, waste management, water conservation, energy efficiency, and the use of green materials, were the subject of extensive data collection.
3. Software Development:
   - The software was developed using Eclipse and MySQL.
   - Classes were created for different tasks.
   - Jar files and plugins were added for connectivity.
   - Tables were created in MySQL to store data.
   - Java classes were connected to databases for data storage.
4. Manual Assessment: Manual ratings through visual inspections of case studies.

5. Software Assessment: The developed software was used to assess the buildings, and the results were compared with manual ratings to validate the software's accuracy.

The following is an exhaustive breakdown of the approaches used in this work.

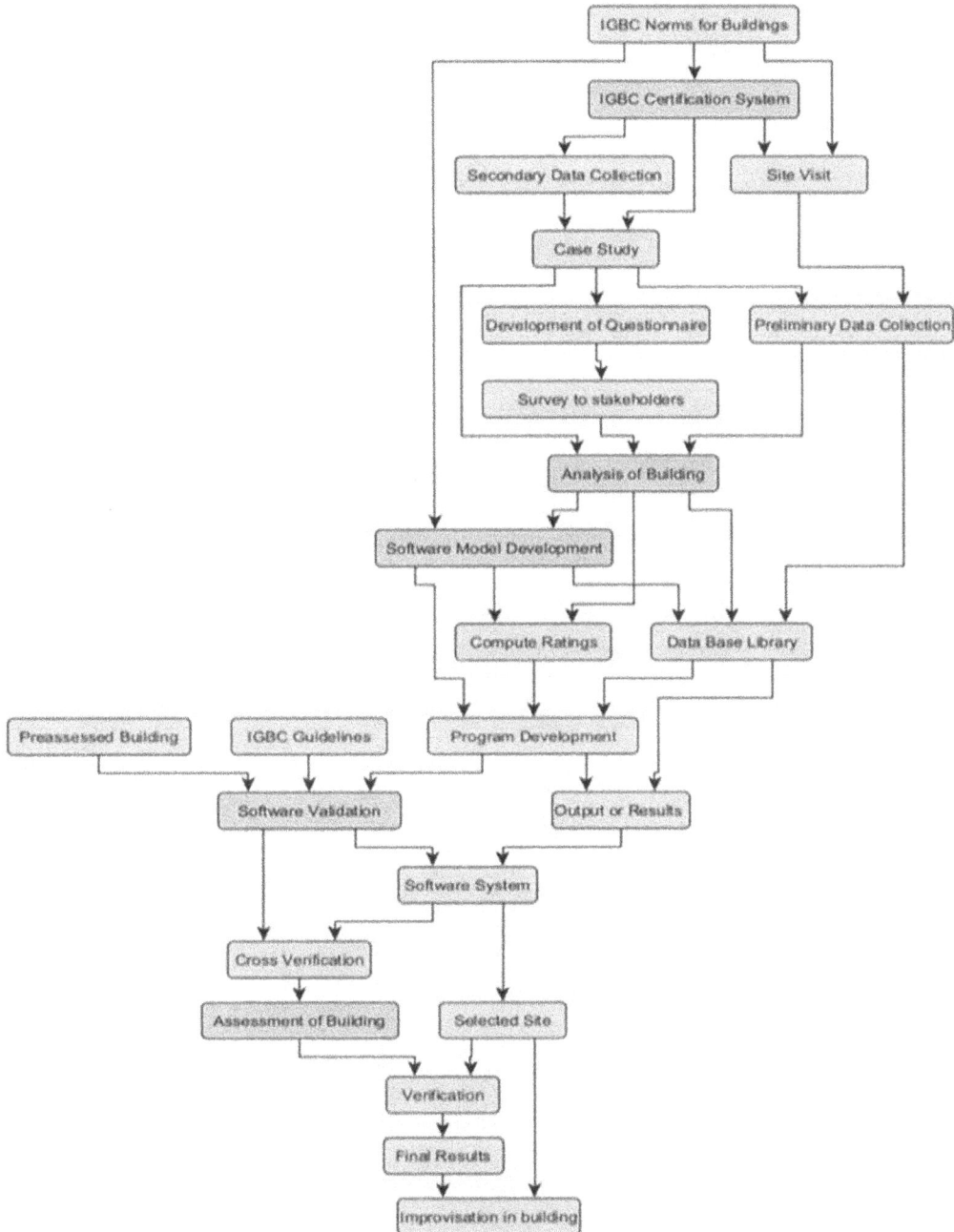

Figure 9.1 Detailed representation of the overall work methodology

Source: Author

3.1 Location of Case Studies

The case studies for both existing and new building are taken from Rajarambapu Institute of Technology, Rajaramnagar, Sangli campus.

Figure 9.2 Location map of RIT

Source: Google Map

Figure 9.3 GIS map survey for RIT campus

Source: Author

3.2 Rating for Existing Building Case Study Methodology

Figure 9.4 Existing building case study-continuing education centre, RIT

Source: Author

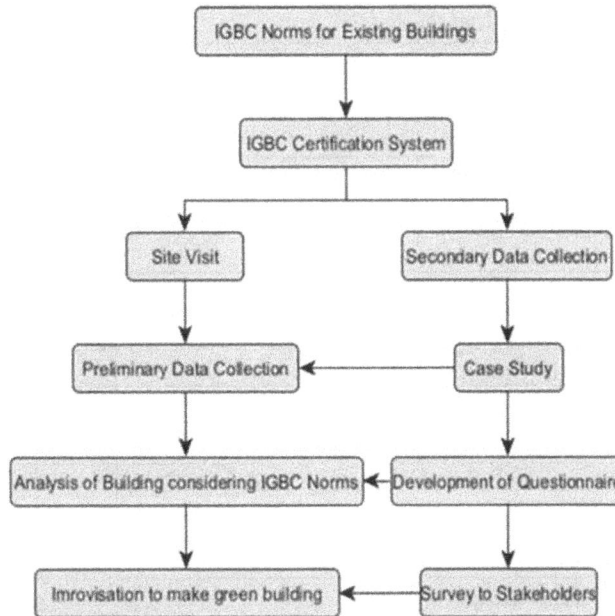

Figure 9.5 Methodology for existing building-continuing education centre at RIT

Source: Author

Table 9.1 Existing building case study IGBC certification details

Site & Facility Management		Credit Points	Selected Site
Required 1	Green Program	0	0
Required 2	Collection and disposal of waste	0	0
Credit 1	Eco Friendly community practices	4	2
Credit 2	Eco Friendly Landscaping practices	2	2
Credit 3, 1	Reduction of heat island – non roof	4	2
Credit 3, 2	Reduction of heat island – roof	4	2
Credit 4	Reduction of outdoor light pollution	2	0
Credit 5	Building maintenance and operations	2	2
	Total	18	10
Water Efficiency		Credit Points	Selected Site
Required 1	Effective fixtures	0	0
Credit 1	Effective fixtures	6	4
Credit 2	Harvesting rainwater	4	2
Credit 3	Treatment of waste water	4	2
Credit 4	Reusing waste water	4	0
Credit 5	Metering of water	4	2
Credit 6	Area of turf	4	4
	Total	26	14

Energy Efficiency		Credit Points	Selected Site
Required 1	Halons and environmentally friendly refrigerants	0	0
Required 2	Lowest possible energy efficiency	0	0
Credit 1	Enhanced energy efficiency	14	6
Credit 2	Renewable energy on-site	6	2
Credit 3	Off-site renewable energy	6	2
Credit 4	Metering of energy	4	4
	Total	30	14
Health & Comfort		**Credit Points**	**Selected Site**
Required 1	Control of tobacco smoke	0	0
Required 2	Ventilation of fresh air	0	0
Credit 1	Monitoring and controlling carbon dioxide	2	1
Credit 2	Pollution-related equipment and systems isolation	2	1
Credit 3	Environmentally friendly cleaning supplies	2	2
Credit 4	RH, indoor temperature, and thermal comfort	2	2
Credit 5	Facilities for individuals with disabilities	4	2
Credit 6	Facilities for occupant well-being	2	2
	Total	14	10
Innovation Category		**Credit Points**	**Selected Site**
Credit 1	Innovation points	10	4
Credit 2	IGBC AP	2	2
	Total	12	6
	Total Points for Existing Building	54/100	

Source: Author's compilation

3.3 Rating for New Building Case Study Methodology

Figure 9.6   New building case study-new classroom building at RIT

Source: Author

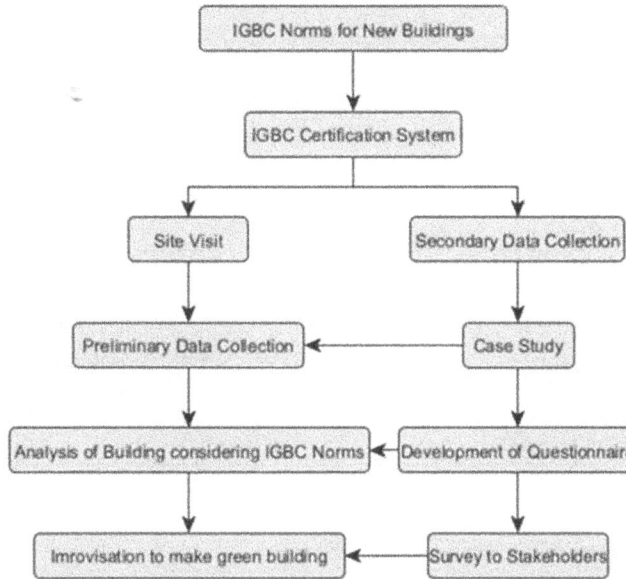

Figure 9.7 Methodology for new building-new classroom building at RIT

Source: Author

Table 9.2 New building case study IGBC certification details

Eco-Friendly Design and Architecture		Credit Points	Selected Site
Credit 1	An Integrated Design Method	1	1
Credit 2	Preservation of Sites	2	2
Credit 3	Architecture That Is Passive	2	2
	Total	5	5
Choosing and Organizing Sites		**Credit Points**	**Selected Site**
Required 1	Local Construction Codes	0	0
Required 2	Control of Soil Erosion	0	0
Credit 1	Essential Facilities	1	1
Credit 2	Close to Public Transportation	1	1
Credit 3	Automobiles with Low Emissions	1	1
Credit 4	Topography or Vegetation in Nature	2	2
Credit 5	Tree conservation or transplantation	1	1
Credit 6	Reduction of Heat Islands, Non-roof	2	1
Credit 7	Reduction of Heat Islands, Roof	2	1
Credit 8	Reduction of Outdoor Light Pollution	1	0
Credit 9	All-purpose Design	1	0
Credit 10	Essential Facilities for Construction Workers	1	1
Credit 11	Green Building Rules	1	0
	Total	14	9

Water Conservation		Credit Points	Selected Site
Required 1	Roof and non-roof rainwater harvesting systems	0	0
Required 2	Water-Saving Plumbing Equipment	0	0
Credit 1	Design of Landscapes	2	1
Credit 2	Controlling Irrigation Systems	1	0
Credit 3	Roof and non-roof rainwater harvesting systems	4	2
Credit 4	Water-Saving Plumbing Equipment	5	3
Credit 5	Reusing and Treating Wastewater	5	3
Credit 6	Measuring Water	1	1
	Total	18	10
Energy Efficiency		**Credit Points**	**Selected Site**
Required 1	Depleting Substances of Ozone	0	0
Required 2	Minimum Efficiency in Energy Use	0	0
Required 3	Plan for Commissioning Building Systems and Equipment	0	0
Credit 1	Sustainable Refrigerants	1	0
Credit 2	Increased Efficiency in Energy Use	15	8
Credit 3	Renewable Energy on-site	6	0
Credit 4	Renewable Energy Off-Site	2	1
Credit 5	Commissioning and Post-Installation Systems and Equipment	2	2
Credit 6	Energy Management and Measuring	2	2
	Total	28	13
Resources and Building Materials		**Credit Points**	**Selected Site**
Required 1	Waste Segregation, Post-Occupancy	0	
Credit 1	Materials for Sustainable Construction	8	4
Credit 2	Management of Organic Waste, After Occupancy	2	1
Credit 3	Managing Waste Materials While Construction	1	1
Credit 4	Utilizing Approved Green Building Supplies, Equipment, and Products	5	2
	Total	16	8
Indoor Environmental Quality		**Credit Points**	**Selected Site**
Required 1	Minimal Ventilation of Fresh Air	0	0
Required 2	Control of Tobacco Smoke	0	0
Credit 1	CO_2 Tracking	1	0
Credit 2	Daylighting	2	2
Credit 3	Outside Views	1	1
Credit 4	Reduce Pollutants Both Indoors and Outside	1	0
Credit 5	Materials with Low Emissions	3	1
Credit 6	Facilities for Occupant Well-Being	1	1
Credit 7	Testing of Indoor Air Quality Before Occupancy and After Construction	2	1
Credit 8	Controlling Indoor Air Quality While Construction	1	1
	Total	12	7

Development and Innovation		Credit Points	Selected Site
Credit 1	Creativeness in the design process	4	2
Credit 2	In structural design, optimization	1	1
Credit 3	Reusing waste water while construction	1	0
Credit 4	Professional accredited by the IGBC	1	0
	Total	7	3
	Total Points for New Building	55/100	

Source: Author's compilation

3.4 Software Model Development

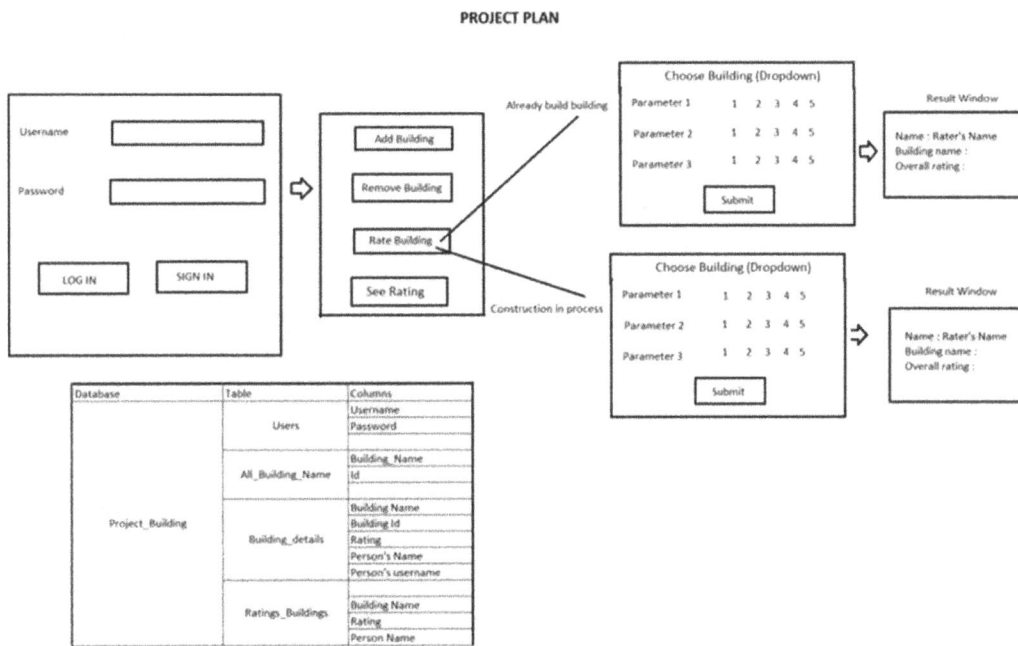

Figure 9.8 Software development flow chart

Source: Author

Methodology used for developing software:

- Downloading Eclipse and MySQL.
- Creating classes for different tasks.
- Adding Jar file and some plug-in settings for connectivity.
- Creating required tables on MySQL for storing of data.
- Connecting java classes with databases for storing the required details.

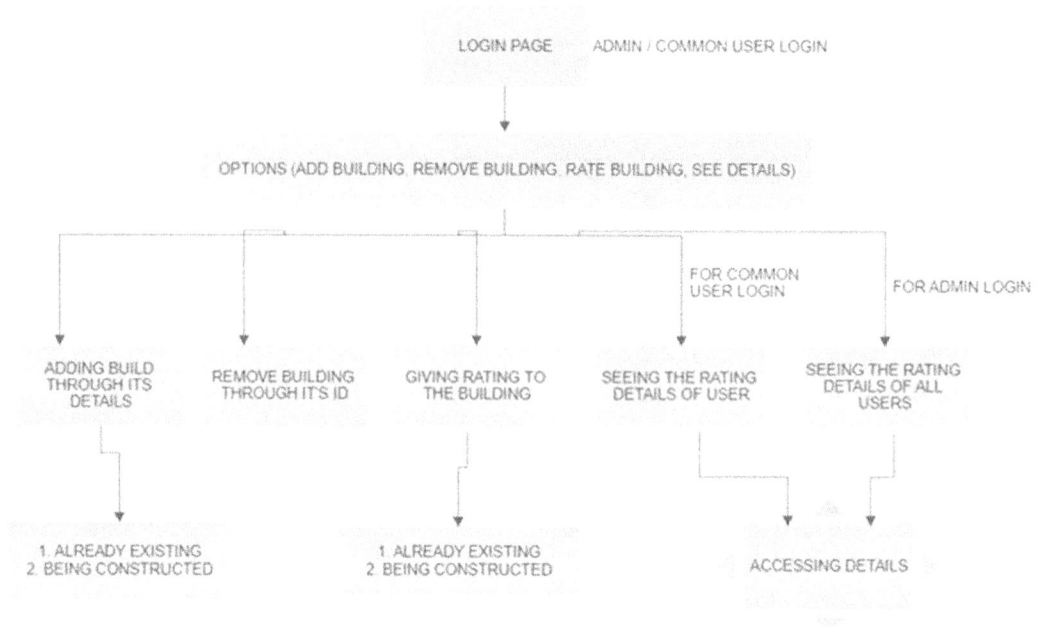

Figure 9.9 Software execution plan

Source: Author

4. RESULTS

On the basis of visual inspection of the case studies, the manual rating has been done for both existing and new building. Then, with the help of software rating has been taken for comparative analysis between manual and software assessment.

Table 9.3 Manually results obtained through building assessment

Building name	Certifications levels	Points	Recognition
Existing	Certified	54	Best Practices
New	Certified	55	Silver (Outstanding Performance)

Source: Author's compilation

Figure 9.10 Validation of rating in developed software-Existing Building case study

Source: Author

building_name	Engineers_na..	ratings	building_status	recognition	day	date	location	building_id	username
Continuing Ed..	Smita Patil	54	Certified	Best Practice	Wednesday	2023-05-31	Islampur	12	smita

Figure 9.11   Overall result in software-Existing building case study

Source: Author

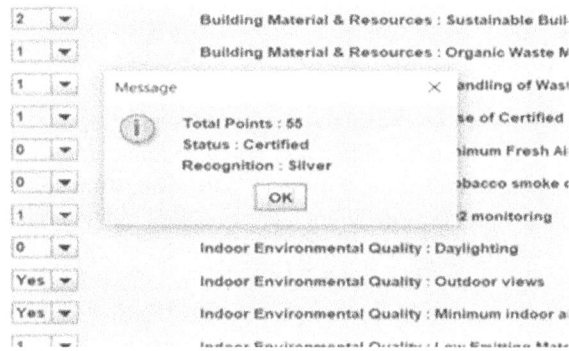

Figure 9.12   Validation of rating in developed software-New building case study

Source: Author

building_name	Engineers_name	ratings	building_status	recognition	day	date	location	building_id	username
Abhi Flats & Aparts	David Harvey	9	Not Certified	Not recognized	Sunday	2023-05-04	Sangli	44	smita
hrnj	rsgkr	0	Null	null	fdkggk	2023-01-04	fogk	567	smita
hgfh	figi	3	Not Certified	Not recognized	fjdg	2023-07-10	dfkgt	754	smita
dgfdg	fuerheru	0	Null	null	udshfu	2023-04-03	jfdgj	780	smita
New Building	Smita Patil	55	Certified	Silver	Wednesday	2023-05-31	Islampur	1234	smita
thk	omkrth	0	Null	null	dkom	2023-04-06	kmok	5666	smita

Figure 9.13   Overall results in software

Source: Author

A comparison between the software-generated ratings and the manual ratings illustrates a considerable level of acceptance, indicating the software model's reliability and efficiency. The rating outputs of the software show a significant relationship with the manual ratings, proving its capacity to accurately assess buildings using the IGBC grading system. The evaluation, involving a visual examination and an in-depth assessment of the building case study, highlighted significant areas where modifications and improvisations can be implemented. The development of the software model for assessing building sustainability compliance with IGBC standards addresses the critical need for standardized and efficient evaluation tools, as highlighted in the literature.

5. Discussion

Mishra and Gour [5] emphasized the importance of managing natural resources efficiently in green building practices. Our software incorporates criteria that assess water conservation measures, energy efficiency, and the use of sustainable materials, aligning with their recommendations for improving existing rating systems. By providing a structured framework for evaluating these aspects, the software helps users implement effective resource management strategies, reducing environmental impact and promoting sustainability. Joseph et al. [2] identified the benefits of incorporating green energy solutions in building projects. The software model addresses the demand for precise and thorough evaluations by assessing the integration of renewable energy sources like solar electricity and wind turbines. In order to create a more sustainable built environment, this feature helps stakeholders find ways to cut back on greenhouse gas emissions and dependency on fossil fuels.

Tang et al. [3] highlighted the importance of waste management in green building practices. Our software includes criteria for assessing waste management strategies, such as segregation, composting, and recycling, which are often overlooked in other rating systems. By providing detailed evaluations of waste management practices, the software encourages the adoption of circular economy principles, minimizing environmental impact and promoting sustainability. Jadhao et al. [6] analyzed the role of greenery in enhancing building sustainability. The software assesses the provision of green spaces and vegetation within and around building premises, improving air quality, providing shade, and creating a healthier environment. This feature aligns with their findings on the importance of integrating green spaces in building projects, highlighting the software's contribution to promoting holistic sustainability. Sharma [4] compared the effectiveness of various rating systems in promoting energy optimization. The software includes criteria for evaluating advanced energy systems and technologies, such as energy-efficient lighting and smart controls, ensuring buildings optimize energy use. By providing a standardized approach to assess energy optimization measures, the software supports stakeholders in implementing effective strategies to reduce energy consumption and improve building performance. The software evaluates the adequacy of ventilation and central cooling systems, addressing indoor air quality concerns. The conclusions of Tang et al. [3], who stressed the significance of these aspects in building sustainability assessments, are supported by this feature. The program assists users in improving indoor air quality, which benefits occupant health and well-being, by offering thorough assessments of ventilation and cooling systems.

The use of clay-based construction materials, as discussed in the literature, is assessed in the software. This feature aligns with the findings of Jadhao et al. [6], who highlighted the benefits of using stabilized compressed earth blocks in reducing energy consumption and environmental impact. The software facilitates the adoption of green building practices by encouraging the use of sustainable building materials. The software addresses the significance of water conservation emphasized by Mishra and Gour [5] by providing criteria for assessing rainwater harvesting systems, graywater recycling, and water-saving fixtures. By providing comprehensive assessments of water management practices, the software helps stakeholders implement effective strategies to conserve water resources and support natural ecosystems.

6. CONCLUSION

This research presents a comprehensive and user-friendly software model that provides an efficient method for assessing buildings' sustainability compliance with IGBC standards. The software facilitates the evaluation process by integrating critical sustainability factors, offering a standardized and systematic approach for quick and accurate assessments. The model's development involved extensive research and collaboration with industry experts to ensure adherence to IGBC guidelines. The developed software provides a practical answer to the problems connected with manual ratings, such as high costs, complexity, and time consumption. Its application simplifies and speeds up the rating process. The validation of the software against manual ratings demonstrated its reliability and efficiency in assessing building sustainability. This innovation promotes sustainable building practices by providing stakeholders with essential information to make informed decisions, ultimately contributing to the creation of environmentally responsible built environments.

REFERENCES

[1] Tarde, G. B., & Binayake, R. A. (2022). A review paper on study of practical aspect of IGBC rating system. *International Journal of Creative Research Thoughts (IJCRT)*, 10(04), 119–127. ISSN:2320-2882.

[2] Joseph, K., Jose, V., Dinesh Kumar, A. N., & Sunny, S. M. (2018). A review on various green building rating systems in India. *International Journal of Scientific and Engineering Research (IJSER)*, 09(05), 1851–1859. ISSN:2229-5518.

[3] Tang, K. H. D., Foo, C. Y. H., & Tan, I. S. (2020). A review of the green building rating systems. *Materials Science and Engineering*, doi:10.1088/1757-899X/943/1/012060.

[4] Sharma, M. (2022). A brief overview of GRIHA and IGBC rating systems. *International Research Journal of Engineering and Technology (IRJET)*, 09(06), 21–25. ISSN-2395-0056.

[5] Mishra, P., & Gour, S. (2018). Green building rating system. *International Journal of Research in Engineering, Science and Management (IJRESM)*, 01(10), 456–461. ISSN:2581-5782.

[6] Jadhao, P. R., Chaudhari, R. S., & Mali, S. T. (2022). Analysis of building sustainability for LEED, GRIHA & IGBC rating systems. YMER, 21(11), 2467–2474. ISSN:0044-0477.

[7] Sankpal, S. J., & Patil, A. B. (2021). A review paper on study and analysis of green building rating system to improve performance of residential buildings in Sangli and Kolhapur region. *International Research Journal of Engineering and Technology (IRJET)*, 08(09), 197–200. ISSN: 2395-0056.

Advances in Construction, Real Estate, Infrastructure and Project Management – Anil Kashyap et al. (eds)
© 2026 Taylor & Francis Group, London, ISBN 978-1-041-13433-6

10 | Feasibility Analysis of Maintenance and Management of Buildings

Pruthvi M.[1],
Sathwik B. C.[2],
H. Saarthak[3], Arkith R. Rai[4]
Student,
NICMAR University Pune

Shashank B. S.[5]
Assistant Professor,
NICMAR University Pune

■ **ABSTRACT:** Building maintenance is a crucial for ensuring a building's durability, security, and efficiency. Without proper building maintenance, structure can deteriorate and which in turn leading to many structural component issues, safety related issues and their hazards, and reduces the service life of the building. Building maintenance is a source of contention between management and occupants. It is found that the some of the factors that are responsible for the building maintenance are age of building, building users, financial and technical parameters. Finding the causes of these issues is the main goal of this study. The primary objective of the questionnaires, which were developed and distributed to users and stakeholders, were Fund for maintenance, efficiency of maintenance management, and satisfaction with maintenance management. Secondary information was gathered through the thorough literature review. Obtained element is then prioritized based on their ranking by suitable methods. Eight significant elements are considered from related literatures were scrutinised by users and their effects on the upkeep of such facilities were assessed. Based on the respondents evaluations, the Relative Importance Index was used to calculate relevance of each aspect. Study would then give insights upon the Current Building maintenance practices and the analysis results would help to explore or confirm the relationships between various factors considered as survey.

■ **KEYWORDS:** Preventive maintenance, Lifetime engineering, Maintenance manuals

[1]P2370181@student.nicmar.ac.in, [2]P2370184@student.nicmar.ac.in, [3]P2370210@student.nicmar.ac.in, [4]P2370236@student.nicmar.ac.in, [5]shashankbs@nicmar.ac.in

DOI: 10.1201/9781003669814-10

1. INTRODUCTION

For infrastructure development to be viable for the long run, building maintenance is very essential. It plays a important role in the operational efficiency and long life of buildings [2, 3]. The goal of maintenance is to preserve or return building components to their functional state through a variety of technical operations. Preventive maintenance involves proactive measures to prevent failures, while corrective maintenance addresses issues post-failure [11]. The goal of Building maintenance's major goal is to keep buildings in their original structural, functional, and aesthetic states so that their investment value is protected over time [13]. Peeling paint, dampness, surface cracking, fungus or small plant growth, floor or tile breaking, plumbing and w/s, toilet, bathroom and sanitary, electrical problems, and inadequate maintenance. and procedures are the most important building maintenance issues. Damage to a building's contents and occupants can result from the fabric and finishing failing, deteriorating, and decaying due to poor care [1]. To keep structures operating at the desired levels, effective maintenance management include organizing, carrying out, assessing, and regulating various tasks. This requires technical expertise and adherence to standards to avoid failures that can lead to financial losses, safety hazards, and legal implications. Additionally, adapting to evolving construction techniques and user demands fosters greater efficiency and quality in building maintenance practices.

A structured maintenance plan offers several benefits. It ensures systematic care and organization of properties, monitors building services for efficient use, and maintains consistent presentation standards [21]. By reducing subjective decision-making and emergency repairs, it minimizes potential damages and legal liabilities arising from neglected maintenance. Furthermore, building maintenance is essential for maintaining operational efficiency and mitigating future risks and financial burdens associated with building deterioration [4, 5]. It involves various stages, starting from the planning and design phase, where efforts are made to minimize future maintenance needs through efficient design and construction practices. During the construction phase, high-quality workmanship is crucial to minimizing maintenance requirements during the building's lifespan [14]. Maintenance activities grouped as planned and not planned types. Planned maintenance includes preventive, scheduled, corrective and condition-based tasks, which are performed following defined schedules or specific conditions to enhance the efficiency and lifespan of a building. In contrast, unplanned maintenance, which includes emergency, corrective, and unpredictable maintenance, addresses immediate issues that arise unexpectedly [13].

The decision to perform maintenance work is influenced by the many factors, like design complexity, the selection of suitable materials, the availability of skilled labor, and conformity with local standards and specifications. Neglecting maintenance can lead to accelerated building deterioration, increased repair costs, and compromised safety, underscoring the importance of proactive maintenance strategies throughout a building's lifecycle [6].

2. LITERATURE REVIEW

According to the study, public buildings are extremely precious assets in any country, and developing nations like Nepal are finding it difficult to keep them in a regularly operable shape. The most important building maintenance difficulties include plumbing and water supply, toilet, bathroom, and sanitary issues, electrical issues, poor maintenance management and procedures, paint peeling, dampness, surface cracking, the growth of fungus or small plants, and decay or damaged floor/tile [10, 12]. The use of subpar materials and workmanship, inadequate contractor and construction supervision, a Lack of high-quality materials and skilled labor, inadequate maintenance management, excessive and unforeseen maintenance expenses, the building's age, and natural disasters, earth settlement, and other factors are some of the main causes. To ensure high-quality construction and delivery of all building papers, precautions ought to be done throughout the stages of design and construction.[7–9] The goal of the research was to create mathematical models that simplify the process of building maintenance.

This was accomplished by dividing the building in accordance with division techniques derived from a number of international maintenance guides and earlier research. As a result, a questionnaire covering the majority of building maintenance items was created based on a survey of the literature and conversations with specialists in the field [18–20]. The Social Sciences Statistical Package is then utilized to process questionnaire information collected, and Weighted Sum Model (WSM) technique was employed to identify the most crucial maintenance items. Lastly, this study suggested using the model to quickly assess and properly monitor structures. Additionally, it will assist engineers and architects in making forecasts using scientific approaches rather than relying just on their own judgment [15–17].

2.1 Problem Definition

Most of the buildings constructed today face various problems due to poor design or lack of proper maintenance. Design issues often arise from unclear or incorrect specifications provided by builders or designers who overlooked the importance of maintaining these facilities. Common defects include carpentry issues like broken door handles, faulty door locks, leaking roofs, and damaged kitchen cabinets; electrical problems like malfunctioning light switches and ceiling fans; structural issues like cracks in floors, walls, and structural components; and plumbing faults such as leaking pipes, unstable toilet seats, and poorly fitted valves.

3. RESEARCH METHODOLOGY

The research methodology that is being used is covered in this chapter. First, we have thought about the things that need to be studied in the research technique. Based on the research articles and conference papers we reviewed for our literature study, these parameters were chosen, debated, and finalized. The parameters, which covered general, technical, and financial issues, were primarily centred on the existing situation. As a result, the complete process is covered in full in this chapter.

3.1 Flow Chart for the Entire Methodology Followed

See Figure 10.1.

3.2 Methodology

The research methodology was conducted in four phases:

1. Preparing the research
2. Information gathering
3. Analysis of data
4. Findings and conclusion

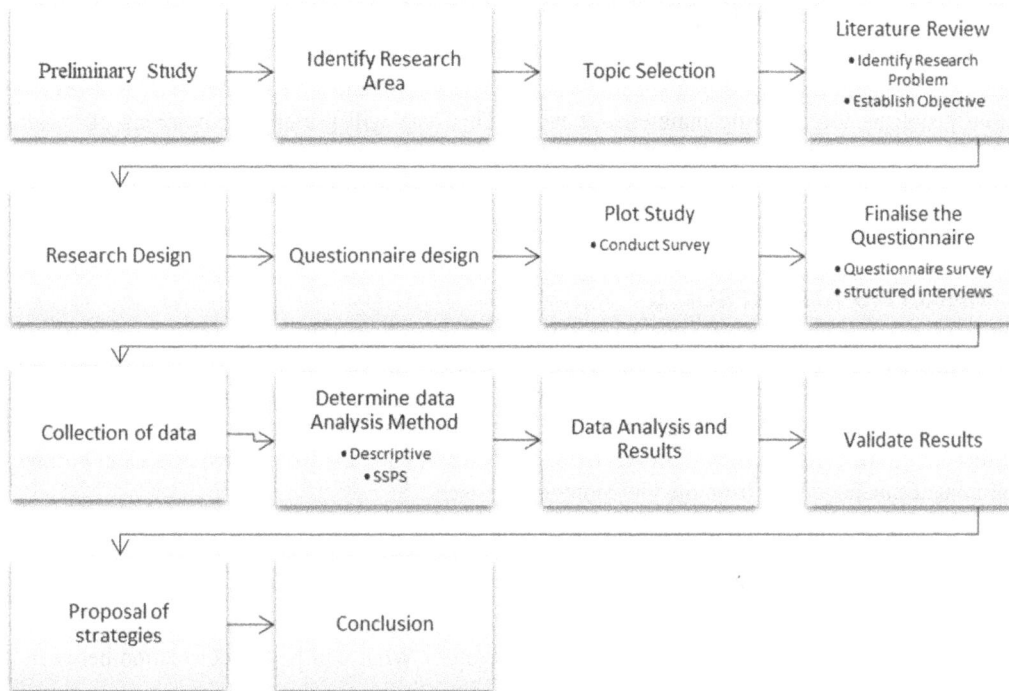

Figure 10.1 Research methodology

Source: Authors

3.3 Preparing the Research

In preparing for this research, the importance of effective building maintenance management was carefully considered from the start. The main objective was to identify the key issues and factors influencing building maintenance. To deepen the understanding of the subject, a comprehensive review of existing literature, including journals, theses, articles, and books, was conducted.

The research followed a quantitative approach, using primary data gathered through a field survey via Google Forms and a structured questionnaire. The management of maintenance in the construction industry was the main focus of the study, focusing on building maintenance management and life cycle engineering. The target audience included private homeowners, apartment residents, and individuals from various housing types, as well as stakeholders such as consultants, contractors, clients, and project management consultants (PMC). These participants were chosen due to their accessibility, expertise in the construction sector, and willingness to share valuable insights on daily maintenance management challenges.

The data collection process involved a well-structured questionnaire with closed-ended questions related to building maintenance issues. The goal was to gather over 200 responses from users and 50 from other stakeholders. So far, approximately 110 user responses have been received, resulting in a response rate of about 55.5%, with data collection still ongoing.

For data analysis, SPSS version 21 will be used to ensure accurate and reliable results based on the responses gathered.

3.4 Information Gathering

Data has been collected by conducting survey for which questionnaire was prepared by research data, visual examination, and plot study. This questionnaire would give the real time data and the actual problems faced during maintenance work. This data will present the opinions of various consumers and stakeholders, which will boost the study information legitimacy.

We have prepared questionnaires for the Users, Developers, Stakeholders, and consultant. This questionnaire is created on google forms and is been circulated to respective people.

4. Analysis and Results

4.1 Analysis of Data

The majority of those surveyed are workers from various fields. Most responders are residents of residential buildings, such as detached homes and apartments. This volume of respondent data is perfect for our research since it allows us to precisely understand the issues encountered in building maintenance management from the viewpoint of the users.

Factor Analysis using IBM SPSS Statistics 21

Our exploratory factor analysis aims to address the following research question:

What are the fundamental aspects of our responses received for factors that contribute more for maintenance of buildings? And how they are interrelated? What will be the correlation between 2 variable data?

4.2 Factor Analysis Results

The Pearson correlation coefficients between The correlation matrix is a square table, shows different variables in a data set we collected.

The A linear link between two variables is measured by the Pearson correlation coefficient. Pearson Correlation coefficient between the required variables is shown in the table below. The variables data set are obtained from the responses that we received from google form. A Factor analysis was run on our data set to simplify the data, in order to decrease the quantity of variables in regression models

Table 10.1 Bartlett's test and KMO

Kaiser-Meyer-Olkin sampling adequacy measure.		.695
Bartlett's test of sphericity	Approximate Chi-Square	643.518
	Df	210
	Sig.	.00

Source: Compiled by authors

Our model has value of 0.695, which is close to 0.7, indicating that the sampling adequacy is moderately sufficient.

Communalities show The extent to which the factors account for the variance in each variable. Initial communalities estimate the variance in each variable explained by all the components or factors. In this study, which uses principal components extraction for the analysis, the initial communalities are always 1.0 for correlation analyses.

A variable's inclusion or exclusion from the factor analysis is also determined by the communality value. An ideal value is one that is more than 0.5. However, even if a variable has a low communality value (less than 0.5), it can still contribute to a well-defined factor, despite having a low loading. In communality table, two variables have communality less than 0.5, meaning these variables should be excluded from the factor analysis.

Table 10.2 Explained total variance

Com-ponent	Initial Eigen values			Extraction sums of squared loadings			Rotation Sums of Squared loadings		
	Total	Percentage of Variance	Cumulative %	Total	Percentage of Variance	Cumulative %	Total	Percentage of Variance	Cumulative %
1	3.895	18.549	18.549	3.895	18.549	18.549	3.367	16.033	16.033
2	1.830	8.712	27.261	1.830	8.712	27.261	1.959	9.326	25.360
3	1.421	6.766	34.027	1.421	6.766	34.027	1.447	6.889	32.249
4	1.319	6.280	40.307	1.319	6.280	40.307	1.302	6.198	38.446
5	1.281	6.100	46.407	1.281	6.100	46.407	1.294	6.163	44.609
6	1.102	5.248	51.655	1.102	5.248	51.655	1.284	6.115	50.725
7	1.074	5.113	56.768	1.074	5.113	56.768	1.179	5.614	56.338
8	1.023	4.872	61.640	1.023	4.872	61.640	1.113	5.302	61.640
9	.989	4.709	66.349						
10	.938	4.464	70.813						
11	.837	3.984	74.798						
12	.794	3.782	78.579						
13	.770	3.668	82.248						
14	.700	3.333	85.580						
15	.602	2.864	88.445						
16	.547	2.604	91.049						
17	.489	2.330	93.379						
18	.415	1.976	95.355						
19	.387	1.843	97.198						
20	.299	1.426	98.624						
21	.289	1.376	100.000						
Analysis of Principal Components is the extraction method.									

Source: Compiled by authors

Eight of the components in the table above are extracted factors loaded under the extracted sums of squared loadings column since their eigenvalues are greater than 1.

The scree plot in the figure shows the eigenvalues of factors or principal components in an analysis as a line plot. This plot helps determine how many components to keep in principal component analysis (PCA). The process of using the scree plot to identify significant components is known as the scree test, which was introduced by Raymond B. Cattell in 1966.

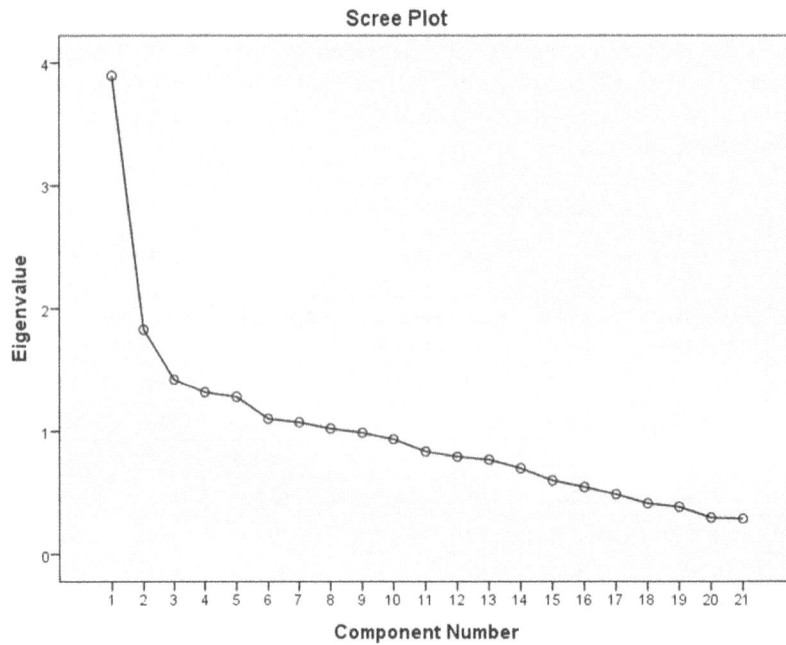

Figure 10.2 Scree plot

Source: Authors

In the plot, the eigenvalues are shown in descending order. The "elbow" point, where the slope levels off, is identified through the scree test, and factors or components to the left of this point are considered significant. The plot shows that 8 factors have eigenvalues greater than 1, meaning these 8 variables should be retained and are considered highly influential for our study.

Table 10.3 Component matrix

	Component							
	1	2	3	4	5	6	7	8
Wear and tear	.712	-.190	-.144				.148	
Biological agents Fungi Insects	.687	-.343			.107		.139	-.181
Geographic allocation and site condition	.671		.116	-.125	-.164	.182	-.226	
Faulty construction practices	.665	-.224			.128			
Quality of materials	.655	.404	.189			.159	-.114	
Unused building after construction	.652	.200	-.182				.183	.271
Heavy rains	.628	-.256			-.133			.145
Design Complexity	.546	.428	.345		-.167			
Age of building	.467	.150	.216	.460	.270			-.219
What is the age of building auto	-.222	.480	-.118	-.185	.296			.201
What type of building are you staying		.431	.182	.356	.399	-.176	-.159	-.186
Which climate condition affect most auto	.251	.342	.142	-.230		-.262	-.134	

	Component							
	1	2	3	4	5	6	7	8
How often is maintenance work carried out in your building auto		.281	-.460	.410		.275	-.364	.159
How long have you been staying in building or apartment	-.175	-.118	.454	.393	-.342	.201	.172	-.183
Do you feel maintenances could be done on regular basis auto	.169	-.186	-.318	.169	.541	-.124		.150
Give your rating for agency providing maintenance auto	-.184	-.403	.257		.499	.160	.166	.363
How do you recognize the need of maintenance auto	.135	-.269	.377	.169		-.577	-.143	
Do you find building maintenance affordable auto			.463	-.371	.364	.488		
Are you aware of building maintenance auto		.264	-.115	.426		.208	.620	
Which service in your premises needs more maintenance auto		.422		-.269		-.376	.524	.117
How much do you spend on building maintenance auto	-.170		.344	.281	-.200			.714

Source: Compiled by authors

Table 10.4 Rotated component matrix

	Component							
	1	2	3	4	5	6	7	8
Wear and tear	.748			-.134				
Biological agents Fungi Insects	.746				-.141			-.273
Faulty construction practices	.698			.131		.124		
Heavy rains	.685							.104
Unused building after construction	.591	.268		-.134	.243	.170	.227	.142
Geographical location and site condition	.585	.435					-.252	
Design Complexity	.278	.703	.162			-.116		.149
Quality of materials	.368	.660	.273		.127			-.109
Which climate condition affect most auto		.467			-.186	.238	.164	
What type of building are you staying in auto	-.221	.183	.704			.117		
Age of building	.308	.153	.695			-.114		
Do you find building maintenece affordable auto		.179		.815				
Give your rating for agency providing maintenance auto		-.443		.597		.104		.343
How do you recognize the need of maintenance auto	.108		.226		-.720			.163

	Component							
	1	2	3	4	5	6	7	8
How often is maintenance work carried out in your building auto			.174	-.416	.537	.167	-.388	.168
Are you aware of building maintenance auto		-.236	.364		.497	-.267	.430	
How long have you been staying in building or apartment	-.110		.116			-.754		.137
Do you feel maintenances could be done on regular basis auto	.245	-.333	.294			.510	-.109	
What is the age of building auto	-.363	.171		.132	.275	.414	.163	
Which service in your premises needs more maintenance auto		.188				.114	.789	
How much do you spend on building maintenance auto	-.111					-.107		.869

Source: Compiled by authors

To better match the data, SPSS can rotate the factors after they have been extracted. Varimax is the most widely utilized rotation technique. Varimax is an orthogonal rotation approach that usually results in very high or very low loadings and tries to make the factor loadings more distinct. This facilitates the process of assigning a certain factor to each object. Varimax rotation, also known as Kaiser-Varimax rotation, basically maximizes the variance of the squared loadings, or the correlations between factors and variables. Therefore, some variables will have weak loadings or no loadings at all on a single component, while others will have substantial loadings.

Table 10.5 Component transformation matrix

Compo-nent	1	2	3	4	5	6	7	8
1	.878	.418	.198	-.056	-.024	.074	-.001	-.081
2	-.388	.637	.328	-.172	.406	.146	.347	.015
3	-.125	.361	.188	.534	-.462	-.491	.021	.281
4	.047	-.332	.676	-.365	.146	-.341	-.200	.346
5	-.027	-.256	.537	.535	-.051	.584	.015	-.123
6	.073	.035	-.105	.474	.731	-.257	-.393	-.060
7	.201	-.336	-.013	.177	.211	-.303	.823	-.051
8	.122	-.010	-.247	.088	.136	.342	.089	.879

Source: Compiled by authors

It is evident which things are connected to each particular component by looking at the rotational Component Matrix above, which displays the loadings of each individual item on the rotational components. This aids in determining which variables have the most influence on each element. Furthermore, the Component Transformation Matrix helps to comprehend how the components have been redefined and related to one another throughout the rotation process by showing the correlations between them both before and after rotation.

Table 10.6 Correlation matrix

Correlation Matrix[a, b]		Moisture	Quality of materials	Design complexity	Chemical agents Alkalis and Chlorides	Heavy rains	Wear and tear	Biological agents Fungi Insects	Building Archi-tecture
Correlation	Moisture	1.000	.408	.489	.503	.136	.586	.560	.102
	Quality of materials	.408	1.000	.614	.282	.045	.478	.150	136
	Design complexity	.489	.614	1.000	.440	-.143	.631	.060	.520
	Chemical agents Alkalis and Chlorides	.503	.282	.440	1.000	.280	.627	.503	.176
	Heavyrains	.136	.045	-.143	.280	1.000	.158	.082	-.077
	Wear and tear	.586	.478	.631	.627	.158	1.000	.230	.687
	Biological agents Fungi Insects	.560	.150	.060	.503	.082	.230	1.000	-.251
	Building Architecture	.102	136	.520	.176	-.077	.687	-.251	1.000
	Geographical location and site location	273	.301	.196	.189	.409	213	.278	-.255
	Age of the building	.483	.750	.308	.243	.396	.206	.269	-.321
	Non use of building after construction	.081	219	.614	.056	-.363	.478	-.452	.686
	Faulty construction practices	.105	.408	.225	.073	.160	.123	-.065	-.060

Source: Compiled by authors

a. Determinant = .000
This matrix is itive definita

4.3 Findings from Factor Analysis for Consultants, Developers, and Stakeholders

Table 10.7 Finding of factor analysis

Geographical Location and Site Location	Age of the building	Nonuse of building after construction	Faulty construction practices
273	.483	.081	.105
.301	.750	219	.408
.196	.308	.614	.225
.189	243	.056	.073
.409	.396	-.363	.160
.213	206	.478	.123
.278	.269	-.452	-.065
-.255	-.321	.686	-.060
1.000	.575	-.028	.524
.575	1.000	-.276	.658
-.028	-.276	1.000	-.094
.524	.658	-.094	1.000

Source: Compiled by authors

Table 10.8 Extraction method: Principal component analysis

Table Communalities		
Communalities	**Initial**	**Extraction**
Moisture	1.000	.738
Quality of materials	1.000	.771
Design complexity	1.000	.839
Chemical agents Alkali sand Chlorides	1.000	.724
Heavy rains	1.000	.901
Wear and tear	1.000	.901
Biological agents Fungi Insects	1.000	.883
Building Architecture	1.000	.872
Geographical location and site location	1.000	.609
Age of the building	1.000	.935
Non use of building after construction	1.000	.860
Faulty construction practices	1.000	.723

Source: Compiled by authors

Each variable's degree of variance is displayed by the Communalities. The initial communalities are used to calculate the variance in each variable that explains all of the components or factorsIt is always equal to 1.0 for correlation studies using principal components extraction as the extraction method.

Another consideration when deciding whether to include a variable in a factor analysis is the importance of community. Ideally, several larger than 0.5 are considered. A variable with a low communality value, i.e., 0.5, is found to contribute to a well-defined factor in the study despite the low loading.

The communal value of each variable in the communality table above is more than 0.5.

Table 10.9 Total variance explained

Compo-nent	Initial Eigenvalues			Extraction Sums of Squared Loadings			Rotation Sums of Squared Loadings		
	Total	**% of Variance**	**Cumulative %**	**Total**	**% of Variance**	**Cumulative %**	**Total**	**% of Variance**	**Cumulative %**
1	4.102	34.183	34.183	4.102	34.183	34.183	3.034	25.280	25.280
2	2.901	24.176	58.359	2.901	24.176	58.359	2.835	23.622	48.902
3	1.671	13.924	72.284	1.671	13.924	72.284	2.580	21.497	70.400
4	1.081	9.011	81.295	1.081	9.011	81.295	1.307	10.896	81.295
5	.657	5.472	86.767						
6	.526	4.386	91.154						
7	.438	3.653	94.807						
8	.339	2.829	97.635						
9	.220	1.836	99.472						
10	.062	.518	99.990						
11	.001	.010	100.000						
12	$-1.002E-013$	$-1.017E-013$	100.000						
Extraction method: Principal Component Analysis.									

Source: Compiled by authors

In the table shown above, it is observed that 4 factors have eigenvalues greater than 1, which are the extracted factors. These factors are listed under the "extracted sums of squared loadings" column. Since the factor analysis is based on the correlation matrix, the variables are standardized, meaning each variable starts with a variance of 1. With 12 variables, the total variance is 12, as each variable has a variance of 1.

This scree plot from figure shows the importance of different factors or components in an analysis. It helps decide which components to keep in principal component analysis (PCA). The

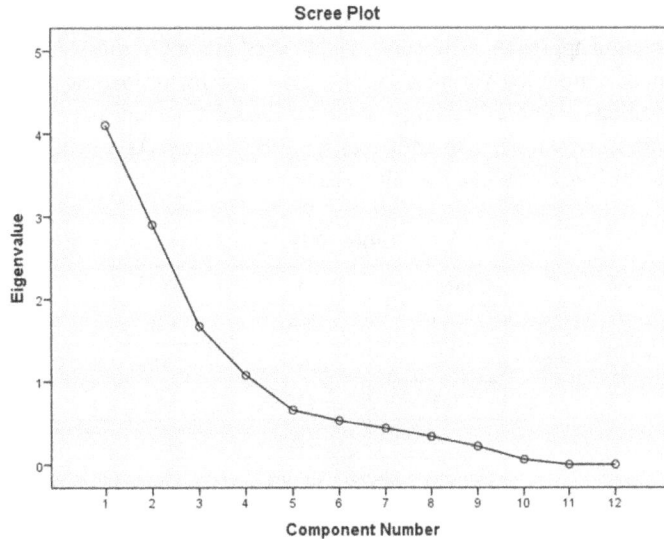

Scree Plot

Figure 10.3   Scree plot

Source: Authors

screen test, introduced by Raymond B. Cattell in 1966, uses this plot to find the important components.

In the plot, the eigenvalues (which show the importance) are listed from highest to lowest. The "elbow" is the point where the graph flattens, and the components to the left of this point are considered important and should be kept.

Table 10.10   Rotated component matrix[*a]

	Component			
	1	2	3	4
Building Architecture	.908	-.206		
Non use of building after construction	.874		-.167	-.258
Wear and tear	.747	.159	.522	.210
Design complexity	.716	.394	.335	-.242
Age of the building	-.155	.901	.293	.112
Faulty construction practices		.827	-.147	.133
Quality of materials	.316	.737	.290	-.209
Geographical location and site location		.668	.158	.364
Biological agents Fungi Insects	-.336		.877	
Moisture	.194	.281	.788	
Chemical agents Alkali sand Chlorides	.287		.740	.300
Heavy rains	-.105	.207	.106	.914
Extraction Method: Principal Component Analysis. Rotation Method: Varimax with kaiser normalization.				
a. Rotation converged in 5 iterations.				

Source: Compiled by authors

After extraction, SPSS can modify the components to better fit the data. Varimax, an orthogonal rotation approach, is the most widely used method for this. Assigning each item to a certain factor is made simpler by Varimax's very high or low factor loadings. Increasing the variance of the squared loadings—where "loadings" stand for the relationships between variables and factors—is the aim of Varimax rotation, also known as Kaiser-Varimax. As a result, some variables frequently have high factor loadings while the others have low ones.

Table 10.11 Component transformation matrix

Component	1	2	3	4
1	.460	.630	.613	.128
2	.845	-.415	-.145	-.304
3	.145	.642	-.744	-.118
4	.230	-.140	-.224	.937
Extraction Method: Principal component analysis. Rotation Method: Varimax with Kaiser Normalization.				

Source: Compiled by authors

Table 10.12 Relative importance index of the factors

Factors	very high	high	medium	low	very low	total	total number	A*N	RII	rank
Quality of Materials	280	224	99	30	10	643	170	850	0.756471	2
Age of the Building	300	220	93	30	9	652	170	850	0.767059	1
Design Complexity	110	136	228	54	11	539	170	850	0.634118	5
Faulty Construction Practices	90	184	213	46	12	545	170	850	0.641176	4
Wear and Tear	55	168	225	60	12	520	170	850	0.611765	6
Weather	150	188	144	50	20	552	170	850	0.649412	3
Unused Buildings After Construction	40	120	45	58	88	351	170	850	0.412941	7
Other Factors	15	76	75	14	116	296	170	850	0.348235	8

Source: Compiled by authors

Eight important factors identified from relevant studies were examined by users to assess their impact on facility maintenance. Based on respondents' evaluations, the RII was used to calculate each factor's significance. The age of the building, quality of materials, and environmental conditions (especially during the rainy season) were found to be the most important elements, with corresponding RII values of 0.77, 0.76, and 0.65.

5. CONCLUSION

Analysis of user data using descriptive and factor analysis shows that respondents perceive several key factors as the main reasons for increased building maintenance and challenges in effective maintenance. These factors include the building's age, material quality, rainy season and climatic conditions, poor construction practices, complex designs, wear and tear, and biological agents.

Correlation between variables was analyzed using factor analysis. Similarly, data from stakeholders, developers, and consultants indicate that the building's age, material quality, poor construction practices, moisture, and wear and tear were the most important elements influencing effective maintenance of building.

Suggested strategies is to minimize the impact of these factors on building maintenance management are as follows. As buildings get older, maintenance management is often impacted by wear and tear. To prevent this, it is crucial to prioritize preventive maintenance to ensure the building remains functional throughout its designed lifespan. Buildings older than 10 years should be regularly inspected, with a thorough inspection every 5 years. Respondents also feel that assessing the quality of materials used during construction is important to reduce future maintenance and repair costs. High-quality materials are more durable, weather-resistant, and offer better safety assurance over time. The overall design's integrity is improved by using the right building materials. Maintenance manuals must include standard checklists for checking the quality of materials, and proper records must be maintained about the type of materials used and the tests done on them. Based on the survey, extended warranties will increase the quality of materials, so users should pay the extra charges to attain the warranty.

Through the analysis, it was observed that most respondents answered that the rainy season affects buildings the most and increases maintenance costs. Rain can mainly affect external finishes, causing moisture to result in leakages of walls and ceilings, corrosion of steel, and weakening of the structure. A pre-monsoon examination should be done before the onset of the monsoon season to get cracks filled and damaged tiles repaired to avoid seepage. The roof and terrace of the building should be checked, and waterproofing coatings should be applied if any cracks or leakage are present. Applying waterproof paints and using water sealants can act as a protective coating on concrete walls. Faulty construction practices often stem from design flaws, especially in structural design, like not accounting for space needed for expansion and contraction, which leads to cracks and, eventually, issues like broken pipes or joint failure. It's crucial to maintain quality during construction, which relies on effective teamwork, and contractors' performance should be closely monitored to prevent defects, errors, or the need for frequent inspections.

To make buildings easier to maintain, design complexity should be reduced. Complex designs make maintenance more challenging, time-consuming, and costly. By regularly cleaning and carrying out small repairs, major replacements can often be prevented. execution of maintenance necessitates the usage of some tools, and designers should provide sufficient clearance for the tools to enter and exit. Additionally, designers should avoid permanently fixing features that require ongoing maintenance. Observations related to the financial aspect of maintenance of buildings indicate that a high number of users are ready to pay extra charges for maintenance services and feel that the charges for maintenance services are fair. Users are willing to pay service providers if they provide extra warranty for their service, which will increase the efficiency of building maintenance. Further, service providers should provide a certificate for their services.

REFERENCES

[1] Dahal, R. C., & Dahal, K. R. (2020). A review on problems of the public building maintenance works with special reference to Nepal. *American Journal of Construction and Building Materials*, 4(2), 39.

[2] Lacasse, M. A., Gaur, A., & Moore, T. V. (2020). Durability and climate change—implications for service life prediction and the maintainability of buildings. *Buildings*, 10(3), 53.

[3] Avci, O., Abdeljaber, O., Kiranyaz, S., Hussein, M., Gabbouj, M., & Inman, D. J. (2021). A review of vibration-based damage detection in civil structures: from traditional methods to machine learning and deep learning applications. *Mechanical Systems and Signal Processing*, 147, 107077.

[4] Olanrewaju, A. L., Khamidi, M. F., & Idrus, A. (2010). Building maintenance management in a Malaysian university campuses: a case study. *Australasian Journal of Construction Economics and Building*, 10(1/2), 101–114.

[5] Shaikh, Y., Patel, V. K., Patel, D., & Patel, V. D. (n.d.). A case study of existing institutional building for assessment and repair.

[6] Aliyu, A., Bello, A., Muhammad, S., Singhry, M., & Bukar, G. (2016). An assessment of building maintenance management practice for abubakar tafawa balewa university teaching hospital, Bauchi. In Proceedings/Abstracts and Programmes of the Academic Conference on Transformation Assessment, (Vol. 5, no. 1, pp. 12–20).

[7] Ahzahar, N., Karim, N. A., Hassan, S. H., & Eman, J. (2011). A study of contribution factors to building failures and defects in construction industry. *Procedia Engineering*, 20, 249–255.

[8] Chua, S. J. L., Au-Yong, C. P., Ali, A. S., & Hasim, M. S. (2018). Building maintenance practices towards the common defects and resident's satisfaction of elderly homes. *Journal of Design and Built Environment*, 62–71.

[9] Kiong, N. B., & Akasah, Z. B. (2012). Analysis building maintenance factors for IBS precast concrete system: a review. *System*, 2(6).

[10] Ofori, I., Duodu, P. M., & Bonney, S. O. (2015). Establishing factors influencing building maintenance practices: Ghanaian perspective. *Journal of Economics and Sustainable Development*, 6(24), 184–193.

[11] Allotey, S. E. (2014). An evaluation of the impact of defects in public residential buildings in Ghana. *Civil and Environmental Research*, 6(11), 58–64.

[12] Waziri, B. S., & Vanduhe, B. A. (2013). Evaluation of factors affecting residential building maintenance in Nigeria: user's perspective. *Civil and Environmental Research*, 3(8).

[13] Pontan, D., Surjokusumo, S., Johan, J., Hasyim, C., Setiawan, M. I., Ahmar, A. S., et al. (2018). Effect of the building maintenance and resource management through user satisfaction of maintenance. *International Journal of Engineering and Technology*.

[14] Khalid, E. I., Abdullah, S., Hanafi, M. H., Said, S. Y., & Hasim, M. S. (2019). The consideration of building maintenance at design stage in public buildings: the current scenario in Malaysia. *Facilities*, 37(13/14), 942–960.

[15] Singh, V. K. (2013). Structural repair and rehabilitation of 3 no.(G+ 8) multi-storeyed residential buildings, at ONGC colony at Chandkheda, Ahmedabad, Gujrat. *Procedia Engineering*, 51, 55–64.

[16] Shrivastava, A., & Shukla, V. K. (2019). Rehabilation and maintainance of ancient building-a case study of Surguja district. *International Journal of Advanced Research in Engineering and Technology*, 10(1).

[17] Bakri, N. N. O., & Mydin, M. A. O. (2014). General building defects: causes, symptoms and remedial work. *European Journal of Technology and Design*, 3(1), 4–17.

[18] Khode, H. (2019) Advanced retrofitting techniques for reinforced concrete structures: a state of an art technical review. *International Journal of Innovative Research in Science, Engineering and Technology*, 8(10).

[19] Sarja, A. (2002). Lifetime engineering of buildings and civil infrastructures; centre scient fique et technique du bâtiment (CSTB) France CT – G1RT-CT-2002-05082.

[20] Gáspár, L. (2008). Lifetime engineering for roads. *Acta Technica Jaurinensis*, 1(1), 37–46.

[21] Waziri, B. S. (2016). Design and construction defects influencing residential building maintenance in Nigeria. *Jordan Journal of Civil Engineering*, 10(3).

Advances in Construction, Real Estate, Infrastructure and Project Management – Anil Kashyap et al. (eds)
© 2026 Taylor & Francis Group, London, ISBN 978-1-041-13433-6

11

Determining the Extent of the Environmental, Social, and Governance (ESG) Compliance of Organizations in the Engineering, Procurement, and Construction (EPC) Industry in India

Devang Desai[1]
Associate Professor,
NICMAR University, Pune

Pranav Ramnathkar[2] and Sakshi Jeurkar[3]
MBA Student,
NICMAR University, Pune

■ **ABSTRACT:** The paper investigates the extent of the Environmental, Social, and Governance (ESG) compliance of organizations in the Engineering, Procurement, and Construction (EPC) industry in India. With an increasing emphasis on sustainable investing, investors, companies, and stakeholders must understand the impact of ESG criteria. This study seeks to give a thorough analysis of ESG performance indicators to analyse their association within the EPC business.

The methodology adopted in the study has the following approach. The first step is to evaluate EPC firms using ESG characteristics from the CRISIL Report, with an emphasis on environmental sustainability, social responsibility, and governance structure. The study has used a score system for ESG performance, with weights allocated to each category indicating its importance.

The study acknowledges various problems, such as the heterogeneity of ESG frameworks, the difficulty of quantifying intangible ESG elements, and data quality issues. Despite these hurdles, the research reveals tremendous opportunity for EPC businesses to improve their resilience by incorporating ESG concepts into their fundamental business strategies.

The research finds that ESG compliance is important for EPC companies. It provides investors, legislators, and business leaders with valuable insights into the importance of implementing sustainable practices. By understanding the level of ESG compliance, the study contributes to a better understanding of the value of sustainability in business, encouraging more EPC companies to incorporate ESG criteria into their operational and strategic decision-making processes.

[1]ddesai@nicmar.ac.in, [2]P2271003@student.nicmar.ac.in, [3]P2271054@student.nicmar.ac.in

DOI: 10.1201/9781003669814-11

Currently, few EPC businesses in India strictly adhere to ESG norms. The research contributes to understanding the benefits of ESG practices and the impact they will have on the growth of EPC enterprises. The research is expected to shed light on the EPC businesses that adopt ESG standards, as well as their level of implementation.

■ **KEYWORDS:** Corporate governance, Corporate social responsibility, EPC company performance, ESG compliance, ESG practices, Sustainable development goals

1. INTRODUCTION

The use of and interest in Environmental, Social, and Governance (ESG) investment strategies has significantly increased in recent years. While making investment judgements, investors are increasingly seeing ESG components in addition to traditional financial performance indicators. Numerous causes are driving this trend, such as the increased public awareness of social and environmental issues, regulatory restrictions, and the realization that businesses that implement solid ESG practices may have a greater chance of long-term success.

Investors utilize a set of standards known as ESG criteria to appraise a company's governance structure, societal and environmental impact, and operational performance. These criteria cover a wide range of elements that are used to assess a company's conduct in relation to environmental (E), social (S), and governance (G) issues. The carbon footprint, energy efficiency, and waste management strategies of a business are a few examples of environmental criteria. Social factors may include things like community relations, product safety, staff diversity, and labour standards. The governance criteria generally evaluate the company's ethical standards compliance, executive compensation, leadership structure, and shareholder rights.

Despite mounting evidence that links great ESG performance to financial success, there are still obstacles and constraints to consider. These could include problems with data condition and availability, the variety of ESG frameworks along with the standards, the challenge of quantifying intangible ESG elements, and the difficulty of proving a link between ESG practices and financial results. By addressing these issues, the knowledge of possible interlinkages between ESG and financial performance, ways to improve ESG disclosure and reporting standards can be enhanced. The study focuses on the level of ESG compliance and ESG practices in the Engineering, Procurement, and Construction (EPC) industry in India.

2. LITERATURE REVIEW

This literature review is divided into two parts. The first part covers all the factors which are related to ESG compliance drawn from analysing research papers on related topics. The second part describes a summary of all the ESG indicators in a tabular form.

ESG compliance, though ardently established in international institutions, is relatively new and catching up in the Indian Market. A variety of factors are responsible for an organisation or state or national institution to follow ESG compliance norms. Factors include but are not limited to size of the firm, the governance structure, the firm value, corporate controversies and many more. The ESG research has been developed expeditiously in recent years, with the percentage of publications increasing by almost 200% from 2019 to 2021.It is also found that ESG research is highly interdisciplinary, with contributions from a wide range of fields, including business, management, economics, and environmental science [22].

Following Paragraphs tackle each of these parameters and illustrate references from relevant papers found during out literature survey.

2.1 Corporate Controversies and ESG Practices

Nirino et al. [18] directed a study to address the aftermath of corporatecontroversies on financial performance of a company. The result obtained states that corporate controversies have a compelling unfavourable effect on the financial conduct of an organisation. This is even more prevalent in EPC companies as they are subject to more unprecedented risks and face environmental and social challenges due to the nature of their operations. According to this study, ESG practises acted as a moderating factor between the exchangeof corporate controversies and financial conduct of an organisation by soothing the harmful consequences of corporate controversies. Effect of ESG performance on firm risk is stronger for larger firms, firms in more regulated industries, and firms in countries with stronger investor protection [19]. Robust ESG practises ensure that the goodwill and image of the company stay intact, the reputation is less hampered and investors' confidence is maintained in the organisation [14].

2.2 Determinants of ESG Disclosure in Indian Companies

ESG disclosure is dependent upon cascading list of factors that affect the disclosure positively as well as negatively [23]. In one of the studies conducted by [23], 100 Indian companies were studied to identify the relationship between factors like firm magnitude, firm leverage, firm profitability, board size, board liberty, and ESG disclosure. Firm size, firm profitability, board size and board independence were associated positively with ESG disclosure that is as the firm's size and profitability increases, the organisation is more prone to disclose its ESG practices. The same runs true with organisation's board size and independence as this leads to increased number of decision-making stakeholders and to ensure transparency across the board, disclosures become of paramount importance. Firm leverage is negatively associated with ESG disclosure. With the increase in debt-to-equity ratio, the need to disclose ESG data to current and potential shareholders decrease as the organisation focuses more on debt mode for financing.

2.3 Portfolio Optimisation and ESG Investing

ESG investing has been picking up scale in the last few years in the international markets. Investors are looking for companies having ESG compliances to ensure that their investments are well protected in the invested company. Schmidt [21] examined the construction of optimal ESG portfolios for the Dow Jones Index, using mean-variance optimisation on data from 2018 to 2020. This led to formulation of a portfolio having higher returns and lower risks compared to other companies listed in the Dow Jones Index. This study concluded that optimal ESG portfolios can be constructed which can help investors achieve stable risk adjusted returns from investments in stock markets. ESG considerations play a critical role in investor decisions, as companies with powerful ESG performance are anticipated as more sustainable and less perilous [26]. ESG integration can be achieved by incorporating ESG factors into all stages of the investment management process, including investment strategy, portfolio construction, risk management, and engagement [25]. Most common motivation for using ESG information is to improve investment performance. Other motivations include meeting client demand, complying with regulations, and promoting responsible investment [3].

2.4 Financial Development and ESG Performance in Asia

Financial development of a country refers to the process of growth, expansion, and sophistication of its financial system. It encompasses various aspects of the financial sector, including the depth, breadth, efficiency, stability, and inclusiveness of financial markets, institutions, and infrastructure.

This defines the overall development of the country and reflects on the country's GDP. Ng et al. [17] contribute by examining the positive effect of financial advancement on ESG performance in the Asian context, providing insights into the interconnectedness of economic and sustainability metrics for companies. Financial developments have an impact on the ESG performance of a country. As the financial development increases countries are prone to approach sustainable approaches in its operations which include ESG parameters as well. This leads to an increase in ESG performance of a country. This is more profound andstronger in countries with higher levels of income and institutional quality.

2.5 ESG Factors in Credit Ratings and Finance Cost

ESG ratings are granted to companies depending upon their performance in the threefold outlooks of ESG that is Environmental, Social and Governance. These ratings have implications on credit ratings and finance cost of companies. In the context of ESG ratings, companies respond to these ratings in a variety of ways, depending on their size, industry, and culture. Some companies are proactive and use ESG ratings to drive their ESG strategies, while others are more reactive and only respond when they feel pressure from their stakeholders [6]. Hence a heterogeneity in response to ESG ratings are observed in companies across different sectors and geographical locations. ESG rating affects the credit rating of organisations. Companies having better ESG performance lead to better ESG rating ultimately leads to improved credit ratings, which in turn reduces the financing cost for the organisation [15]. Profitability, credit management, leverage, and sustainable growth rate were found to have a significant impact on ESG ratings. Profitability has a positive impact on all three ESG factors, while leverage has a negativeimpact on all three ESG factors [2]. This correlation is more profound in certain industries namely financial services industry and energy related firms. Overall, there is an affirmative interaction between ESG performance andcredit Rating and negative correlation with financing cost. Organisation with prevalent ESG conduct tend to have higher gains on assets, return on equity, and market values. It was also found that companies with better ESG conduct tend to have lower hazard and cost of capital [11].

2.6 Firm Size and Ownership Structure: Implications for ESG

Firm size refers to the size of the operations of an organisation and its market presence, number of people employed along with other functional aspects. As the firm size increases, companies are prevalent to act on sustainability aspects leading to improved ESG performance. Hence firms with larger size have generally higher ESG ratings [9]. Firm size is complemented with the ownership structure of the organisation that determines how the policies of the company are made, power and responsibility distribution and the overall workings of the company. The interaction between firm magnitude and ESG conduct is more prevalent in sectors like financial services and energy sector.

Ownership structure points to the composition and dissemination of ownership stake in a company, organization, or entity. It identifies the individuals, groups, or entities that hold ownership rights or equity stakes in the organization and delineates their respective ownership positions and voting rights. Ownership structures can be overseas, public, state and family ownerships. Overseas and publicownership affirmativelyaffect ESG disclosure whereas state and family ownership doesn't affect ESG disclosure [12]. There is an affirmative interaction between ESG practises and ownership structure on the organisation's conduct in emerging and developing markets as well [5]. Shareholder engagement has a significant positive impact on ESG performance. Better engagement leads to an improvement in the ESG scores. The effects are strongest for engagements that focus on environmental and social issues, and for firms with poor ex-ante ESG performance [4]. Audit committees play a mitigating character in the interaction of ESG disclosure and organisation value across all forms of ownership. Hence inclusion of diverse stakeholders in ownership structure is important to facilitate ESG disclosure promoting better ESG performance and better firm value and presence of audit committee is important to moderate adverse effects.

2.7 Challenges in ESG Reporting: Greenwashing and ESG Responsibilities

Environmental, Social, and Corporate Governance (ESG) greenwashing points to the practice of misleadingly portraying a company's environmental, social, and governance policies as more favourable or sustainable than they actually are. It involves using deceptive or superficial measures to create a positive public image of the company's ESG performance without implementing substantive changes or improvements in its practices. The characteristics of ESG greenwashing include: Misleading or exaggerated claims, selective disclosure, token measures, lack of transparency, green marketing or lack of accountability. Firm-level variables, such as company size, financial performance, and industry type, have a significant impact on ESG greenwashing. As the firm size increase along with improved financial performance, companies tend to engage more in ESG green washing [10]. Country level variables also have impact on ESG greenwashing. These include country's level of economic development and the quality of its institutional environment. Firms in nations with reduced proportion of economic development and weaker institutional surroundings are also more likely to engage in ESG greenwashing. However, firm level factors are more important than country level factors in discouraging green washing practices of a firm [27].

2.8 External Factors and ESG Performance

External environment determines the functioning of an organisation. An organisation working with same set of standard operating procedures, policies and practices in different nations may have different outcomes in each one of them, this is because factors like public visibility, industry sensitivity and stability of government may either promote or hamper an organisations growth. ESG ratings determined by ESG performance play an important role in financial performance of companies that are more visible to the public eye and/or are operating in industries sensitive to ESG performance [1]. Companies having better sustainability aspects and ESG performance are fare better than their counterparts in stabilising turmoil caused by political volatility. The country governance environment has affirmative and important impact on corporate ESG conduct [16]. There exists an affirmative interaction between economic development and ESG conduct of companies operating in the nation. Companies operating in countries having better economic development have an overall improved ESG performance. Hence there is an affirmative interaction between country governance and ESG performance.

2.9 ESG and Sustainable Development Goals (SDGs) in Companies

ESG which stands for Environmental, Social & Governance are the non-financial parameters that define the functioning of an organisation. These parameters directly lead to achievement of SDGs (Sustainable Development Goals) [13]. This is done as the company achieves sustainability which is the scope to gratify ongoing demands without endangering the capacity of future generations to satisfy their own needs. ESG aspects are factors that companies consider when making decisions, such as their environmental impact, their relationships with employees and stakeholders, and their corporate governance practices. ESG aspects are relevant to sustainability, grouped into 10 categories namely: environmental, social, governance, economic, ethical, cultural, technological, political, legal, and regulatory. Even in the BRICS countries (Brazil, Russia, India, China, and South Africa) ESG responsibilities and economic development have an affirmative and important impact on achieving the SDGs in the BRICS countries. The impact of ESG obligations and economic advancement on achieving the SDGs is stronger in countries with higher levels of institutional development. Industry 4.0 technologies, such as the Internet of Things (IoT), artificial intelligence (AI), and big data, have the potential to empower ESG initiatives and help companies achieve their sustainability goals [20]. ESG responsibilities and economic development can help to achieve the SDGs in the BRICS countries. All the above ESG aspects are interrelated and that they contribute to continued sustainability of the organisation leading to achievement of SDGs.

2.10 ESG Ratings, Transparency, and Tax Practices

Transparency in tax practices pertains to the clarity, accessibility, and candidness of information regarding a company's tax matters, encompassing its strategies for tax planning, obligations, and payments. This entails the disclosure of pertinent tax-related details to stakeholders—such as investors, regulators, and the public—in a manner that is both comprehensible and unambiguous. The transparency in taxation matters aims to cultivate accountability, instil trust, and inspire confidence in the company's tax-related operations. This is especially important in public companies which are listed and source finance in the form of equity from investors. In this case transparency helps the investors better gauge the company and help in taking investment decisions. A study was conducted by Thiart, [24] among companies listed in Johannesburg Stock Exchange which showed an affirmative and important interaction between ESG ratings and tax disclosure. Companies that are good in ESG performance tend to be more transparent in their tax practises Hence companies having ESG ratings follow standard and ethical tax practises to ensure transparency.

The Table 11.1 presents the summary of all the important indicators for ESG compliance.

Table 11.1 Important indicators for ESG compliance

Sr. No	Factors	Contributing Authors
1	Corporate controversies	[18]
2	Financial Performance	[19]
3	ESG Practices	[14]
4	Firm Size	[9, 19]
5	Firm Profitability	[2, 23]
6	Board Size	[23]
7	Board Independence	[23]
8	Firm Leverage	[2, 23]
9	Financial Development	[17]
10	ESG Performance	[15, 17]
11	Level of Income	[17]
12	Institutional Quality	[17]
13	Industry Sector	[6]
14	Organization Culture	[6]
15	ESG Rating	[2, 6, 9, 24]
16	ESG Response	[6]
17	Credit Rating	[15]
18	Financing Cost	[15]
19	Ownership Structure	[12]
20	ESG Disclosure	[12]
21	Firm Value	[12]
22	Country governance structure	[2, 10]
23	ESG greenwashing	[10]
24	Company Governance Structure	[10]
25	Company Visibility (to public)	[2]
26	Resilience to Political Volatility	[1]
27	Sustainability Development Goals	[13]
28	Transparency in Tax Practices	[24]
29	Company Ownership Structure	[5]
30	ESG Responsibility	[17]
31	Economic Development	[17]
32	Financial Sustainability	[2]
33	Firm's Marketing Expense	[2]
34	Credit Management	[2]
35	Sustainable growth rate	[2]

Sr. No	Factors	Contributing Authors
36	Firm Risk	[19]
37	ESG Integration	[25]
38	Investment Strategy	[25]
39	Portfolio Construction	[25]
40	ESG Information	[3]
41	Investment Performance	[3]
42	Technological Advancement	[20]
43	Shareholder Engagement	[4]

Source: Authors' compilation

3. RESEARCH METHODOLOGY

The research methodology for the study was divided into two parts. The first part consisted identification of ESG parameters from literature review and second part consisted studying ESG ratings for different EPC companies from CRISIL Report.

With an emphasis on the (EPC) sector specifically, the ESG scoring system adopted by CRISIL aims to offer a thorough assessment framework for evaluating how well businesses perform in Environmental, Social, and Governance (ESG) areas that have a direct impact on financial performance. Establishing open standards and metrics will help stakeholders, consumers, and regulators evaluate the ESG conduct of businesses in the EPC industry. The scoring system aims to encourage businesses to embody ESG considerations into their firm strategies, procedures, and managerial mechanisms by rating elements like environmental sustainability practices, social obligation programmes, and administration structures. In the end, the objective is to support lifelong value generation, reduce risks, encourage sustainable advancement, and advance accountability and transparency within the ecosystem of the EPC industry.

In order to ensure a thorough evaluation of the EPC industry, the scope of companies to be assessed should include those directly involved in the Engineering, Procurement, and Construction (EPC) industry across various sectors such as infrastructure, energy, utilities, transportation, and real estate development. The study included companies of diverse sizes, from multinational corporations to small and medium-sized enterprises (SMEs), and from public to privately held firms.

In the Engineering, Procurement, and Construction (EPC) sector, the CRISIL ESG framework offers an organised method for assessing Environmental, Social, and Governance (ESG) factors. Assimilation of industry-specific ESG issues found in the literature review will improve its applicability and relevance. This covers things like following technical standards, involving the community, and using sustainable building techniques. Furthermore, it is imperative to consider variables such as project execution proficiency, inventiveness, and market volatility in all areas to guarantee a thorough evaluation of EPC firms' achievements and their influence on financial results.

Secondary data obtained from CRISIL report on ESG performance of fifteen companies chosen from Indian EPC sector. These companies are Tata Projects Limited, KEC International Limited, Godrej Properties Limited, Adani Power Limited, Welspun Enterprises Limited, Gail India Limited, JSW Energy Limited, Larsen and Toubro Limited, Macrotech Developers Limited, Mahindra Lifespace

Developers Limited, H.G. Infra Engineering Limited, J. Kumar Infra projects Limited, Reliance Power Limited, Tata Realty and Infrastructure Limited, VA Tech Wabag Limited.

In order to reflect the components' relative importance within the ESG scoring system adopted by CRISIL [8], weights must be assigned to the E, S, and G components. This weighting gives environmental factors a 35% weight, social factors a 25% weight, and governance factors a 40% weight. The highest weight is given to governance, which is recognised for its fundamental role in effectively advancing social and environmental agendas within EPC industry companies. Robust governance protocols establish a foundation for the execution of enduring policies, guaranteeing responsibility, and controlling hazards, all of which have an impact on the ecological and societal outcomes. The scoring system seeks to provide a fair evaluation of ESG performance by assigning weights based on relative importance, emphasising the crucial role that governance plays in encouraging sustainability and ethical business practices in the EPC industry.

Higher scores indicate stronger adherence to ESG principles and practices within the Engineering, Procurement, and Construction (EPC) industry. A scoring scale with a range of 0–100 was used to quantify ESG performance. Based on predetermined criteria and metrics, each E, S, and G component was assigned a score; 100 denotes excellent performance, and 0 denotes serious shortcomings or non-compliance.

Pilot testing with a small number of businesses enables the methodology to be improved and applied in the real world. In order to ensure that the scoring model is accurate and effective in evaluating ESG performance within the Engineering, Procurement, and Construction (EPC) industry, the framework was tested in ten local real estate companies from Pune, Maharashtra, India to identify potential biases or inconsistencies.

4. RESULTS AND DISCUSSIONS

The study consisted of case studies covering fifteen Engineering, Procurement, and Construction (EPC) firms that work in a variety of industries, such as Waste Management, Real Estate, Oil & Gas, Infrastructure, and Water. Every case study outlines the interaction between Environmental, Social & Governance (ESG) aspects and the extent of their application by EPC companies.

The Table 11.2 below displays various EPC companies with their ESG Scores derived from CRISIL report [7].

The above CRISIL ratings, are given based upon the parameters that each company satisfy in the Environment, Social and Governance spectrum. Companies that have strong ESG scores have practices such as energy management system, energy saving procedures, use of smart technology in daily processes. They were found to be incorporating ESG and sustainability approach in their value chain and technical processes. Companies like GAIL and JSW are pioneers in this. Remaining companies which have adequate performance lack some of these practices or do not enforce them to that strong level. The practices which they incorporate are four pillars of sustainability (environmental, social, economic, and cultural), agenda for sustainable development, use of solar energy, sewage treatment along with decarbonization and water positive approach. The companies which have weak rating are still in R&D aspects of these above-mentioned practices or are just trying to reduce their traditional footprint. HG Infra and J Kumar fall into these categories.

Table 11.2 ESG scores for various EPC companies

Company Name	Sector	Environment Score	Social Score	Governance Score	ESG Score	Category of ESG compliance	Scoring Period
Tata Projects Limited	Construction EPC	41	48	60	50	Adequate	Mar-22
KEC International Limited	Heavy Engineering	34	49	67	51	Adequate	Mar-22
Godrej Properties Limited	Real Estate	54	44	71	58	Adequate	Mar-22
Adani Power Limited	Power Thermal	52	61	57	56	Adequate	Mar-22
Welspun Enterprises Limited	Construction EPC	62	39	69	47	Adequate	Mar-22
Gail India Ltd.	Oil & Gas	57	73	62	63	Strong	Mar-22
JSW Energy Limited	Power Thermal	49	61	75	62	Strong	Mar-22
Larsen and Toubro Limited	Diversified	58	60	68	63	Strong	Mar-22
Macrotech Developers Limited	Real Estate	59	54	74	64	Strong	Mar-22
Mahindra Lifespace Developers Limited	Real Estate	52	53	73	61	Strong	Mar-22
H.G. Infra Engineering Limited	Transport Infra – Roads	26	37	66	45	Below Average	Mar-22
J. Kumar Infra projects Limited	Construction EPC	25	41	61	43	Below Average	Mar-22
Reliance Power Limited	Power Thermal	19	25	53	34	Below Average	Mar-22
Tata Realty and Infrastructure Limited	Real Estate	25	29	60	40	Below Average	Mar-22
VA Tech Wabag Limited	Water and Waste Management	30	38	61	44	Below Average	Mar-22

Source: Authors' compilation

In the social category strongly rated companies align their CSR policies with national welfare policies, like Atal innovation mission, provision of medical healthcare etc. Adequately rated companies are presently changing their CSR norms to adapt to the upcoming trends and focus on occupational health and safety, community development and happiness. KEC falls into this category. The weakly rated companies are still following age old CSR approach and are trying to initiate proactive efforts to challenge the social aspects.

Companies that are strongly rated on the governance aspect, emphasize "Better every day tagline" to improve on their core principle i.e. integrity, accountability and transparency. The companies that are rated adequately lack the above approach but emphasize being corporate citizen and achieving goals while maintaining accountability. They also incorporate independent directors, and utilize business level sustainability roadmap. The companies which are weakly rated lack this proactive approach and just ensure compliance.

5. CONCLUSIONS & FUTURE DIRECTIONS

By analysing fifteen EPC companies on their ESG compliance, it is derived that there is a diverse level of ESG compliance across different companies. The criteria vary from Strong, Adequate and Below Average. This level depends on multitude of factors mentioned in the literature review.

5.1 Unpacking the Benefits of ESG Integration

The study highlights several key benefits associated with robust ESG practices.

- Route to Capital: Stockholders are progressively integrating ESG factors into their investment choices. Organisations with strong ESG conduct attract a wider pool of environmentally and socially conscious investors, expanding their access to capital and potentially lowering the cost of financing. This facilitates long-term growth and investment in sustainable initiatives.

- Improved Brand Reputation: More and more customers are weighing ethical issues while making purchases. Businesses that show a sincere dedication to sustainability by using open and honest ESG practices build a strong brand and win over customers' confidence and loyalty. This converts into a distinctive brand, which may result in a gain in market share and a competitive edge.

- Operational Efficiencies: Reducing waste and optimizing resources are common components of implementing sustainable practices. This reduces the negative effects on the environment and saves money by using resources and energy more effectively. Optimized processes also lead to increased output and earnings.

5.2 Beyond the Tangible: Additional Advantages of ESG

The positive impact of ESG extends beyond immediate financial gains. These case studies suggest that:

- Risk Mitigation: Businesses are protected from possible liability and reputational harm by taking proactive measures to control environmental and social risks, such as labour practices scrutiny and climate change rules. This proactive strategy promotes sustainability over the long run and resilience.

- Engagement of Employees: Putting ethical and sustainable policies into practice creates a happy workplace, which may boost morale and productivity. An engaged and driven staff makes a big difference in the success of a firm.

- Innovation and Competitive Advantage: Adopting ESG principles frequently calls for innovation in technology and processes, which results in the creation of sustainable goods and solutions. As a result, businesses are positioned at the forefront of the changing market landscape and are encouraged to continuously grow in that manner.

5.3 Addressing Challenges and Moving Forward

Even though the case studies emphasize the many advantages of ESG compliance, putting these principles into reality is not without difficulties. Businesses might encounter:

- Upfront Costs: Making large upfront financial investments in social activities, green infrastructure, and new technology can be expensive. On the other hand, these investments may be seen as long-term strategic choices with high prospective rewards.
- Data Management and Transparency: Reliable data collecting and analysis tools are essential for monitoring and reporting ESG performance. Establishing trust and openness with stakeholders necessitates detailed reporting of ESG activities and their results as well as transparent communication.
- Integration with Current Operations: Careful planning and implementation are necessary for the successful integration of ESG concepts into current business models. Businesses must make sure that sustainability goals and primary business objectives are aligned and overcome any possible opposition to change.
- Financial Instability: Although certain businesses may not be doing well financially even if they are in the strong ESG score category, they will gain in the long run because of their increased market value and the advantages that come with adopting these practices. Few of the reasons are stated below:
- Time Horizon: Long-term results are frequently seen from ESG practices. It's possible that investments in ethical labour practices, responsible sourcing, and sustainable technology won't pay off right away. But these expenditures may also create the foundation for future cost reductions, enhanced brand recognition, and entry into new markets, which will eventually result in long-term financial stability.
- Industry-Specific Difficulties: Converting ESG principles into instant cash rewards presents particular difficulties for some sectors. For instance, adopting renewable technology may come with hefty upfront expenses for businesses in transitional industries like energy, which might negatively affect their short-term financial performance.
- Market Inefficiencies: Although the trend is changing, not all investors presently give ESG ratings a high priority when making decisions. Strong ESG standards may cause a temporary undervaluation of corporations. But, as demand and awareness for sustainable investments increase, the market is probably going to adjust as investors realize the lifelong perks of contributing in ESG-compliant businesses.
- External variables: Regardless of an organization's ESG score, macroeconomic variables, industry upheavals, and unanticipated occurrences can have a negative financial performance impact. But by encouraging responsible risk management, attracting ethical investors, and building stakeholder confidence, robust ESG practices may provide businesses the resilience they need to overcome these obstacles.

Therefore, while short-term financial instability might exist for some companies with strong ESG scores, it's crucial to consider the following parameters. By adopting sustainable practices, these businesses are probably establishing the foundation for future financial success. The timescales for obtaining the financial advantages of ESG differ throughout businesses. The market is expected to realize the long-term benefits of sustainable businesses as ESG knowledge rises. A company's capacity to withstand economic downturns can be strengthened by implementing strong ESG policies.

In Conclusion, the EPC industry can unleash the enormous potential of ESG compliance by working together to create a supportive and collaborative environment. This will make a way for a moreviable andsuccessful future. The case examples, taken together, make a strong argument for seeing ESG compliance as a strategic investment rather than a cost, one that has the possibility to have a major affirmative impact on the economy, the environment, and society. EPC firms may position themselves as leaders in a fast-changing global market, secure long-term success, and contribute to a more sustainable future by embracing this paradigm change and conquering the related obstacles.

REFERENCES

[1] Aboud, A., & Diab, A. (2018). The impact of social, environmental and corporate governance disclosures on firm value: evidence from Egypt. *Journal of Accounting in Emerging Economies*, 8(4), 442–458.

[2] Alam, M. M., Tahir, Y. M., Saif-Alyousfi, A. Y., Ali, W. B., Muda, R., & Nordin, S. (2022). Financial factors influencing environmental, social and governance ratings of public listed companies in Bursa Malaysia. *Cogent Business and Management*, 9(1), 2118207.

[3] Amel-Zadeh, A., & Serafeim, G. (2018). Why and how investors use ESG information: evidence from a global survey. *Financial Analysts Journal*, 74(3), 87–103.

[4] Barko, T., Cremers, M., & Renneboog, L. (2022). Shareholder engagement on environmental, social, and governance performance. *Journal of Business Ethics*, 180(2), 777–812.

[5] Bilyay-Erdogan, S., & Öztürkkal, B. (2023). The role of environmental, social, governance (ESG) practices and ownership on firm performance in emerging markets. *Emerging Markets Finance and Trade*, 59(12), 3776–3797.

[6] Clementino, E., & Perkins, R. (2021). How do companies respond to environmental, social and governance (ESG) ratings? evidence from Italy. *Journal of Business Ethics*, 171(2), 379–397.

[7] CRISIL (2022). CRISIL ESG Report 2022. https://www.crisil.com/content/dam/crisil/investors/annual-reports/2022/sustainability/sustainibility-esg-report.pdf.

[8] CRISIL (2023). CRISIL's ESG Scoring Methodology. https://www.crisil.com/content/dam/crisil/our-businesses/india_research/sustainability-yearbook-2022/methodology/crisil-esg-methodology.pdf.

[9] Drempetic, S., Klein, C., & Zwergel, B. (2020). The influence of firm size on the ESG score: corporate sustainability ratings under review. *Journal of Business Ethics*, 167(2), 333–360.

[10] Erol, D., & Çankaya, S. (2023). The impacts of firm-level and country-level variables on environmental, social and corporate governance greenwashing. (Doctoral dissertation, Istanbul Ticaret University).

[11] Friede, G., Busch, T., & Bassen, A. (2015). ESG and financial performance: aggregated evidence from more than 2000 empirical studies. *Journal of Sustainable Finance and Investment*, 5(4), 210–233.

[12] Fuadah, L. L., Mukhtaruddin, M., Andriana, I., & Arisman, A. (2022). The ownership structure, and the environmental, social, and governance (ESG) disclosure, firm value and firm performance: the audit committee as moderating variable. *Economies*, 10(12), 314.

[13] Jámbor, A., & Zanócz, A. (2023). The diversity of environmental, social, and governance aspects in sustainability: a systematic literature review. *Sustainability*, 15(18), 13958.

[14] Khandelwal, V., Sharma, P., & Chotia, V. (2023). ESG disclosure and firm performance: an asset-pricing approach. *Risks*, 11(6), 112.

[15] Kiesel, F., & Lücke, F. (2019). ESG in credit ratings and the impact on financial markets. *Financial Markets, Institutions and Instruments*, 28(3), 263–290.

[16] Mooneeapen, O., Abhayawansa, S., & Mamode Khan, N. (2022). The influence of the country governance environment on corporate environmental, social and governance (ESG) performance. *Sustainability Accounting, Management and Policy Journal*, 13(4), 953–985.

[17] Ng, T. H., Lye, C. T., Chan, K. H., Lim, Y. Z., & Lim, Y. S. (2020). Sustainability in Asia: the roles of financial development in environmental, social and governance (ESG) performance. *Social Indicators Research*, 150, 17–44.

[18] Nirino, N., Santoro, G., Miglietta, N., & Quaglia, R. (2021). Corporate controversies and company's financial performance: exploring the moderating role of ESG practices. *Technological Forecasting and Social Change*, 162, 120341.

[19] Sassen, R., Hinze, A. K., & Hardeck, I. (2016). Impact of ESG factors on firm risk in Europe. *Journal of Business Economics*, 86, 867–904.

[20] Saxena, P. K., Seetharaman, A., & Shawarikar, G. (2024). Factors that influence sustainable innovation in organizations: a systematic literature review. *Sustainability*, 16(12), 4978.

[21] Schmidt, A. B. (2022). Optimal ESG portfolios: an example for the dow jones index. *Journal of Sustainable Finance and Investment*, 12(2), 529–535.

[22] Senadheera, S. S., Gregory, R., Rinklebe, J., Farrukh, M., Rhee, J. H., & Ok, Y. S. (2022). The development of research on environmental, social, and governance (ESG): a bibliometric analysis. *Sustainable Environment*, 8(1), 2125869.

[23] Sharma, P., Panday, P., & Dangwal, R. C. (2020). Determinants of environmental, social and corporate governance (ESG) disclosure: a study of Indian companies. *International Journal of Disclosure and Governance*, 17(4), 208–217.

[24] Thiart, C. (2023). The correlation between environmental, social and governance ratings and the transparency in Johannesburg stock exchange companies' tax practices. *South African Journal of Economic and Management Sciences*, 26(1), 4886.

[25] Van Duuren, E., Plantinga, A., & Scholtens, B. (2016). ESG integration and the investment management process: fundamental investing reinvented. *Journal of Business Ethics*, 138, 525–533.

[26] Velte, P. (2017). Does ESG performance have an impact on financial performance? evidence from Germany. *Journal of Global Responsibility*, 8(2), 169–178.

[27] Yu, E. P. Y., Van Luu, B., & Chen, C. H. (2020). Greenwashing in environmental, social and governance disclosures. *Research in International Business and Finance*, 52, 101192.

Advances in Construction, Real Estate, Infrastructure and Project Management – Anil Kashyap et al. (eds)
© 2026 Taylor & Francis Group, London, ISBN 978-1-041-13433-6

12

Predicting Climate Impact on Ghoramara Coastal Structures: Using Deep Learning

Neha Priyadarshini,
Vikas Prasad
Research Scholar and Associate Professor
(NICMAR UNIVERSITY)

■ **ABSTRACT:** Climate change is a pressing global issue that is significantly impacting coastal regions, leading to accelerated sea-level rise, increased storm intensity, and coastal erosion. Ghoramara Island, situated in the Sundarbans delta of India, is particularly vulnerable to these climate-induced changes. This study employs deep learning techniques to predict and assess the impact of climate change on coastal structures and land degradation in Ghoramara Island. The research methodology integrates a multidisciplinary approach, incorporating satellite imagery data, climate data, and historical records of coastal changes. Deep learning algorithms, including Convolutional Neural Networks (CNNs) and Recurrent Neural Networks (RNNs), are utilized to analyze and model the intricate relationships between climate variables, coastal infrastructure, and land degradation patterns over time. By leveraging satellite imagery data, the study captures the spatial and temporal dynamics of coastal erosion, shoreline retreat, and infrastructure vulnerability on Ghoramara Island. The deep learning models are trained to recognize and predict patterns of coastal changes, considering factors such as sea-level rise, storm surge events, sediment transport, and human interventions.

The results of this research provide valuable insights into the future scenarios of coastal degradation and infrastructure vulnerability in Ghoramara Island under different climate change scenarios. These predictions are essential for informing decision-makers, urban planners, and coastal communities about the potential risks and challenges posed by climate change in the region.

■ **KEYWORDS:** Multidisciplinary approach, Satellite imagery data, CNNs (Convolutional neural networks), RNNs (Recurrent neural networks), Sea-level rise, Storm surge events, Sediment transport

[1]neha.phd2@pune.nicmar.ac.in, [2]vprasad@nicmar.ac.in

DOI: 10.1201/9781003669814-12

1. INTRODUCTION

One of the most pressing issues of the 21st century is climate change because its effects are being felt most strongly in coastal areas worldwide. Ghoramara Island, a tiny deltaic island in West Bengal in the Indian Sundarbans, is a striking example of how vulnerable coastal regions are to the negative effects of climate change. On the island, significant land degradation, coastal erosion, and submersion have occurred due to rising sea levels, an increase in the frequency of cyclonic storms, and other weather conditions. These progressions altogether affect the island's foundation, economy, and lifestyle. Advanced techniques for anticipating and mitigating climate change's impact on coastal buildings and land degradation are required to address these issues. Albeit significant, conventional strategies for appraisal as often as possible neglect to catch the perplexing, non-direct connections between various climatic components and their limited ramifications. This necessitates the use of increasingly sophisticated and trustworthy forecasting models. With the silence of man-made brainpower and Artificial Neural Networks, profound learning has been demonstrated to be a helpful device in this present circumstance. Its capacity to dissect huge datasets and distinguish designs that are not promptly clear utilizing ordinary measurable strategies makes it especially appropriate for examinations of the impacts of environmental change. To make projections of changes to the coast that are both more precise and more comprehensive, deep learning models can make use of historical climate data, satellite imagery, and other relevant datasets. Using profound learning procedures, we mean to foresee what environmental change would mean for the framework along Ghoramara Island's shoreline and land debasement.

2. LITERATURE REVIEW

In the Ganges-Brahmaputra-Meghna delta using a deep learning technique has evaluated coastline changes and land degradation, underscoring the significance of cutting-edge technologies in coastal environmental studies [1]. Another study focused on, using deep learning models to predict the impact of climate change on coastal structures and emphasized the significance of predictive modeling in climate change adaptation techniques [4]. A study investigated the advantages and drawbacks of implementing new modeling techniques in climate research, particularly focusing on the use of deep-learning methods [9]. This study examines, how sea level rise's effects on coastal regions were evaluated using deep learning models in research that provided insightful information for prospective future methods of coastal planning and adaptation [7]. Another study used remote sensing to illustrate environmental monitoring and evaluate land deterioration on Ghoramara Island, utilizing geospatial technologies and deep learning algorithms [2]. The study covered current advancements and difficulties in applying deep learning techniques to studies of the coastal environment, as well as the use of these techniques for monitoring coastal zones [5]. The use of sophisticated modeling techniques in environmental research was demonstrated through a case study of Ghoramara Island, where deep learning models were employed to evaluate the effects of climate change [6]. The author of this research emphasized the possibility of utilizing cutting-edge technology in coastal environmental management and covered the present trends and prospects of deep learning applications in coastal zone management [11]. In this study, deep learning models were utilized to evaluate land deterioration in coastal locations, with a particular focus on Ghoramara Island, after the study carefully examined the degradation patterns seen on the island [8].

The study highlighted the accuracy and effectiveness of deep learning models in predicting changes in the environment, offering insightful information for managing coastal regions [12]. A variety of case studies, approaches, and results were examined, emphasizing how crucial these strategies are becoming for coastal planning and management. They also covered the most current developments in deep learning methods for coastal zone prediction [3].

2.1 Background

This paper proposes to make use of certain Machine learning (ML) algorithms as described in the introduction, so that Deep Learning can be used for predicting climate impact on Ghoramara coastal structures.

Important Machine Learning algorithms/techniques are listed below:

i. Artificial Neural Networks (ANNs)
- ANNs are computer networks that draw inspiration from the organic neural networks seen in animal brains. They are composed of artificial neurons (nodes), which are connected groupings.
- An input layer, one or more hidden layers, and an output layer are the three layers that typically make up an artificial neural network (ANN). All neurons within a layer can communicate with all other neurons within that layer.
- Artificial neural networks (ANNs) acquire task-specific programming typically by analyzing examples. They are employed in applications where pattern recognition and prediction are required, such as speech and picture recognition.

ii. Convolutional Neural Networks (CNNs)
- CNNs are a particular kind of ANN built for handling structured grid data, like pictures.
- They utilize convolutional layers, which apply a convolution operation to the input and send the output to the following layer. These layers are capable of identifying shapes, textures, and edges in photos.
- CNNs are good at evaluating satellite photos and spotting changes to the coast because they work especially well with spatial data.

iii. Recurrent Neural Networks (RNNs)
- An RNN, or Recurrent Neural Network, is an artificial neural network type where the connections between nodes form a directed network following a temporal sequence.
- Unlike feedforward neural networks, RNNs incorporate loops, allowing information to persist.
- This makes them suitable for processing sequential data.

iv. Long Short-Term Memory Networks (LSTMs)
- A specific type of RNN known as an LSTM is designed to retain long-term dependencies.
- They have a more intricate structure than regular RNNs, with cells, input, output, and forget gates that control the information flow.
- LSTMs are effective for predicting long-term climate changes and their effects on coastal structures because they excel at capturing long-term dependencies in data sequences.

Application to Ghoramara Island

For Ghoramara Island, a comprehensive approach might involve:

- CNN: These systems analyze satellite photos to identify physical alterations in coastal landforms and structures over time.
- RNNs and LSTMs: Utilized to forecast future modifications by utilizing past data. To project future effects, these models can evaluate time-series data, including sea levels, temperatures, and rates of erosion.

3. PROBLEM STATEMENT

By 2030, the world temperature needs to cease rising steadily; if we fail to do so and miss the deadline by seven years, the rise might then become irreversible. The worldwide mean temperature map will climb, according to information from the National Aeronautics and Space Administration (NASA). By 2030, the United Nations has set a goal of 1.5 degrees Celsius for the world temperature average to be reached. This can be accomplished by cutting carbon emissions; the countries that emit the most carbon dioxide, such as China, the US, India, and Russia, should cut their emissions first to maximize the efficacy of the plan. Since the effects will extend beyond their nations, we ought to encourage them to take those actions. By enhancing their standard of living, they should help the developed countries in the future. Although handling large datasets is a significant challenge in climate science research, we might be able to solve the system of equations for these properties and all of these grid points, especially since the majority of real-world time series datasets are multivariate and contain a wealth of dynamic system information. However, due to computation and other limitations, this will involve millions and millions of calculations. To approximate the description of specific processes and climate models, including cloud models, parameterizations are representations used. There is considerable ambiguity about the climate due to these parameterizations. Few climate researchers have developed and used complex, comprehensive models. To educate international policymakers on climate change, one must be aware of the existing and expected climate. By simulating the average monthly temperature, the standard models can be used to experience a modeling of temperature change. However, because weather and climatic variations are more visible at global scales, using these models becomes more difficult. With the use of our artificial intelligence (AI)-driven climate model, scientists will have more trustworthy resources to analyze past climate change and predict future changes for many years to come. The Indian Sundarbans consist of 102 islands in total. Of these, 48 are covered with mangrove forests, and the remaining 54 are home to people. Situated in the South 24 Parganas district, Ghoramara is the most endangered island in India. This island is located 1.91 kilometers from Sagar Island and around 4 km from Kakdwip. This island's precise coordinates are $88° 7' 2''E$ to $88° 8' 26''E$ and $21° 54' 5''N$ to $21° 55' 29''N$. There used to be 40, 000 people living on Ghoramara Island, but only 3000 remain now because of soil degradation and the island's fragility. Most of the locals moved elsewhere in search of better opportunities. Fishing, farming, and cultivation are the major industries for the people that live on this island.

3.1 Research Objective

a. The initial stage is to establish CNN and ANN models to extract spatial features from different datasets, such as satellite photos and data from coastal monitoring.

b. Using RNNs to analyze storm surge strength and sea level rise while looking for temporal trends in time-series climate data.

c. To use LSTM networks to mimic the intricate and protracted interactions of environmental factors influencing coastal conditions.

d. To recommend a trustworthy forecasting instrument that makes use of time series analysis and can convert data into readily usable patterns for upcoming projections.

4. PROPOSED METHODOLOGY

Shoreline changes between 1990 and 2020 may be studied with the use of Landsat satellite photographs. Climate-related data, including sea-level rise and wave patterns, are provided by meteorological departments. Anecdotes from the area and historical documents can also provide interesting context. Long-term shoreline change rates may be calculated with the Digital Shoreline Analysis System (DSAS) software, whilst transitory changes can be measured using the CEDAS model's RMAP. The link between shoreline changes and contributing variables can be examined using statistical methods. By 2050, building and training a Convolutional Long Short-Term Memory (LSTM) model could enable shoreline locations to be predicted. The results of the research and forecast can then be used to develop mitigation strategies, such as coastal management initiatives.

The proposed methods/ techniques are briefly outlined in Table 12.1.

Data Collection: Obtaining necessary datasets such as satellite photos and historical climate data.

Pre-processing: Cleaning, arranging, and preparing the pictures for analysis, together with the time-series data.

Training of Models:

CNN: Utilizing satellite imagery for training, patterns of coastal erosion and structural damage are identified.

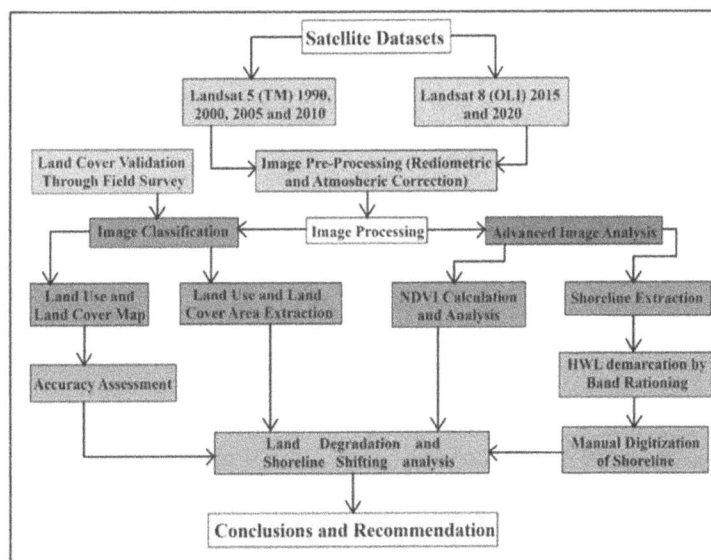

Figure 12.1 Proposed methodology flowchart

Source: Authors

Table 12.1 Methods/Techniques with brief outline

Method/Technique	Brief Outline
Remote Sensing (RS)	Uses satellite imagery to monitor and analyze changes in coastal areas and land cover over time.
Geographic Information System (GIS)	Utilizing spatial data to map, visualize, and analyze changes and vulnerabilities in coastal regions.
Landsat Datasets	Offers access to historical satellite imagery data for evaluating long-term changes in coastal structures.
Deep Learning (DL)	Uses neural networks, specifically convolutional neural networks (CNNs) and recurrent neural networks (RNNs), to model and predict climate impacts.
Convolutional Neural Networks (CNNs)	It involves extracting spatial features from satellite imagery to identify patterns and changes in coastal structures.
Recurrent Neural Networks (RNNs)	Captures temporal dependencies and sequences in data for predicting future coastal changes.
Long Short-Term Memory (LSTM)	A type of RNN that effectively handles long-term dependencies for accurate climate impact predictions.
Digital Shoreline Analysis System (DSAS)	Analyzes shoreline changes and computes rates of change using statistical methods.
Machine Learning (ML) Models	Includes algorithms like Random Forest and Support Vector Machines to support prediction tasks.
Climate Data Analysis	Involves the examination of temperature, precipitation, and sea level rise data to assess their impacts.
Historical Records Review	Utilizes historical documents and records to understand past climate events and their impacts on coastal regions.
Scenario Analysis	Develop and analyze several future scenarios to evaluate potential climate impacts on coastal structures.
Statistical Analysis	Employs statistical methods to quantify and interpret data trends and relationships.
Field Surveys and Ground Truthing	Conducts on-site surveys to validate satellite data and model predictions.
Integrated Coastal Zone Management (ICZM)	Applies an interdisciplinary approach to managing coastal resources sustainably in light of predicted changes.

Source: Compiled by authors

RNN/LSTM: RNN/LSTM training is used to predict future trends and impacts using time-series climate data.

Prediction: Projecting upcoming changes and their consequences on coastal infrastructure using well-trained models.

Evaluation: Models are evaluated for performance using metrics like as accuracy, precision, recall, and F1 score, and any required improvements are performed.

5. Research Contribution

This study presents a novel deep learning-based framework to predict and mitigate the impacts of climate change on coastal infrastructure and land degradation, specifically focusing on Ghoramara Island in the Sundarbans region of India. The contributions of this research are threefold:

1. **Novelty of the Study:**

   Unlike conventional statistical and physical models, this study integrates advanced deep learning algorithms—**CNN**, **RNN**, and **LSTM**—to capture both spatial and temporal patterns in coastal changes. By analyzing satellite imagery and historical climate data, the research develops a **comprehensive predictive model** capable of addressing the intricate and nonlinear dynamics of shoreline erosion and infrastructure degradation.

2. **Significance in Climate Impact Prediction:**

   This study provides a robust tool for accurately forecasting the future effects of climate change, including sea-level rise, land loss, and coastal erosion. The proposed model is tailored to the specific vulnerabilities of Ghoramara Island, an area highly susceptible to climatic variations. The ability to make precise predictions enables proactive measures to be undertaken, addressing both short-term risks and long-term impacts.

3. **Practical Applications for Sustainable Coastal Management:**

   The research outputs will provide actionable insights for policymakers, urban planners, and environmental scientists. By simulating future scenarios of shoreline shifts and land degradation, the model can inform **sustainable coastal management strategies** such as:

   • Designing protective coastal infrastructure.

   • Implementing integrated coastal zone management (ICZM).

   • Prioritizing relocation plans for at-risk communities.

   This work advances the integration of **AI-driven models** into climate adaptation strategies and contributes to global efforts in combating coastal vulnerabilities. The methodologies proposed in this research can also serve as a blueprint for similar regions facing climate-induced changes.

   Limitations:

   • Accurate forecasting may be hampered by incomplete or inaccurate data on climate, coastal structures, and land degradation.

   • Lack of long-term datasets may make it more challenging to recognize historical patterns and produce reliable projections for the future.

6. CONCLUSION

The use of Artificial Neural Networks (ANN), Convolutional Neural Networks (CNN), Recurrent Neural Networks (RNN), and Long Short-Term Memory networks (LSTM) to predict the effects of climate change on the land degradation and coastal structures of Ghoramara Island is a viable direction for future research. Nevertheless, there are benefits and drawbacks to employing these cutting-edge deep learning models. In conclusion, despite certain difficulties in predicting the effects of climate change, ANN, CNN, RNN, and LSTM are essential tools for fostering resilience and sustainable development on Ghoramara Island due to their in-depth comprehension of the interrelated dynamics of land degradation and coastal structures.

REFERENCES

[1] Halder, B. (2020). Land degradation and shoreline changes in the Ganges-Brahmaputra-Meghna delta: a deep learning approach. *Journal of Coastal Research*, 36(3), 512–517.

[2] Chen, Z. (2022). Remote sensing-based assessment of land degradation using deep learning algorithms in Ghoramara Island. *International Journal of Applied Earth Observation and Geoinformation.*

[3] Das, S. (2021). Predicting coastal zone changes using deep learning techniques: a comprehensive review. *Journal of Coastal Research.*

[4] Johnson, S. (2021). Predicting climate change impacts on coastal structures using deep learning models. *Environmental Modelling and Software.*

[5] Kim, L. (2021). Deep learning techniques for coastal zone monitoring: a review. *Ocean Engineering.*

[6] Kumar, P. (2020). Deep learning models for climate change impact assessment: a case study of Ghoramara Island. *Journal of Hydrology.*

[7] Li, W. (2021). Predicting sea level rise impacts on coastal areas using deep learning models. *Journal of Geophysical Research: Oceans.*

[8] Liu, C. (2020). Assessing land degradation in coastal regions using deep learning models: Ghoramara Island case study. *Journal of Environmental Management.*

[9] Martinez, G. (2021). Deep learning applications in climate change research: challenges and opportunities. *Climate Dynamics*, 57, 1021–1036.

[10] Tran, N. (2021). Deep learning applications in coastal zone management: current trends and future prospects. *Ocean and Coastal Management.*

[11] Wang, L. (2021). Deep learning-based assessment of shoreline changes and land degradation in Ghoramara Island. *Journal of Applied Remote Sensing.*

Advances in Construction, Real Estate, Infrastructure and Project Management – Anil Kashyap et al. (eds)
© 2026 Taylor & Francis Group, London, ISBN 978-1-041-13433-6

13 Predicting *Turnover Intention*: A Study on Indian Construction Professionals Working on Infrastructure Projects

Rumjhum Bandyopadhyay[1]
Junior Research Fellow,
Department of Management Symbiosis International
(Deemed University), Pune, India

Amal Bhattacharya[2]
Assistant Vice President,
Tata Projects Limited Pune, India

Seema R. Singh[3]
Professor & Director,
SCCE Symbiosis International (Deemed University),
Pune, India

■ **ABSTRACT:** This study explored the predictability of Perceived Organizational Support (POS), Perceived Demotivational Managerial Practices (PDMP), and Perceived Fairness in Human Resource Management Practices (PFHRMP) on the Turnover Intention (TI) among Indian Construction professionals. An online survey form was created and distributed to various online forums to reach to the Indian construction professionals working on infrastructure projects across India. The findings revealed that the independent variables can predict 69% of the variance in the criterion variable. POS and PFHRMP are negatively related to turnover intention. In contrast, PDMP is positively related to the TI of Indian construction professionals working on various infrastructure projects in India. The findings contribute to the current literature.

■ **KEYWORDS:** Indian construction professionals, Turnover intention, Perceived organisational support, Perceived demotivational managerial practices, Perceived fairness in human resource management practices

1. INTRODUCTION

The construction sector significantly contributes to the economic growth of different countries, including India [22]. India's construction industry has a worth of over three lakhs crores in 2022

[1]phdgrad.rumjhum.bandyopadhyay@siu.edu.in, [2]amalb@tataprojects.com, [3]director@scce.edu.in

DOI: 10.1201/9781003669814-13

[21]. However, people working in construction projects face various challenges due to adverse working conditions, highly demanding jobs, and poor control over the assigned task [28]. Construction professionals are most vulnerable to stress. They face the challenge of job burnout and other negative consequences [40]. The construction industry faces challenges regarding work-life balance and high turnover [24, 39]. A high turnover rate causes significant economic loss for the organization regarding hiring and training, decreased team performance, and disrupted service delivery [26, 35].

Research on turnover in the construction industry primarily examines why employees leave their jobs. Researchers found various causes, such as work-life balance, ethical leadership, stress, and job satisfaction, which are predictors of turnover intention [19, 26]. However, a gap is identified in the existing knowledge body concerning the predictive values of all three variables: Perceived Organizational Support (POS), Perceived Demotivational Managerial Practices (PDMP), Perceived Fairness in Human Resource Management Practices (PFHRMP) on turnover intention (TI) among construction professionals working on various infrastructure projects in India.

Social support at work is perceived as organizational support [5]. The POS is operationally defined as the perception of employees about the organization's acknowledgement of their contribution to how the organization takes care of their well-being and supports their professional development and success.

Demotivated employees show decreased productivity at work and higher turnover rates, which leads to economic loss for the organization [4]. While various studies have explored employee demotivation in the construction industry, there needs to be more focus on demotivational managerial practices specifically [2]. PDMP is operationally defined as the employees' viewpoint about supervisors' behaviour or actions as obstacles to their motivation and engagement at work. This perception can be measured through surveys or interviews that record employees' beliefs about specific managerial practices that they find discouraging, demoralizing, or detrimental to their motivational levels, potentially leading to increased turnover intention.

Studies show that if employees perceive human resource management (HRM) practices as transparent, then employees will exhibit more positive work attitudes such as organizational citizenship behaviour, loyalty, satisfaction, and commitment. This perception will help to increase employees' willingness to continue with the organization. However, if the human resource department lacks fairness, it can lead to absenteeism, counterproductive work behaviour, and turnover [17]. Perceived fairness in Human Resource Management (HRM) practices is operationally defined as the subjective evaluation by the employees of the performance appraisal, transparency, consistency, and impartiality in implementing HR policies and procedures within an organization. This perceptual evaluation of the HRM practices by the employees can be measured through surveys or interviews that record employees' views about the factors that define fairness in HRM practices.

Turnover intention (TI) is an employee's readiness to leave the organization within a certain period. This readiness or intention to leave is a vital indicator for understanding actual turnover. Thus, exploring the factors related to TI may help to avoid the potential negative impacts of actual employee turnover on business [25]. TI can be operationally defined as the degree to which employees express their willingness or desire to leave their current jobs and seek alternative employment opportunities. This intention is measured through self-reported surveys. High turnover intention often indicates dissatisfaction with the current job or work environment.

This study is the first to understand the predictive value of all three variables: PDMP, PFHRMP, and POS on TI of Indian construction professionals. The study's findings help frame Human resource policies to improve employee retention. The following research questions are pertinent to the study:

1. Does POS predict the TI of Indian construction professionals who work on various infrastructure projects?
2. Does PDMP predict the TI of Indian construction professionals who work on various infrastructure projects?
3. Can PFHRMP predict the TI of Indian construction professionals who work on various infrastructure projects?

2. LITERATURE REVIEW

2.1 Perceived Organizational Support (POS) & Turnover Intention (TI) of Construction Professionals

Jiang and Li [22] established that POS has a moderating effect on the relationship between TI and job embeddedness among project managers of the construction industry. Furthermore, they noted that insufficient person-environment fit reduces job engagement and raises TI in this group. Job engagement serves as a mediator between an individual's environment fit and TI. This study was done on the Chinese population and should be tested in another cultural context. Person-environment fit is a broad concept that includes various aspects of a person and the environment. Understanding the specific factors is of utmost importance in deploying employee retention strategies. However, employees' perception of the workplace environment as supportive, fair, and motivating significantly influences turnover intention. Individuals decide to leave a company based on these fundamental aspects of work.

2.2 Supervisors' Support & TI of Construction Professionals

Kissi et al. [23] examined how engagement at work serves as a mediator between the supervisors' support and employee turnover among construction workers. The result indicates that work engagement and supervisors' support positively impact the TI. Conversely, TI has a detrimental effect on project and organizational performance. Dodanwala and Santoso [8] discovered that satisfaction about the supervisor's behaviour and job security directly reduce stress levels, resulting in low TI.

2.3 PFHRMP & TI of Construction Professionals

Hazeen and Umarani [18] identified a positive correlation between engineers' desire to continue with an organization and fairness in HRM practices. Employees' satisfaction with human resource practices influences their decision to continue with the organization.

2.4 Other Factors Related to TI

Through a qualitative study, Borg and Scott-Young [6] found significant factors affecting construction managers' intention to leave. Work culture relates to factors like bad behaviour at the workplace, work-related stress and poor work-life balance. Despite existing research on TI, there is a dearth of research to understand the relevant factors related to TI within the construction sector. Uğural and Giritli [37] addressed this gap by examining how identification with the organization and acknowledgement of organizational prestige affect TI among construction professionals differently across genders. Uğural et al. [38] identified important factors related to voluntary TI of construction professionals, focusing on individual value orientations and their mediating roles in external prestige and organizational identification. Mohd Kasmuri et al. [30], through a qualitative study, identified the factors related to employee turnover in the construction sector of Malaysia. Findings revealed three main themes of turnover determinants: organization, personal, and external factors. Some discrepancies in research findings were identified, particularly in areas like compensation, training, stress, and external influences. Dodanwala et al. [9] found that satisfaction with job and job stress directly affect TI. Additionally, ambiguous roles, conflicting roles, and work-family conflict were linked to TI. Satisfaction and stress on the job were mediating factors in this relationship.

Choi [7] explored the characteristics of the construction job which impact TI. The key factors identified are variety in skill and justification for the assigned job location are two important factors which can significantly affect TI of employees working in construction industry.

The authors identified a gap in the literature as no study has been found that considered all three factors, perceived organizational support, perceived demotivational managerial practices and perceived fairness in human resource practices, to predict the turnover intention of the construction professionals working on infrastructure projects in India. The present study explores how much predictability these three variables have on the turnover intention of Indian construction professionals.

2.5 Theoretical Background & Hypotheses Development

An employee's decision to continue or to leave the organization is not made in isolation. It is a thoughtful decision employees make, influenced by various factors [34]. The theoretical base of the present study is grounded in the "Social Exchange Theory". As per the theory, if employers extend their support to employees, they feel obligated due to social exchange. In response, employees strive to perform effectively and maintain a positive attitude, as reflected in their organizational commitment. Studies indicate that similar exchanges or interactions among co-workers can foster this sense of obligation and encourage positive behaviours [20]. Perception of organizational support, fairness in HRM practices and motivational aspects of supervisors' behaviours help reduce TI [1, 14–16, 29]. Researchers used social exchange theory concepts to develop hypotheses for the present study. According to Organizational Support theory, employees form beliefs about their organization's values and how it contributes to and cares about their well-being over time. This POS triggers a sense of obligation in employees. Thus, employees reciprocate this feeling by significantly contributing to the organization's welfare and success. The POS will help to enhance commitment and willingness to contribute towards its success [13]. There must be a significant negative correlation between POS and TI. Therefore, it is hypothesized that POS can predict TI. Supervisors establish strategies to inspire employees and assign tasks to individuals or teams to accomplish objectives. The retention rate would increase in such cases [32]. When senior management creates organizational policies and practices that fall short of meeting employees' expectations, it can lead to employee demotivation. Demotivation leads to less productivity and higher TI [2, 36]. Employees experiencing demotivation are prone to being dissatisfied with their roles. This dissatisfaction can affect their performance and TI [31]. The findings serve as background for the second hypothesis. Fairness in an organization's policies refers to the organization's fair practices towards its employees. A significant relationship exists between HRM practices and employee attitudes, especially when an employee perceives the strength of HRM practice [10]. Based on this concept, a third hypothesis is formed.

H1: *POS can predict TI of Indian construction professionals working on various infrastructure projects.*

H2: *PDMP can predict the TI of Indian construction professionals working on various infrastructure projects.*

H3: *PFHRMP can predict the TI of Indian construction professionals working on various infrastructure projects.*

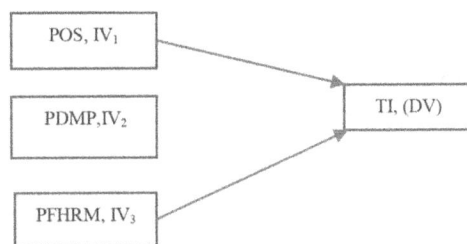

Figure 13.1   Proposed model on POS, PDMP, PFHRMP & TI

Source: Authors

3. METHODOLOGY

3.1 Sample and Procedure

Data is collected through a questionnaire survey method. The sample for this cross-sectional study is drawn following snowball and purposive sampling methods. The subjects are all civil engineers working on various infrastructure projects across India. All of them are working in MNCs. Google forms are created and circulated through social media groups. Data collection is done within one month. Nearly 200 persons were contacted, and 104 responses were obtained (above fifty per cent). The response rate is acceptable as the variables chosen are sensitive to the response.

3.2 Instruments

The variables chosen for the study are as follows:

Perceived Organizational Support (POS, IV1); Perceived Demotivational Managerial Practice (PDMP, IV2); Perceived Fairness in HRM Practices (PFHRM, IV3); Turnover Intention (TI, DV). Eight items are selected from the scale developed to measure POS [12]. The scale's reliability (Coefficient alpha) ranges from .74 to .95 [11]. Validity is established for POS by finding a positive correlation between organizational commitment and job satisfaction [11]. Eight items are selected from an already standardised scale to measure PDMP [2]. Three items are adapted from a study by Lentz [27] to measure turnover intention [33]. The Cronbach alpha for the scale is 0.88. Seven items are adapted from the scale with a composite reliability of 0.86 for fair performance management, 0.82 for fair compensation and benefits and 0.90 for fair employee relations [17].

4. RESULTS

The demographic details of the sample are presented in tabular form (Table 13.1).

Table 13.1 Demographic details

Age	Gender	Marital Status	Educational experience	Years of Experience
56.3% falls in the 25–40 years range	97.1% Male	85.6% Married	Diploma holder 16.3%	1–10 years 27.9%
43.75% falls in the 40–60 years range	2.9% Female	14.4% Unmarried	Graduate 55.8%	11–20 years 36.5% 21–30 years 24%
			Postgraduate 27.9%	Above 30 years 11.5%

Source: Compiled by authors

5. DISCUSSION

Regression analysis is done by using the MS Excel application. The adjusted R square is 0.69 (approx.), which indicates that the three predictors: Perceived Organizational support (POS), Perceived Demotivational Managerial Practice (PDMP) and Perceived Fairness in Human Resource Management Practice (PFHRMP) can explain 69% of the variance in the turnover intention (Table 13.2). All these three variables are organizational factors related to turnover

Table 13.2 The multiple regression output by MS Excel

Regression Statistics	
Multiple R	0.833698
R Square	0.695052
Adjusted R Square	0.685904
Standard Error	2.003247
Observations	104
ANOVA	

	Df	SS	MS	F	Significance F	
Regression	3	914.6617	304.8872395	75.97493	1.09479E-25	
Residual	100	401.2998	4.012998199			
Total	103	1315.962				

	Coefficients	Standard Error	t Stat	P-value	Lower 95%	Upper 95%
Intercept	15.21611	1.596747	9.529446269	1.07E-15	12.04821067	18.38400997
POS	−0.21181	0.036152	−5.858999366	5.99E-08	−0.283537786	−0.140089366
PDMP	0.104535	0.032985	3.169125608	0.002029	0.039092687	0.169976557
PFHRMP	−0.14663	0.048291	−3.036427034	0.003053	−0.242440841	−0.050824292

Source: Compiled by authors

intention as perceived by the employees. These variables contribute to the formation of work culture. Research evidence shows that work culture-related factors like organizational support, HRM strategies and practices, and workplace behaviours can influence employee retention in the construction industry [6]. The three independent variables of the present study act on the turnover intention in various ways. According to the social exchange perspective, employees' perception of their exchange relationship with the organization contributes to their motivation and well-being. The exchange relationship works in various ways, but POS is the most influential [3]. The result shows that POS has a strong negative association (Pearson r = −0.80) with TI. POS might have played a role in enhancing motivation, thus decreasing the turnover intention among Indian construction professionals. Demotivational managerial practice has a significant positive relation with TI (Pearson r = 0.57). PDMP leads to job dissatisfaction among Indian construction professionals eventually affecting turnover intention [2]. Perceived fairness in HRM practices leads to a sense of attachment and obligation among employees. This perception helps to decrease turnover intention [18]. The present study shows that PFHRMP and TI are significantly negative correlation (Pearson r = −0.69). It can be said that the Indian construction professional's positive perception of HRM practices contributes to lowered TI. The F value is well below the 0.05 confidence level; hence, the model is statistically significant (Table 13.2, Figure 13.1). The P values for PDMP, POS, and PFHRMP are well below the 95% confidence level (5.99E-08, 0.002029 and 0.003053, respectively) and hence have significant predictive values for the turnover intention of Indian construction professionals (Table 13.2). All three research hypotheses are accepted. The result shows that POS and PFHRMP have a significant negative correlation with TI, and PDMP has a positive correlation with TI (Table 13.3).

Table 13.3 Correlation coefficient of the variables (Pearson r) generated by MS Excel

	POS	PDMP	PFHRMP	TI
POS	1			
PDMP	−0.53376	1		
PFHRMP	0.72441626	−0.391665746	1	
TI	−0.79737	0.574819575	−0.694258756	1

Source: Compiled by authors

6. LIMITATIONS AND FUTURE SCOPE

The sample is drawn using the non-probability sampling method. Therefore, this study opens the scope for future research using probability sampling to increase the generality of the findings. Only quantitative correlational design does not explain the causality of the variables. A mixed-method approach can be used to deal with this limitation in future studies. The sample size is small (N=104) due to the limited time frame and the difficulty of reaching professionals because of their tight work schedules. In the future, a larger sample will increase the validity of the study.

7. IMPLICATIONS

This study contributes by studying the predictive value of three critical variables on the turnover intention of construction professionals, which has yet to be studied in India. The findings may help in developing HR strategies to enhance employee retention.

8. CONCLUSION

The findings of the present study indicate that three variables, namely, perceived organizational support (POS), perceived demotivational managerial practices (PDMP), and perceived fairness in Human Resource Management practices (PFHRMP), have good predictive value for the turnover intention of the Indian construction professionals working in various infrastructure projects.

REFERENCES

[1] Ali, H., Li, M. & Qiu, X. (2024). Examination of HRM practices in relation to the retention of Chinese Gen Z employees. *Humanities and Social Sciences Communications*, 11(1). https://doi.org/10.1057/s41599-023-02472-6

[2] Aung, Z. M., Santoso, D. S., & Dodanwala, T. C. (2022). Effects of demotivational managerial practices on job satisfaction and job performance: Empirical evidence from Myanmar's construction industry. *Journal of Engineering and Technology Management*, 67, 101730. https://doi.org/10.1016/j.jengtecman.2022.101730

[3] Avanzi, L., Fraccaroli, F., Sarchielli, G., Ullrich, J., & van Dick, R. (2014). Staying or leaving: A combined social identity and social exchange approach to predicting employee turnover intentions. *International Journal of Productivity and Performance Management*, 63(3), 272–289.

[4] Azar, M., & Shafighi, A. A. (2013). The effect of work motivation on employeesâ job performance (Case study: Employees of Isfahan Islamic Revolution Housing Foundation). *International Journal of Academic Research in Business and Social Sciences*, 3(9). https://doi.org/10.6007/ijarbss/v3-i9/2

[5] Bonaiuto, F., Fantinelli, S., Milani, A., Cortini, M., Vitiello, M.C. & Bonaiuto, M. (2022). Perceived organizational support and work engagement: the role of psychosocial variables. *Journal of Workplace Learning*, 34 (5), 418–436. https://doi.org/10.1108/JWL-11-2021-0140

[6] Borg, J., & Scott-Young, C. M. (2022). Contributing factors to turnover intentions of early career project management professionals in construction. *Construction Management and Economics*, 40(10), 835–853. https://doi.org/10.1080/01446193.2022.2110602

[7] Choi, S. (2024). Identifying the job characteristics affecting construction firm employee turnover intention. *Construction Research Congress 2022*, 591–600. https://doi.org/10.1061/9780784485286.059

[8] Dodanwala, T. C., & Santoso, D. S. (2022). The mediating role of job stress on the relationship between job satisfaction facets and turnover intention of the construction professionals. *Engineering, Construction and Architectural Management*, 29(4), 1777–1796. https://doi.org/10.1108/ecam-12-2020-1048

[9] Dodanwala, T. C., Santoso, D. S., & Yukongdi, V. (2022). Examining work role stressors, job satisfaction, job stress, and turnover intention of Sri Lanka's construction industry. *International Journal of Construction Management*, 23(15), 2583–2592. https://doi.org/10.1080/15623599.2022.2080931

[10] Edgar, F., & Geare, A. (2005). HRM practice and employee attitudes: different measures – different results. *Personnel Review*, 34(5), 534–49. https://doi.org/10.1108/00483480510612503

[11] Eisenberger, R., Fasolo, P., & Davis-LaMastro, V. (1990). Perceived organizational support and employee diligence, commitment, and innovation. *Journal of Applied Psychology*, 75(1), 51–59. https://doi.org/10.1037/0021-9010.75.1.51

[12] Eisenberger, R., Huntington, R., Hutchison, S., & Sowa, D. (1986). Perceived organizational support. *Journal of Applied Psychology*, 71(3), 500–507. https://doi.org/10.1037/0021-9010.71.3.500

[13] Eisenberger, R., Armeli, S., Rexwinkel, B., Lynch, P., & Rhoades, L. (2001). Reciprocation of perceived organizational support. *The Journal of Applied Psychology*, 86(1), 42–51. https://doi.org/10.1037//0021-9010.86.1.42

[14] Fulmore, J. A., Fulmore, A. L., Mull, M., & Cooper, J. N. (2022). Reducing employee turnover intentions in the service sector: the connection between human resource development practices and employee engagement. *Human Resource Development Quarterly*, 34(2), 127–153. https://doi.org/10.1002/hrdq.21471

[15] Georgiadou, A., Amari, A., Swalhi, A., & Hofaidhllaoui, M. (2024). How does perceived organizational support improve expatriates' outcomes during global crises? The mediating role of the ethical organizational climate in global organizations. *Journal of International Management*, 30(3), 101131–101231. https://doi.org/10.1016/j.intman.2024.101131

[16] Halid, H., Kee, D. M. H., & Rahim, N. F. A. (2020). Perceived Human Resource Management Practices and Intention to Stay in Private Higher Education Institutions in Malaysia: The Role of Organizational Citizenship Behaviour. *Global Business Review*, 25(1), 162–179. https://doi.org/10.1177/0972150920950906

[17] Hazeen, F. M., & Umarani, C. (2022a). A study on the impact of role stress on engineer intention to leave in Indian construction firms. *Scientific Reports*, 12(1). https://doi.org/10.1038/s41598-022-21730-2

[18] Hazeen, F. M., & Umarani, C. (2022b). Fairness in human resource management practices and engineers' intention to stay in Indian construction firms. *Employee Relations: The International Journal*, 45 (1). https://doi.org/10.1108/er-07-2021-0308

[19] Hom, P. W., Lee, T. W., Shaw, and Hausknecht, J. P. (2017). One hundred years of employee turnover theory and research. *Journal of Applied Psychology*, 102(3), 530–45.

[20] Harden, G., Boakye, K. G., &Ryan, S. (2016). Turnover intention of technology professionals: a social exchange theory perspective. *Journal of Computer Information Systems*, 58(4), 291–300. https://doi.org/10.1080/08874417.2016.1236356

[21] India: GDP from Construction 2023 | Statista, 2023. *Statista*. https://doi.org/1065000/1069818-blank-355

[22] Jiang, S., & Li, J. (2023). Impact of person-environment fit on construction project managers' turnover intention: a moderated mediation model. *Engineering, Construction and Architectural Management*. https://doi.org/10.1108/ecam-04-2023-0384

[23] Kissi, E., Ikuabe, M., Aigbavboa, C., Smith, E. D., & Babon-Ayeng, P. (2023). Mediating role of work engagement in the relationship between supervisor support and turnover intention among construction workers. *Engineering, Construction and Architectural Management*. https://doi.org/10.1108/ecam-06-2023-0556

[24] Koc, K., Kunkcu, H., Asli, G. (2022). Organizational commitment of construction professionals in the construction industry. In Proceedings of International Structural Engineering and Construction, pp., 9. 10.14455/ISEC.2022.9(1). CPM-10

[25] Lazzari, M., Alvarez, J. M., & Ruggieri, S. (2022). Predicting and explaining employee turnover intention. *International Journal of Data Science and Analytics*, 14(3), 279–292. https://link.springer.com/article/10.1007/s41060-022-00329-w

[26] Le, H., Lee, J., Nielsen, I., & Nguyen, T. L. A. (2022). Turnover intentions: the roles of job satisfaction and family support. *Personnel Review*, 52(9), 2209–2228.

[27] Lentz, E. (2004). The link between the career plateau and mentoring: Addressing the empirical gap [Doctoral dissertation, University of South Florida]. *Scholar Commons*. http://scholarcommons.usf.edu/etd/1129

[28] Leung, M.-Y., Bowen, P., Liang, q., & Famakin, I. (2015). Development of a job-stress model for construction professionals in South Africa and Hong Kong. *Journal of Construction Engineering and Management*, 141(2), 04014077. https://doi.org/10.1061/(asce)co.1943-7862.0000934

[29] Dzansi, L. W., Chipunza, C., & Dzansi, D. Y. (2016). Impact of municipal employees' perceptions of fairness in human resources management practices on motivation: evidence from a South African Province. *Problems and Perspectives in Management*, 14(1), 138–149. https://doi.org/10.21511/ppm.14(1-1).2016.01

[30] Kasmuri, M., Hawa, S., Ismail, Z., Nordin, R. M., & Hashim, N. (2022). Analysis of the Malaysian construction industry professional employee turnover antecedents. *Jurnal Kejuruteraan*, 34(5), 871–86. https://doi.org/10.17576/jkukm-2022-34(5)-15.

[31] Pang, K. & Lu, C-S. (2018). Organizational motivation, employee job satisfaction and organizational performance. *Maritime Business Review*, 3(1), 36–52. https://doi.org/10.1108/mabr-03-2018-0007

[32] Pattali, S., Sankar, J. P., Qahtani, H. Al., Menon, N., & Faizal, S. (2024). Effect of leadership styles on turnover intention among staff nurses in private hospitals: the moderating effect of perceived organizational support. *BMC Health Services Research*, 24(1). https://doi.org/10.1186/s12913-024-10674-0

[33] Priyashantha, K.G., & Hunnes, J. A. (2011). The impact of job satisfaction on perceived desirability of leaving: a study in cable manufacturing organizations in Sri Lanka. https://doi.org/10.13140/RG.2.2.30101.45282

[34] Rahman, W., & Nas, Z. (2013). Employee development and turnover intention: theory validation. *European Journal of Training and Development*, 37(6), 564–79. https://doi.org/10.1108/ejtd-may-2012-0015

[35] Rumawas, W. (2021). Employees turnover intention in the construction industry in Indonesia. *Journal of Construction in Developing Countries*, 1–41. https://doi.org/10.21315/jcdc-03-21-0050

[36] Shareef, R. A., & Atan, T. (2019). The influence of ethical leadership on academic employees' organizational citizenship behavior and turnover intention. Management Decision, 57(3), 583–605. https://doi.org/10.1108/md-08-2017-0721

[37] Ugural, M., & Giritli, H. (2021). Three-way interaction model for turnover intention of construction professionals: some evidence from Turkey. *Journal of Construction Engineering and Management*, 147(4), 04021008. https://doi.org/10.1061/(asce)co.1943-7862.0002012

[38] Uğural, M. N., Giritli, H., & Urbański, M. (2020). Determinants of the turnover intention of construction professionals: a mediation analysis. *Sustainability*, 12(3), 954. https://doi.org/10.3390/su12030954

[39] Wang, Y., Hu, N., Zuo, J., & Rameezdeen, R. (2020). Project management personnel turnover in public sector construction organizations in China. *Journal of Management in Engineering*, 36(2), 05019009. https://doi.org/10.1061/(asce)me.1943-5479.0000735

[40] Wu, Z., Wang, Y., & Liu, M. (2023). Job stress and burnout among construction professionals: the moderating role of online emotions. *Engineering, Construction and Architectural Management*, 31(12), 4831–4851 https://doi.org/10.1108/ecam-09-2022-0868

Advances in Construction, Real Estate, Infrastructure and Project Management – Anil Kashyap et al. (eds)
© 2026 Taylor & Francis Group, London, ISBN 978-1-041-13433-6

14 Enhancing Recycled Concrete Aggregate Properties: A Systematic Review of Treatment Methods

Yuvraj R. Patil*

PhD Scholar,
Department of Civil Engineering,
Visvesvaraya National Institute of Technology,
Nagpur

Vaidehi A. Dakwale and Rahul V. Ralegaonkar

Assistant Professor,
Department of Civil Engineering,
Visvesvaraya National Institute of Technology,
Nagpur

■ **ABSTRACT:** Recycled concrete aggregates (RCA), derived from old and demolished buildings, offer a sustainable alternative to natural aggregates, reducing carbon emissions by 20–35% and saving 15–40% of natural resources. However, the old adhered mortar, due to its porosity and water absorption, adversely impacts the properties of RCA. Treatments like mechanical, chemical, or thermal methods can enhance RCA strength by up to 22% by removing or strengthening the mortar. Despite their benefits, treatments such as carbonation and nanomaterial intrusion add costs, making the method selection crucial. This article reviews the treatment methods and proposes a systematic approach for choosing cost-effective and scalable solutions.

■ **KEYWORDS:** Recycled concrete aggregate, Multi-criteria decision-making tool, Sustainable construction, Physical properties

1. INTRODUCTION

In recent past construction industry is actively seeking methods to enhance sustainability by conserving natural resources. Sustainability can be achieved through maximum utilization of

---
*Corresponding author: yuvrajpatil14@gmail.com

DOI: 10.1201/9781003669814-14

construction and demolition (C&D) waste [1]; however, the composition of C&D waste is complex, which is a major constraint for replacement of natural aggregates [18]. Major components of C&D waste include concrete, bricks, mortar, and steel. Recycled concrete aggregate (RCA) forms about 50–60 % of C&D waste [36], and its use in concrete is a way to promote sustainability in construction [24], and reduces the amount of C&D waste sent to landfills [2], contributing to environmental conservation. Landfilling is often considered as a convenient disposal method for C&D waste [41]. Alternatively, recycling C&D waste serves two purposes: conserves landfilling space and produces alterative for virgin resources [24]. According to previous studies, substituting RCA with natural aggregates in both structural and non-structural applications can reduce carbon emissions by 20–35 percent [29] and conserve natural resources by 15–40 % [36]. During the crushing of RCA, some part of the old mortar remains on its surface, known as adhered mortar [30, 31]. As it is porous by virtue [42, 49], It is noted that the water absorption, aggregate crushing value, and aggregate impact value, are increased, and the apparent specific gravity, bulk density, and specific gravity are reduced [19, 48]. This restricts the use of RCA in concrete [8]. Concrete is a three-phase material, viz bulk cement paste, aggregate, and interfacial transition zone (ITZ). Previous studies have shown that the properties of concrete are significantly influenced by the adhered mortar [49]. Thus, improvements in the adhered mortar directly improve the properties of the concrete [54]. Several treatments have been investigated in the past, which are classified into groups: first, removal of adhered mortar, which includes mechanical grinding [15], horizontal and vertical impact crushing, acidic soaking and rubbing [46], and soaking and washing with high-pressure water [3]; second, strengthening of the adhered mortar, which includes nano-silica intrusion and, carbonation and wrapping [10, 13], and cementitious material slurry coating [43]. However, all these methods involve the use of machinery, chemicals and dedicated setup. Success and acceptability of a treatment method in industrial practices depends upon several factors viz are cost of treatment, quality of RCA after treatment [2], environmental impact, suitability with the existing treatment process, weather dependency, adaptability for scaling up [6], and ease of the system. This study offers a framework for selecting an appropriate treatment method based on these factors.

2. TREATMENT METHODS

The treatment methods reviewed in this article are classified into two groups based on their approach to improving the strength of RCA viz are adhered mortar removal and strengthening of adhered mortar.

2.1 Adhered Mortar Removal

Pre-soaking and Washing with Water

Pre-soaking and washing with water are commonly adopted treatment method [16]. Because of its ease of application, this method was proposed by [22], in which RCA is cleaned with low to moderate water pressure to remove adhered mortar and other impurities. This method has two main limitations: it can only remove a small amount of adhered mortar, and the water used for washing the RCA requires recycling due to contamination [31]. However, ultrasonic cleaning of RCA can remove more adhered mortar, resulting in a 7 percent increase in the compressive strength of concrete made with RCA [22]. Although cleaning with low to moderate pressure does not have a significant effect on strength enhancement [58], due to its simplicity and ease of operation, it is preferred over other methods in field applications [44].

Mechanical Abrasion Method

It is the simplest machine treatment method for adhered mortar removal compared to other methods [51], but it is weather-dependent; hence, it is operated for a limited period [32]. Adhered mortar is removed using a Los Angeles abrasion machine, horizontal and vertical impact crushing [2], and additional water washing is necessary. However, this method cannot completely remove the adhered mortar. Additionally, it damages the surface of RCA, causing fissure cracks and partial destruction of its shape [24]. Using this method to treat a large quantity of RCA is feasible; however, it requires special equipment and extra space, leading to increased treatment costs. [24] used acetic acid immersion before treating RCA via mechanical grinding. Compared with normal mechanical grinding, the results obtained show improved performance in the physical and mechanical properties of RCA, and the preconditioning of RCA reduces the cost of treatment.

Thermal Treatment

Thermal treatment is another method to remove adhered mortar because it can weaken the bond between the old mortar and RCA [2]. In this method, RCA is heated to temperature ranging from 250–500°C for a period of half hour [43], followed by cooling for a period of 10–30 minutes and rubbing it by hand or by an abrasion machine; however if steel balls are used in the abrasion process, it causes more damage to the surface and shape of the RCA, and a similar effect is observed by [2] when the temperature is above 600°C. Precautions should be taken to maintain the temperature within a desirable range, as temperature increases can negatively affect the surface texture of RCA, causing more pores, elongated cracks, and higher water absorption. Previous studies have not extensively highlighted the effects of temperature on the properties of RCA. In addition, this method is expensive, requires specialised equipment, and is nonsuitable for large-scale projects [4]. Also, it is challenging to maintain uniform heating for large amounts of RCA. Hence, this method is rarely adopted to treat the RCA in the field.

Chemical Treatment

Conditioning of RCA is necessary in all chemical treatments, which involves washing and drying the RCA at 100°C–105°C for period of one day. RCA is pre-soaked in acetic acid and hydrochloric acid to remove impurities [43]. Next, the RCA was washed in deionized water to remove acidic residues from the surface. At this point, the RCA is prepared for further treatment. To remove the adhered mortar [51] followed RCA soaking in hydrochloric acid solution for 1, 3, and 7 days at molarities of 0.1–0.7 Molar. The aggregates were washed with deionized water to eliminate the mortar from the RCA surface. The results obtained were satisfactory; however, as the concentration and timing of treatment increased, the RCA shape was significantly disturbed. Another method for removing adhered mortar is the use of a chemical and calcium metasilicate slurry [43]. RCA was treated in two stages. Initially, RCA was immersed in hydrochloric acid 0.1 mol for 24 hours. Subsequently, the aggregate was rinsed with deionized water to remove any residual acid. The application of calcium metasilicate slurry, which appears as a white powder similar to cement particles, is the second stage of the surface treatment. For 24 hours, the RCA was submerged in a calcium metasilicate solution with a 5% concentration. The use of chemical treatment methods is limited because they are complex and require special attention to maintain a specific concentration during their operation.

2.2 Adhered Mortar Strengthens Methods

Pozzolana Slurry

Pozzolana materials are by-products of industrial waste that effectively mitigate major drawback of RCA, mainly its durability [3]. The quality of concrete made with RCA can be improved by appropriately altering the design mix through the addition of a suitable percentage of fly ash, silica fume, GGBS, and other materials. [3] demonstrated that by incorporating a pozzolanic material for the modification of RCA, the aggregate crushing value is improved by 13 percent. The pore refinement of the adhering mortar and secondary hydration between the added mineral admixture and cement hydration by products CaOH2 are the fundamental principles of employing mineral admixtures [37]. The pores and internal fissures that are present in the composite directly affect the strength of the concrete. [38] investigated the effects of coating RCA with pozzolana. Different mixtures of pozzolan were used, and the combination of Portland cement, silica fume and fly ash was found to be effective for high-strength RCA with a higher packing density and a higher ITZ. This could be because of the better packing in terms of pore filling. Pozzolana treatment is easy and effective compared to other adhered mortar strengthening methods; however, the presence of fine particles in the mixture may decrease workability. This can be mitigated by employing a suitable concrete design mix and appropriate admixture.

Carbonation

Carbonation is the most advanced method for treating RCA. However, the process is time consuming [11]. Preconditioning of the sample is necessary for both carbonation and accelerated carbonation treatment methods [43]. A special chamber was required to maintain stable temperature of $20 \pm 2°C$ with a humidity level of $70 \pm 5\%$ and a carbon dioxide concentration of $20 \pm 3\%$. Followed by air-drying, the aggregate was immersed in a solution of calcium hydroxide and mixed intermittently every five minutes. Subsequently, it was dried at 50°C before carbonation. The carbonation process involved exposing the aggregate to a carbon dioxide concentration of $10 \pm 2\%$ [24]. The performance of RCA improved marginally, the compressive strength improvement is in the range of 15–25%, but owing to its high initial investment, its adoptability is not common. [49] carried out the combination of carbonation followed by the slurry wrapping method and obtained an 8 percent higher performance of RCA. Conversely, the wrapping method involved immersing and wrapping the aggregate for a duration of 60 minutes, with mixing intervals of 5 minutes. After the coating treatment, the aggregate was allowed to air in a controlled environment for 24 hours, followed by immersion in a curing solution for seven days, and finally dried at 50 °C. The treatment of RCA with this method is difficult because of its ability to replicate in the field, and this method is laboratory limited.

Cement Slurry and Nanomaterials

In past [3] studied the effect of cement slurry coating on RCA surfaces and compared the properties of treated RCA and non-treated RCA, as well as the performance of concrete made with a controlled mix and RCA mix; the aggregate impact value was enhanced by 12 percent while water absorption was reduced by 54 percent. The porosity of the coated RCA (0.16–0.23 mm) cement slurry layer on the outside has been cut down by 55%, which means it can't hold as much water. When treated, the coating layer enhances the physical, mechanical, and durability properties of the concrete [26]. This treatment needs to be explored further, as the cement slurry is more porous in nature than the attached mortar of RCA. Nanomaterials are more effective than cement slurry methods [34], direct intrusion of nano-based material is performed to reduce the surface pores and fissure cracks of RCA, but the cost and limited field applicability restrict its use. Hence, to select an appropriate treatment method, a user-friendly framework is essential, as discussed in the following section.

3. FRAMEWORK FOR SELECTION OF APPROPRIATE METHOD

The selection of an appropriate method to treat RCA is crucial for the success of the treatment operation [46]; hence, to select a suitable method, a well-designed strategy is required. There are several methods available for the framework design, but in the present study multi criteria decision making framework (MCDM) weighted average method is adopted, due to its reliability of results and ease.

3.1 MCDM Framework Tool
Define Alternatives and Criteria

Alternative available are the treatment methods, hence all the methods discussed in section 2 are considered as alternative available which are: Pre-soaking and washing in water, Mechanical grinding method, Thermal treatment, Chemical treatment, Pozzolana slurry, Carbonation and wrapping, Cement slurry, Nano silica. While designing the MCDM framework tool the following criterias were considered: (a) treatment economy, (b) on site adaptability, (c) upscaling potential, (d) RCA quality after treatment, and (e) treatment duration. The criteria for designing an MCDM were further classified according to previous studies. The treatment economy is classified as low, moderate, and high, where low indicates that it is less cost-effective, similarly moderate, and high indicates higher cost-effectiveness. Site adaptability and upscaling potential are classified as either yes or no based on their true or false conditions related to their adaptability and potential for upscaling, as indicated by [16]. RCA quality after treatment is broadly classified because the effectiveness of the treatment is directly associated with it; hence, according to [39], it is classified as low, medium, and good. A low indicates that the average physical and mechanical properties, such as water absorption, specific gravity, aggregate crushing value, and aggregate impact value are improved by 0.1 to 5 percent. A medium indicates 5 to 10 percent improvement, and a good indicates more than 10 percent improvement. Treatment duration is classified into three classes, where low indicates a one to two days' time requirement, moderate indicates three to four days, and high indicates five or more days' time.

Allocation of Weight

The weight assigned to each criterion indicated its importance. The weightage calculations are based on the parameters considered by each researcher in the past studies, and Table 14.1 indicates their significance indicated by each researcher based on the percentage weightage and fractional percent weightage is calculated in Table 14.2.

Table 14.1 Significance of each criteria

Criteria for MCDM	Studies
Treatment Economy	[1, 20, 33, 45, 47, 50, 55]
On Site Adaptability	[17, 21, 35, 45, 52, 59]
Upscaling Potential	[5, 14, 25, 55, 57]
RCA Quality after Treatment	[7, 12, 23, 27, 28, 40, 56]
Treatment Duration	[9, 27, 45, 53]

Source: Compiled by authors

Table 14.2 Weightage calculation

Criteria	Approximate Percent Weight	Fractional Weight	No. of Researcher highlighted importance
Treatment Economy	25	0.25	7
On site Adaptability	20	0.20	6
Upscaling Potential	15	0.15	4
RCA – Quality after treatment	25	0.25	7
Treatment Duration	15	0.15	4

Source: Compiled by authors

Scale of Criteria

The classification of criteria is taken as input for the scale system, the scale varies from 1 to 3, but in cases of true or false conditions, a binary scale system is adopted, where 0 (zero) indicates the false condition and 1 (one) indicates the true condition. Table 14.3 presents the scale system adopted in the MCDM framework.

Table 14.3 Scale system for MCDM framework

Criteria	Rank		
Treatment Economy	1	2	3
On-site Adaptability	0	1	
Upscaling Potential	0	1	
RCA-Quality after Treatment	1	2	3
Treatment Duration	1	2	3

Source: Compiled by authors

4. RESULTS AND DISCUSSION

Based on the alternatives, criteria and scale of criteria, a decision matrix is calculated. The following example shows a detailed calculation of the decision matrix. The alternative available is pre-soaking and washing with water, this method is highly economical hence scale for treatment economy is 3, it is suitable and have potential for upscaling to treat the large quantum of C&D waste hence it has scale of 1, for both on site adoptability and up scaling potential, again the quality of treated waste is low indicating overall physical and mechanical properties are improved in between 0.1 to 5 percent hence it has scale of 1, but the treatment duration is one day hence it has scale of 3 for treatment duration based on that the total score is calculated in as per equation 1, which is shown below.

$$\text{Total Score} = \text{Fractional Weight} * \text{Scale} \tag{1}$$

$$\text{Total score} = (0.25 * 3) + (0.20 * 1) + (0.15 * 1) + (0.25 * 1) + (0.15 * 3) \tag{2}$$

$$\text{Total score} = 1.8 \tag{3}$$

Here Table 14.4 shows the scale of each critera, total score of the different alternatives based on the scale of each criterion and fractional weight assigned.

Table 14.4   Rank of alternatives

Treatment Method	Treatment Economy	On Site Adaptability	Upscaling Potential	RCA Quality after Treatment	Treatment Duration	Total Score	Rank
Pre-soaking and washing	3	1	1	1	3	1.8	2
Mechanical grinding	2	1	1	2	2	1.65	5
Thermal treatment	1	0	1	2	2	1.2	7
Chemical treatment	2	1	0	3	1	1.75	4
Pozzolana slurry	3	1	1	3	2	2.15	1
Carbonation and wrapping	1	0	0	3	1	1.15	8
Cement slurry	2	1	1	2	3	1.8	2
Nano silica	1	0	0	3	2	1.3	6

Source: Compiled by authors

The study reveals that various treatment methods enhance the properties of RCA with varying degrees of effectiveness and feasibility. Mechanical grinding offers modest improvements but may introduce surface damage same in case of thermal treatment, additionally it requires temperature control. Chemical treatments, like acetic and hydrochloric acid, significantly improve RCA properties but raise environmental concerns. Pozzolana slurry and carbonation treatments show promising results, with carbonation yielding the most notable enhancements despite increased expenses and handling challenges. Bio cement slurry and nanomaterial treatments significantly reduce porosity and enhance durability; however, their high costs limit widespread use.

5. Conclusions

Based on the literature review and the MCDM framework tool, following conclusions are drawn:

1. A lot of treatment methods were developed by the researcher, but most of the methods are limited to laboratories; additionally, some of the methods are uneconomical and damage the surface texture of the RCA, which limits their use in the field. Hence, a study should focus on the development of an eco-friendly method that is suitable for large-scale treatment.
2. In comparison with the adhered removal methods, treatment with adhered mortar is more effective and cost-effective, whereas treatment with pozzolana slurry is more cost effective.

3. The MCDM framework tool is a user-friendly approach to selecting a suitable treatment method.

Apart from the alternatives selected in this article, several other methods are also available, such as combinations of two individual methods; however, this framework does not consider these methods, and the study is limited to the above-mentioned methods only.

REFERENCES

[1] Al-Bayati, H. K. A., Das, P. K., Tighe, S. L., & Baaj, H. (2016a). Evaluation of various treatment methods for enhancing the physical and morphological properties of coarse recycled concrete aggregate. *Construction and Building Materials*, 112, 284–298. https://doi.org/10.1016/j.conbuildmat.2016.02.176

[2] Al-Bayati, H. K. A., Das, P. K., Tighe, S. L., & Baaj, H. (2016b). Evaluation of various treatment methods for enhancing the physical and morphological properties of coarse recycled concrete aggregate. *Construction and Building Materials*, 112, 284–298. https://doi.org/10.1016/j.conbuildmat.2016.02.176

[3] Al-Waked, Q., Bai, J., Kinuthia, J., & Davies, P. (2022a). Enhancing the aggregate impact value and water absorption of demolition waste coarse aggregates with various treatment methods. *Case Studies in Construction Materials*, 17. https://doi.org/10.1016/j.cscm.2022.e01267

[4] Al-Waked, Q., Bai, J., Kinuthia, J., & Davies, P. (2022b). Enhancing the aggregate impact value and water absorption of demolition waste coarse aggregates with various treatment methods. *Case Studies in Construction Materials*, 17. https://doi.org/10.1016/j.cscm.2022.e01267

[5] Al-Waked, Q., Bai, J., Kinuthia, J., & Davies, P. (2022c). Enhancing the aggregate impact value and water absorption of demolition waste coarse aggregates with various treatment methods. *Case Studies in Construction Materials*, 17. https://doi.org/10.1016/j.cscm.2022.e01267

[6] Bui, N. K., Satomi, T., & Takahashi, H. (2017a). Improvement of mechanical properties of recycled aggregate concrete basing on a new combination method between recycled aggregate and natural aggregate. *Construction and Building Materials*, 148, 376–385. https://doi.org/10.1016/j.conbuildmat.2017.05.084

[7] Bui, N. K., Satomi, T., & Takahashi, H. (2017b). Improvement of mechanical properties of recycled aggregate concrete basing on a new combination method between recycled aggregate and natural aggregate. *Construction and Building Materials*, 148, 376–385. https://doi.org/10.1016/j.conbuildmat.2017.05.084

[8] de Andrade Salgado, F., & de Andrade Silva, F. (2022). Recycled aggregates from construction and demolition waste towards an application on structural concrete: a review. *Journal of Building Engineering*, 52. https://doi.org/10.1016/j.jobe.2022.104452

[9] Dimitriou, G., Savva, P., & Petrou, M. F. (2018). Enhancing mechanical and durability properties of recycled aggregate concrete. *Construction and Building Materials*, 158, 228–235. https://doi.org/10.1016/j.conbuildmat.2017.09.137

[10] Fang, X., Xuan, D., & Poon, C. S. (2017a). Empirical modelling of CO_2 uptake by recycled concrete aggregates under accelerated carbonation conditions. *Materials and Structures/Materiaux et Constructions*, 50(4). https://doi.org/10.1617/s11527-017-1066-y

[11] Fang, X., Xuan, D., & Poon, C. S. (2017b). Empirical modelling of CO_2 uptake by recycled concrete aggregates under accelerated carbonation conditions. *Materials and Structures/Materiaux et Constructions*, 50(4). https://doi.org/10.1617/s11527-017-1066-y

[12] Fang, X., Xuan, D., & Poon, C. S. (2017c). Empirical modelling of CO_2 uptake by recycled concrete aggregates under accelerated carbonation conditions. *Materials and Structures/Materiaux et Constructions*, 50(4). https://doi.org/10.1617/s11527-017-1066-y

[13] Fang, Y., & Chang, J. (2015). Microstructure changes of waste hydrated cement paste induced by accelerated carbonation. *Construction and Building Materials*, 76, 360–365. https://doi.org/10.1016/j.conbuildmat.2014.12.017

[14] Feng, Z., Zhao, Y., Zeng, W., Lu, Z., & Shah, S. P. (2020). Using microbial carbonate precipitation to improve the properties of recycled fine aggregate and mortar. *Construction and Building Materials*, 230. https://doi.org/10.1016/j.conbuildmat.2019.116949

[15] Forero, J. A., de Brito, J., Evangelista, L., & Pereira, C. (2022a). Improvement of the quality of recycled concrete aggregate subjected to chemical treatments: a review. *Materials*, 15(8). https://doi.org/10.3390/ma15082740

[16] Forero, J. A., de Brito, J., Evangelista, L., & Pereira, C. (2022b). Improvement of the quality of recycled concrete aggregate subjected to chemical treatments: a review. *Materials*, 15(8). https://doi.org/10.3390/ma15082740

[17] Forero, J. A., de Brito, J., Evangelista, L., & Pereira, C. (2022c). Improvement of the quality of recycled concrete aggregate subjected to chemical treatments: a review. *Materials*, 15(8). https://doi.org/10.3390/ma15082740

[18] Gao, Q., Li, X-guang, Jiang, S-qi, Lyu, X-jun., Gao, X., Zhu, X. nan et al. (2023). Review on zero waste strategy for urban construction and demolition waste: Full component resource utilization approach for sustainable and low-carbon. In *Construction and Building Materials*, 395. https://doi.org/10.1016/j.conbuildmat.2023.132354

[19] Guo, H., Shi, C., Guan, X., Zhu, J., Ding, Y., Ling, T. C., et al. (2018). Durability of recycled aggregate concrete – A review. *Cement and Concrete Composites*, 89, 251–259. https://doi.org/10.1016/j.cemconcomp.2018.03.008

[20] Ismail, S., & Ramli, M. (2014). Mechanical strength and drying shrinkage properties of concrete containing treated coarse recycled concrete aggregates. *Construction and Building Materials*, 68, 726–739. https://doi.org/10.1016/j.conbuildmat.2014.06.058

[21] Katkhuda, H., & Shatarat, N. (2017). Improving the mechanical properties of recycled concrete aggregate using chopped basalt fibers and acid treatment. *Construction and Building Materials*, 140, 328–335. https://doi.org/10.1016/j.conbuildmat.2017.02.128

[22] Katz, A. (2004). Treatments for the Improvement of Recycled Aggregate. *Journal of Materials in Civil Engineering*, 16(6), 597–603. https://doi.org/10.1061/(asce)0899-1561(2004)16:6(597)

[23] Kazmi, S. M. S., Munir, M. J., Wu, Y. F., & Patnaikuni, I. (2018). Effect of macro-synthetic fibers on the fracture energy and mechanical behavior of recycled aggregate concrete. *Construction and Building Materials*, 189, 857–868. https://doi.org/10.1016/j.conbuildmat.2018.08.161

[24] Kazmi, S. M. S., Munir, M. J., Wu, Y. F., Patnaikuni, I., Zhou, Y., & Xing, F. (2019a). Influence of different treatment methods on the mechanical behavior of recycled aggregate concrete: a comparative study. *Cement and Concrete Composites*, 104. https://doi.org/10.1016/j.cemconcomp.2019.103398

[25] Kazmi, S. M. S., Munir, M. J., Wu, Y. F., Patnaikuni, I., Zhou, Y., & Xing, F. (2019b). Influence of different treatment methods on the mechanical behavior of recycled aggregate concrete: A comparative study. *Cement and Concrete Composites*, 104. https://doi.org/10.1016/j.cemconcomp.2019.103398

[26] Kazmi, S. M. S., Munir, M. J., Wu, Y. F., Patnaikuni, I., Zhou, Y., & Xing, F. (2020a). Effect of recycled aggregate treatment techniques on the durability of concrete: A comparative evaluation. *Construction and Building Materials*, 264. https://doi.org/10.1016/j.conbuildmat.2020.120284

[27] Kazmi, S. M. S., Munir, M. J., Wu, Y. F., Patnaikuni, I., Zhou, Y., & Xing, F. (2020b). Effect of recycled aggregate treatment techniques on the durability of concrete: A comparative evaluation. *Construction and Building Materials*, 264. https://doi.org/10.1016/j.conbuildmat.2020.120284

[28] Al-Bayati, H. K. A., TIGHE, S. L., & BAAJ, H. (2016). Effect of different treatment methods on the interfacial transition zone microstructure to coarse recycled concrete aggregate. http://www.tac-atc.ca/sites/default/files/conf_papers/al-bayati_.pdf

[29] Liu, J., Huang, Z., & Wang, X. (2020). Economic and environmental assessment of carbon emissions from demolition waste based on LCA and LCC. *Sustainability (Switzerland)*, 12(16). https://doi.org/10.3390/su12166683

[30] Li, W., Long, C., Tam, V. W. Y., Poon, C. S., & Hui Duan, W. (2017). Effects of nano-particles on failure process and microstructural properties of recycled aggregate concrete. *Construction and Building Materials*, 142, 42–50. https://doi.org/10.1016/j.conbuildmat.2017.03.051

[31] Mistri, A., Bhattacharyya, S. K., Dhami, N., Mukherjee, A., & Barai, S. V. (2020a). A review on different treatment methods for enhancing the properties of recycled aggregates for sustainable construction materials. *Construction and Building Materials*, 233. https://doi.org/10.1016/j.conbuildmat.2019.117894

[32] Mistri, A., Bhattacharyya, S. K., Dhami, N., Mukherjee, A., & Barai, S. V. (2020b). A review on different treatment methods for enhancing the properties of recycled aggregates for sustainable construction materials. In *Construction and Building Materials*, 233. https://doi.org/10.1016/j.conbuildmat.2019.117894

[33] Mistri, A., Bhattacharyya, S. K., Dhami, N., Mukherjee, A., & Barai, S. V. (2020c). A review on different treatment methods for enhancing the properties of recycled aggregates for sustainable construction materials. *Construction and Building Materials*, 233. https://doi.org/10.1016/j.conbuildmat.2019.117894

[34] Mukharjee, B. B., & Barai, S. V. (2014a). Statistical techniques to analyze properties of nano-engineered concrete using recycled coarse aggregates. *Journal of Cleaner Production*, 83, 273–285. https://doi.org/10.1016/j.jclepro.2014.07.045

[35] Mukharjee, B. B., & Barai, S. V. (2014b). Statistical techniques to analyze properties of nano-engineered concrete using recycled coarse aggregates. *Journal of Cleaner Production*, 83, 273–285. https://doi.org/10.1016/j.jclepro.2014.07.045

[36] Neupane, R. P., Imjai, T., Makul, N., Garcia, R., Kim, B., & Chaudhary, S. (2023). Use of recycled aggregate concrete in structural members: a review focused on Southeast Asia. *Journal of Asian Architecture and Building Engineering*. https://doi.org/10.1080/13467581.2023.2270029

[37] Ouyang, K., Shi, C., Chu, H., Guo, H., Song, B., Ding, Y., et al. (2020a). An overview on the efficiency of different pretreatment techniques for recycled concrete aggregate. *Journal of Cleaner Production*, 263. Elsevier Ltd. https://doi.org/10.1016/j.jclepro.2020.121264

[38] Ouyang, K., Shi, C., Chu, H., Guo, H., Song, B., Ding, Y., et al. (2020b). An overview on the efficiency of different pretreatment techniques for recycled concrete aggregate. *Journal of Cleaner Production*, 263. https://doi.org/10.1016/j.jclepro.2020.121264

[39] Qiu, J., Tng, D. Q. S., & Yang, E. H. (2014). Surface treatment of recycled concrete aggregates through microbial carbonate precipitation. *Construction and Building Materials*, 57, 144–150. https://doi.org/10.1016/j.conbuildmat.2014.01.085

[40] Saravanakumar, P., Abhiram, K., & Manoj, B. (2016). Properties of treated recycled aggregates and its influence on concrete strength characteristics. *Construction and Building Materials*, 111, 611–617. https://doi.org/10.1016/j.conbuildmat.2016.02.064

[41] Tavira, J., Jiménez, J. R., Ayuso, J., Sierra, M. J., & Ledesma, E. F. (2018). Functional and structural parameters of a paved road section constructed with mixed recycled aggregates from non-selected construction and demolition waste with excavation soil. *Construction and Building Materials*, 164, 57–69. https://doi.org/10.1016/j.conbuildmat.2017.12.195

[42] Tayeh, B. A., Saffar, D. M. A., & Alyousef, R. (2020). The utilization of recycled aggregate in high performance concrete: a review. *Journal of Materials Research and Technology*, 9(4), 8469–8481. https://doi.org/10.1016/j.jmrt.2020.05.126

[43] Vengadesh Marshall Raman, J., & Ramasamy, V. (2020a). Various treatment techniques involved to enhance the recycled coarse aggregate in concrete: a review. *Materials Today: Proceedings*, 45, 6356–6363. https://doi.org/10.1016/j.matpr.2020.10.935

[44] Vengadesh Marshall Raman, J., & Ramasamy, V. (2020b). Various treatment techniques involved to enhance the recycled coarse aggregate in concrete: a review. *Materials Today: Proceedings*, 45, 6356–6363. https://doi.org/10.1016/j.matpr.2020.10.935

[45] Vengadesh Marshall Raman, J., & Ramasamy, V. (2020c). Various treatment techniques involved to enhance the recycled coarse aggregate in concrete: A review. *Materials Today: Proceedings*, 45, 6356–6363. https://doi.org/10.1016/j.matpr.2020.10.935

[46] Verma, A., Babu, V. S., & Arunachalam, S. (2021a). Characterization of recycled aggregate by the combined method: Acid soaking and mechanical grinding technique. *Materials Today: Proceedings*, 49, 230–238. https://doi.org/10.1016/j.matpr.2021.01.842

[47] Verma, A., Babu, V. S., & Arunachalam, S. (2021b). Characterization of recycled aggregate by the combined method: Acid soaking and mechanical grinding technique. *Materials Today: Proceedings*, 49, 230–238. https://doi.org/10.1016/j.matpr.2021.01.842

[48] Wang, B., Yan, L., Fu, Q., & Kasal, B. (2021). A comprehensive review on recycled aggregate and recycled aggregate concrete. *Resources, Conservation and Recycling*, 171. https://doi.org/10.1016/j.resconrec.2021.105565

[49] Wang, J., Zhang, J., Cao, D., Dang, H., & Ding, B. (2020a). Comparison of recycled aggregate treatment methods on the performance for recycled concrete. *Construction and Building Materials*, 234. https://doi.org/10.1016/j.conbuildmat.2019.117366

[50] Wang, J., Zhang, J., Cao, D., Dang, H., & Ding, B. (2020b). Comparison of recycled aggregate treatment methods on the performance for recycled concrete. *Construction and Building Materials*, 234. https://doi.org/10.1016/j.conbuildmat.2019.117366

[51] Wang, L., Wang, J., Qian, X., Chen, P., Xu, Y., & Guo, J. (2017a). An environmentally friendly method to improve the quality of recycled concrete aggregates. *Construction and Building Materials*, 144, 432–441. https://doi.org/10.1016/j.conbuildmat.2017.03.191

[52] Wang, L., Wang, J., Qian, X., Chen, P., Xu, Y., & Guo, J. (2017b). An environmentally friendly method to improve the quality of recycled concrete aggregates. *Construction and Building Materials*, 144, 432–441. https://doi.org/10.1016/j.conbuildmat.2017.03.191

[53] Wang, R., Yu, N., & Li, Y. (2020). Methods for improving the microstructure of recycled concrete aggregate: A review. *Construction and Building Materials*, 242. https://doi.org/10.1016/j.conbuildmat.2020.118164

[54] Wang, X., Yang, X., Ren, J., Han, N., & Xing, F. (2021a). A novel treatment method for recycled aggregate and the mechanical properties of recycled aggregate concrete. *Journal of Materials Research and Technology*, 10, 1389–1401. https://doi.org/10.1016/j.jmrt.2020.12.095

[55] Wang, X., Yang, X., Ren, J., Han, N., & Xing, F. (2021b). A novel treatment method for recycled aggregate and the mechanical properties of recycled aggregate concrete. *Journal of Materials Research and Technology*, 10, 1389–1401. https://doi.org/10.1016/j.jmrt.2020.12.095

[56] Xuan, D., Zhan, B., & Poon, C. S. (2016). Assessment of mechanical properties of concrete incorporating carbonated recycled concrete aggregates. *Cement and Concrete Composites*, 65, 67–74. https://doi.org/10.1016/j.cemconcomp.2015.10.018

[57] Zhan, B. J., Xuan, D. X., & Poon, C. S. (2018). Enhancement of recycled aggregate properties by accelerated CO2 curing coupled with limewater soaking process. *Cement and Concrete Composites*, 89, 230–237. https://doi.org/10.1016/j.cemconcomp.2018.03.011

[58] Zhang, J., Shi, C., Li, Y., Pan, X., Poon, C. S., & Xie, Z. (2015a). Influence of carbonated recycled concrete aggregate on properties of cement mortar. *Construction and Building Materials*, 98, 1–7. https://doi.org/10.1016/j.conbuildmat.2015.08.087

[59] Zhang, J., Shi, C., Li, Y., Pan, X., Poon, C. S., & Xie, Z. (2015b). Influence of carbonated recycled concrete aggregate on properties of cement mortar. *Construction and Building Materials*, 98, 1–7. https://doi.org/10.1016/j.conbuildmat.2015.08.087

Advances in Construction, Real Estate, Infrastructure and Project Management – Anil Kashyap et al. (eds)
© *2026 Taylor & Francis Group, London, ISBN 978-1-041-13433-6*

15 | Ascertaining Valuation Parameters of a Real Estate Property

Nandkishor Balaji Sonar[1]
Research Scholar,
Civil Engineering Department, MGM University,
India

Vijaya Sitaram Pradhan[2]
Professor,
Civil Engineering Department, MGM University,
India

■ **ABSTRACT:** Valuation parameters contribute in determining the economic value of a real estate property. This study aims at identifying the essential factors affecting real estate property value by conducting interviews with practicing valuers from India. Thirty-four questions of the interview were framed after thorough study of similar research papers and after consulting with various valuation professionals, town planners, architects, civil engineers, chartered accountants, lawyers, professors from the above-mentioned fields. These thirty-four questions were categorized according to building details, location details and service facilities. Data analysis of the raw interview responses was done using its mean score for ranking the parameters, central tendency to identify the most frequent response, range to check the diversity of opinions and standard deviation to verify the dispersion. The demographic analysis regarding the age, experience, location of practice and qualification of valuers was also done. According to the data analysis it was discovered that, property area, locality, proximity to civic amenities, construction quality, availability of public transport facilities in close vicinity were the value indicators with higher preferences followed by water supply facility, electric supply facility, availability of lift, parking facility and security services among others. Appropriate valuation method can thus be used further to determine the property values using the results of these interviews.

■ **KEYWORDS:** Fair market value, Valuation, Valuation parameters, Real estate property

1. INTRODUCTION

Various researchers and scholars have highlighted the complex nature of real estate valuation process and the evolving approaches within the field. Their focus on the need for comprehensive strategies

[1]nsonar@mgmu.ac.in, [2]vizpradhan@gmail.com

DOI: 10.1201/9781003669814-15

and models to assess property values, reflect the challenges associated with accurately determining the worth of real estate assets. There exists multiple factors that may affect the value of a property, positively or adversely. The positive factors like presence of Environment friendly locality, gardens, playgrounds, recreational facilities may increase the property value whereas the negative factors like the presence of mobile towers nearby, unfavourable bye laws, absence of public transit facilities in the close vicinity may decrease it. Since every property is unique with distinct features, their value differs when compared to another similar property in the same locality. Therefore, the crucial factors affecting property value need to be determined first and a technically sound method of calculating property value should be used later. Carefully identifying the crucial factors affecting property value is essential as these factors are the prime indicators which impact the real estate value of any property.

2. LITERATURE REVIEW

Ania and Toosi [1] emphasised about the importance of applying pertinent valuation methods. A real estate property acts as an investment alternative and plays a crucial role in the growth of construction industry as well as the economic development of the country. They addressed the limitations of traditional valuation methods and stressed the necessity for a more inclusive approach that considers both market factors and technical attributes. According to them, the lack of information access and ambiguity of valuation process lead to the difference between the estimated value and the fair market value of a property. Lu and Nora's [3] observations regarding the gaps in the conceptualization of the value and valuation of a built infrastructure focuses on the need for a formalized model that includes the environmental, social, and economic aspects affecting the property value. This draws our attention to the importance of considering a wide range of factors beyond traditional market elements when assessing the value of real estate properties.

Malinowski et al. [4] discovered the challenges associated with property valuation, including the increasing number of properties and transactions. They pointed out the need for an efficient and automated valuation system to support the real estate appraisers. Their insights concentrate on the role of technology in streamlining the valuation process and addressing the limitations of human and time resources. Huanhuan et al. [2] concentrated on the interplay between structural features, transportation accessibility and their impact on the real estate property values in urban areas. They used a Hedonic pricing model to find out the effect of various factors on property values. It was concluded that, structural features of the property and accessibility to nearby places both influence the property value significantly besides location and proximity to public transport services.

Sandbhor and Chaphalkar's [5] approach of employing Principal Component Analysis (PCA) technique to identify the most influential value attributes, demonstrates the significance of data-driven methodologies in predicting the property values. Their efforts to simplify the valuation process and organize the most significant factors affecting value, highlight the demand for an efficient and effective predictive model in the process of real estate valuation. Sukran et al. [7] mentioned that a comprehensive mass valuation system is required to consider the various characteristics of a property and the uniqueness of its location. They noted the judgements of valuation experts and citizens regarding the key parameters affecting property value and employed analytical techniques like Frequency Analysis, Principal Component Analysis, Factor Analysis and Analytical Hierarchy Process to analyze them. Multiple Regression Analysis was used to verify the significance of these parameters under the open market conditions. Yu et al.'s [8] extensive framework for determining property values, incorporating various value attributes and advanced modeling techniques, underscores the requirement of a broad approach for analyzing the housing market. Their use of Geographically Weighted Regression model and Random Forest model reflects the evolving trend in property valuation. They aimed to provide a more detailed and accurate analysis of the determinants of property value thereby contributing to the knowledge of property market dynamics.

Relying solely on the traditional methods of real estate property valuation can lead to a less accurate valuation, especially considering the diverse range of factors that can impact a property's value. Moreover, the role of personal judgment by practicing valuers based on their property surveys, standard references, interviews and past sales transactions, can also lead to determining incorrect market value of the property.

3. METHODOLOGY

The essential attributes of property value were determined by conducting interviews with the renowned practicing valuers in the country. Fifty valuation professionals were approached for the personal visit and conducting interview regarding the research work. Data analysis was done using the received responses in the raw manner. Depending on the results of the interview responses, appropriate valuation method can further be used to determine the property value. The adopted research methodology is shown in Figure 15.1 below:

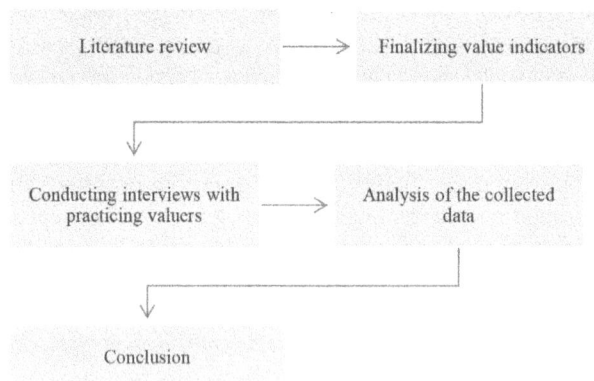

Figure 15.1 Flow of research work

Source: Authors

The mean score technique of quantitative analysis was used to assess the significance of different factors affecting property value. The mean score calculation is essential for determining the average value of responses on a five-point scale and is crucial for ranking options in order of importance. It was done by using the following formula [6]:

$$Mean\ Score = \frac{\sum S \times f}{N}$$

where, s = score given to each factor by respondents, ranking from 1 to 5.

f = frequency of each rating for each factor or option.

N = Total number of responses for that factor or option.

By employing a rating scale that ranges from 1 to 5, with 1 representing very low and 5 denoting the very high rating, respondents provided feedback based on their individual perspectives. The above formula involves summing up the individual scores given to each factor or option and then dividing the sum by the total number of responses. By applying this formula, mean score for each factor was derived, which was then used to rank the options in descending order of importance. The central tendency was checked from the distribution of responses to determine the central point or most frequent response. Also, to find out the diversity of opinions, range of responses was assessed.

4. RESULTS AND DISCUSSION

To accomplish the task of ascertaining property value attributes, personal visits were made to more than 50 valuation professionals. However, only 34 out of them agreed to respond to the questions of the research work. These practicing valuers were asked to rate the value attributes based on the 34 pre-framed questions on a likert scale of 1 to 5. Valuer's opinion concentrates on masses. According to them, what matters most to the majority should be considered as the standard factors affecting property value, based on which the property valuation should be done.

The first question to the valuers was about their age. The youngest of all was 36 years and the oldest was 81 years. Average age of these respondents was 49.65 years.

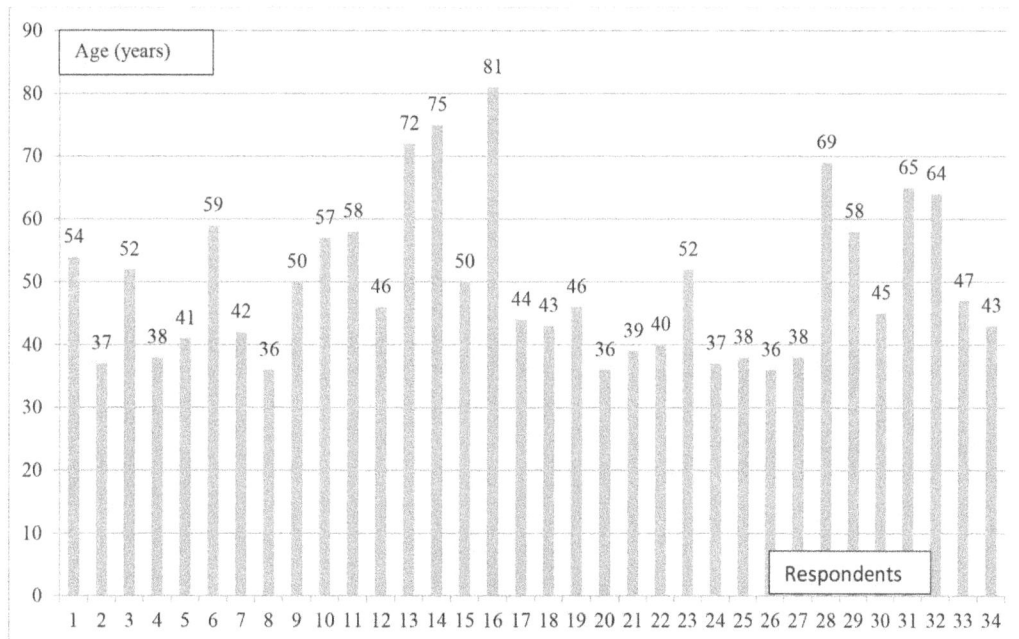

Figure 15.2 Age of valuers

Source: Authors

The next question to the valuers was regarding the city in which they practice valuation. Majority of them i.e. 6 out of 34 were from Pune and Mumbai each.

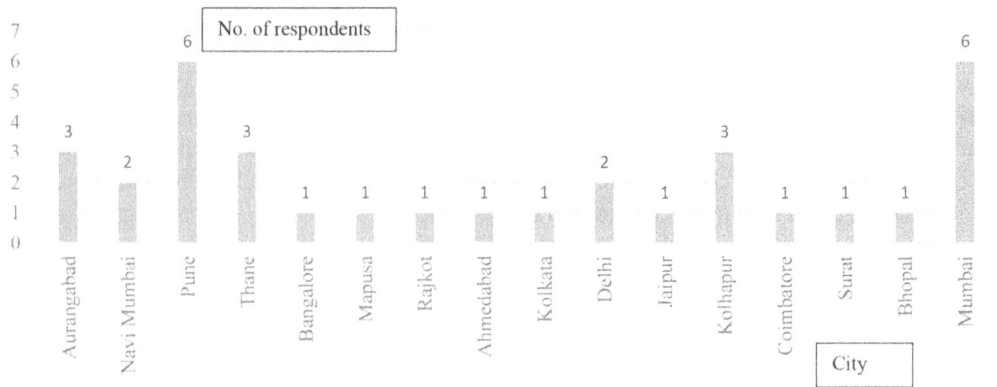

Figure 15.3 City of valuers

Source: Authors

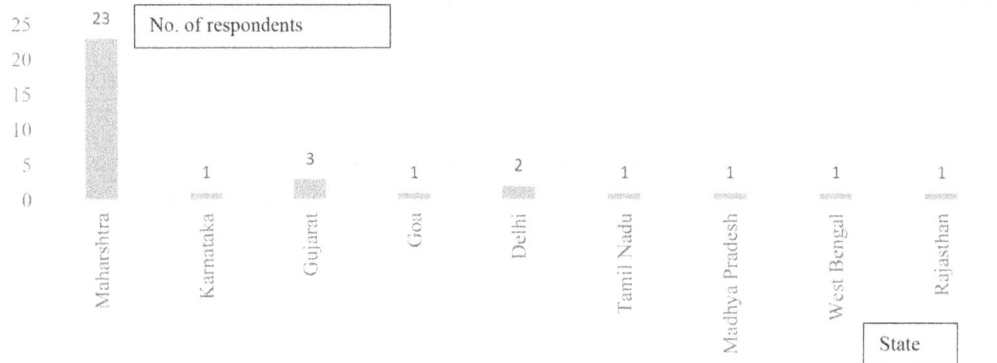

Figure 15.4 State of valuers

Source: Authors

About the qualification of valuers, maximum were from the Civil engineering background. Whereas, the others were from Architecture, Commerce and Law background.

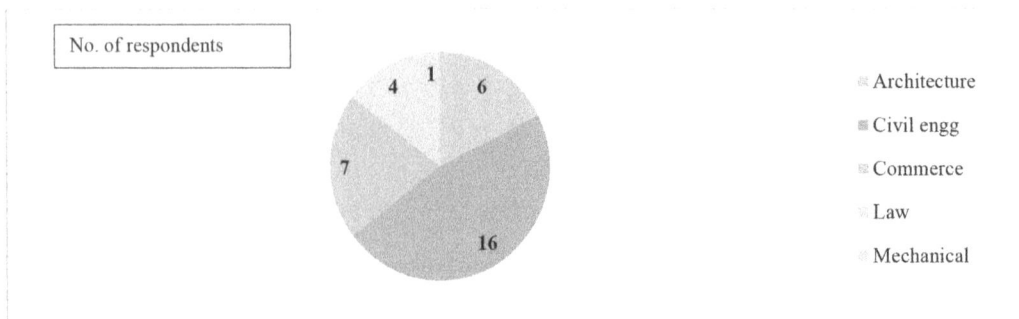

Figure 15.5 Qualification of valuers

Source: Authors

Regarding the experience of all valuers, the lowest was 12 years whereas the highest was 60 years. Average experience of the respondents was 27.56 years.

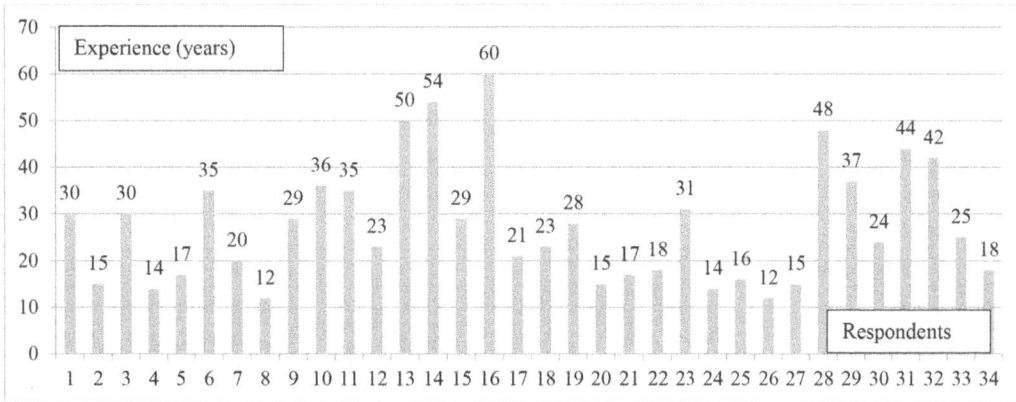

Figure 15.6 Experience of valuers

Source: Authors

Following graph depicts the number of practicing valuers with particular rating for all questions:

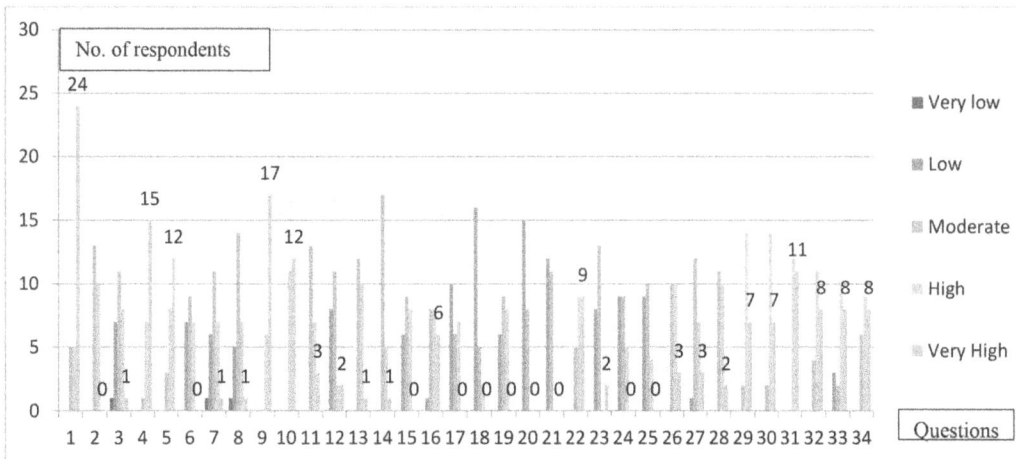

Figure 15.7 Responses to all 34 questions

Source: Authors

The questions about the factors affecting value were categorized according to building details, location details and service facilities. These factors were ranked on the basis of their weighted mean in descending order to highlight the relative importance of each factor in the valuation process. The most frequent response was also calculated for each question, which was observed as 2 to 5 for all the questions. Seven out of the first 12 ranks had the central point as 3 and the remaining five had the central point as 5. Therefore, the central tendency signifies that majority of the respondents preferred these value attributes with a moderate to high level of agreement. The range or spread of responses showed that the responses had relatively low to moderate variability with the range of 1, 2 or 3 for all the questions. This variability among the valuers suggest moderate to high agreement for the value attributes. Table 15.1 shows the rank wise list of valuation parameters.

Table 15.1 Rank wise list of valuation parameters

Question	Central point	Range	Mean	Rank	Category	SD
Size of property	5	1	4.7353	1	Building details	0.7353
Locality around	3	1	4.5588	2	Location details	0.4991
Near market	3	3	4.5294	3	Location details	0.8214
Construction quality	5	2	4.5000	4	Building details	0.6524
Public transport	5	2	4.4706	5	Location details	0.6524
Regular water supply	3	3	4.3529	6	Service facility	0.9935
Continuous electric supply	3	3	4.3235	7	Service facility	0.8650
Particular floor	3	3	4.2941	8	Building details	0.8670
Availability of lift	5	1	4.2941	9	Service facility	0.4412
Maintenance charges	5	1	4.2353	10	Building details	0.5000
Personal parking facility	3	2	4.0882	11	Service facility	0.6471
Security services	3	3	4.0882	12	Service facility	0.8220
Returns from property	3	2	3.7941	13	Building details	0.6424
Legal control of property	3	2	3.7647	14	Building details	0.5549
Good ventilation	4	2	3.6176	15	Building details	0.7957
Age of building	3	3	3.5882	16	Building details	0.9467
Solar panels	4	2	3.5294	17	Building details	0.8570
Free access to property	2	2	3.5000	18	Building details	0.7162
Near transport	3	2	3.4706	19	Location details	0.7579
External view	2	3	3.4706	20	Building details	0.7376
Pollution in area	3	2	3.3235	21	Location details	0.6739
Road conditions	5	2	3.2059	22	Location details	0.7300
Groundwater	3	3	3.2059	23	Location details	0.8381
Near workplace	3	2	3.1765	24	Location details	0.7579
Rain water harvesting	3	3	3.1176	25	Building details	0.8665
Total rooms	4	2	3.1176	26	Building details	0.7186
Depreciation	3	3	3.1176	27	Building details	0.6966
Taxes	4	2	3.0588	28	Building details	0.6885
Future expansion	4	2	3.0294	29	Building details	0.5703
Vastu shastra	4	2	2.9706	30	Building details	0.5703
Insurance of property	4	1	2.8824	31	Building details	0.4779
Total floors	5	2	2.6765	32	Building details	0.6739
Total bathrooms	4	3	2.6765	33	Building details	0.9193
Total bedrooms	4	3	2.5000	34	Building details	0.8177

Source: Compiled by authors

The respective graphs for the category wise questions are shown below for all the three categories namely – building details, location details and service facilities.

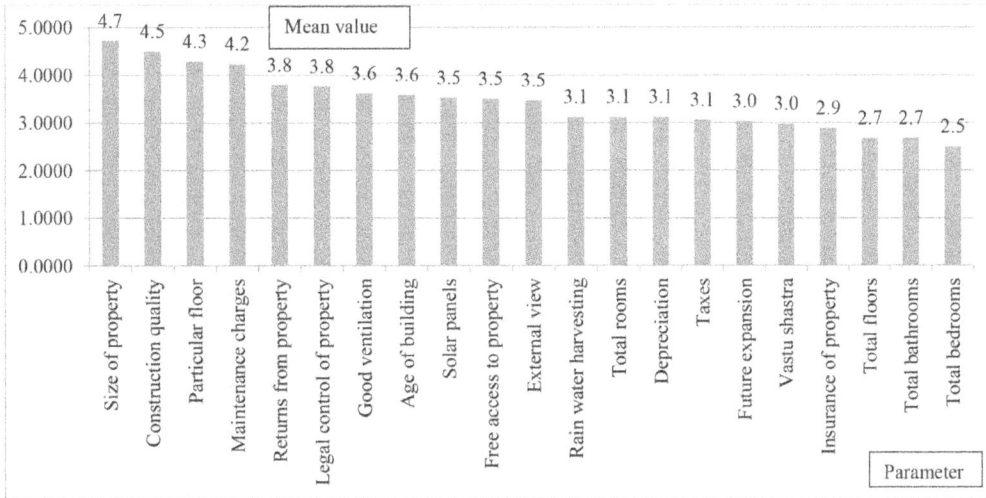

Figure 15.8 Building details parameters

Source: Authors

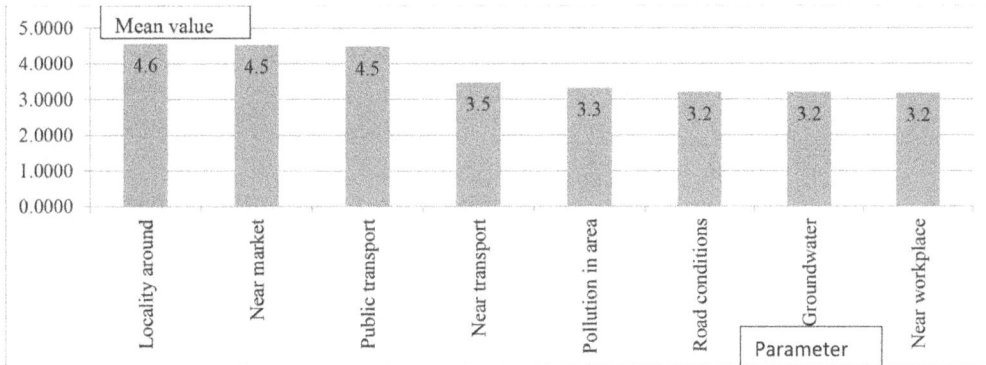

Figure 15.9 Location details parameters

Source: Authors

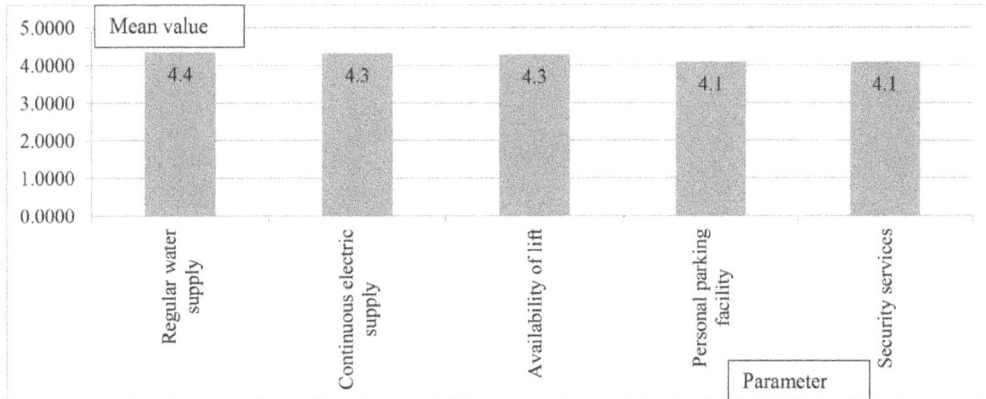

Figure 15.10 Service facility parameters

Source: Authors

5. CONCLUSION

The practicing valuers perception about the essential attributes of a property value was ascertained using personal interviews and its data analysis. It was observed that the prominent factors affecting value were the ones with highest amount of weighted mean including property area (4.7353), surrounding locality (4.5588), proximity to market or civic amenities (4.5294), quality of construction work (4.5000), proximity to public transport (4.4706), availability of regular water supply (4.3529), continuous electric supply (4.3235), property situated on a particular floor in building with provision of lift (4.2941), maintenance charges (4.2353), availability of personal parking facility and security services (4.0882). Among them, five belonged to service facility category – regular water supply, continuous electric supply, provision of lift, personal parking facility and security services, three belonged to the location details category – surrounding locality, proximity to market and proximity to public transport and the remaining four belonged to the building details category – property area, quality of construction work, property situated on a particular floor and maintenance charges. These 12 valuation parameters are crucial factors ascertained from the research work and need to be considered for valuing a real estate property.

REFERENCES

[1] Ania, K., & Toosi, H. (2021). Residential real estate valuation framework based on life cycle cost by building information modeling. *Journal of Architectural Engineering, 27*(3). https://doi.org/10.1061/(ASCE)AE.1943-5568.0000453

[2] Huanhuan, L., Shengchuan, Z., & Ronghan, Y. (2021). Determinants of housing prices in Dalian City, China: Empirical study based on hedonic price model. *Journal of Urban Planning and Development, 147*(2). https://doi.org/10.1061/(ASCE)UP.1943-5444.0000604

[3] Lu, Z., & Nora, M. E. (2016). Discovering stakeholder values for axiology-based value analysis of building projects. *Journal of Construction Engineering and Management, 142*(4). https://doi.org/10.1061/(ASCE)CO.1943-7862.0001103

[4] Malinowski, A., Mateusz, P., Zbigniew, T., Bogdan, T. Olgierd, K., & Tadeusz, L. (2018). An approach to property valuation based on market segmentation with crisp and fuzzy clustering. In N. T. Nguyen et al. (Eds.), *ICCCI 2018, LNAI 11055* (pp. 534–548). Springer Nature Switzerland AG.

[5] Sandbhor, S., & Chaphalkar, N. B. (2019). Effect of training sample and network characteristics in neural network-based real property value prediction. In A. J. Kulkarni et al. (Eds.), *Proceedings of 2nd International Conference on Data Engineering and Communication Technology*, Springer Nature Singapore Pte. Ltd.

[6] Siegel, S., & Castellan, N. J. (2008). *Nonparametric statistics for the behavioural sciences* (2nd ed.). McGraw-Hill.

[7] Sukran, Y., Suleyman, S., Ali, U. A., & Fatma, B. U. (2021). Feature selection applications and model validation for mass real estate valuation systems. *Journal of Land Use Policy, 108*, 105539. https://doi.org/10.1016/j.landusepol.2021.105539

[8] Yu, Z., Dachuan, Z., & Eric, J. M. (2021). Spatial autoregressive analysis and modeling of housing prices in the city of Toronto. *Journal of Urban Planning and Development, 147*(1). https://doi.org/10.1061/(ASCE)UP.1943-5444.0000583

Advances in Construction, Real Estate, Infrastructure and Project Management – Anil Kashyap et al. (eds)
© *2026 Taylor & Francis Group, London, ISBN 978-1-041-13433-6*

16

Challenges in Project Management of Indian Smart City: A Systematic Literature Review

Sarthak Jain,
Vartika Gupta, Utkarsh Tiwari,
Pranav Bawaskar, and Deepak M.D.*
Research Scholar,
Department of Construction Management,
NICMAR University, Pune

■ **ABSTRACT:** The rapid technological advancements and increasing urbanization worldwide have led to the emergence of Smart Cities—urban areas that use digital and IoT technologies to improve quality of life, optimize resource management, and promote sustainability. However, implementing smart city initiatives presents significant project management challenges. This paper examines these challenges by exploring the various factors that influence the successful planning, execution, and operation of smart city projects. The key challenges identified include integrating diverse technological systems and establishing robust data management frameworks to ensure interoperability and security. Coordinating multiple stakeholders, such as government bodies, private sector partners, and community groups, adds complexity, necessitating effective communication and alignment of objectives. Additionally, the dynamic nature of urban environments demands agile project management approaches to adapt to evolving needs and technological advancements. Legal and regulatory issues further complicate the project management landscape, especially regarding data privacy, security standards, and compliance with existing urban policies. Socio-cultural factors, including public resistance to change and the digital divide, can significantly impact the acceptance and sustainability of smart city projects. This paper aims to provide insights into effective project management strategies and frameworks to address these obstacles, ensuring the successful deployment and sustainability of smart city initiatives. By identifying and understanding these challenges, stakeholders can better navigate the complexities of smart city development and foster more resilient, inclusive, and technologically advanced urban communities.

■ **KEYWORDS:** Smart cities, Project management, Challenges, Sustainability, Urbanisation

---

*Corresponding author: deepakmd@nicmar.ac.in

DOI: 10.1201/9781003669814-16

1. INTRODUCTION

Since the middle of the 20th century, urbanisation has impacted the world has in the most unprecedented way in the history of mankind [6]. As economic opportunities become increasingly centred in urban areas and mobility increases between cities and the countryside, billions of people now make their homes within vast agglomerations whose size could scarcely have been imagined as recently as a century ago [2].

At first glance, the integration of smart technologies into urban agendas appears to be a natural process. Millions of individuals are incorporating new smart technologies into their daily lives, and it was only a matter of time until metropolitan regions started to notice this trend more widely. However, the procedure implies that the idea of urban development is somewhat nebulous and poorly defined, which leaves room for varying interpretations and the vested interests of huge information and communication technology (ICT) businesses and municipal governments to be worked out and implemented.

The aim of infrastructure development of smart cities aims is the collective development addressing the social, physical, economical and institutional needs of infrastructure acting as the four pillars of infrastructural development [4].

The goal of the study is to review the existing challenges of smart cities by reviewing important urban literature and addressing the project management loopholes and barriers having a consequential declining impact on the progress of the projects of Smart city.

2. LITERATURE REVIEW

2.1 Smart City and the Indian Context

India's urban population is expected to reach 600 million by 2030 due to the country's fast urbanization and further [3] study estimated that 200 million Indians will move from rural to urban regions over the course of the next 15 years. is expected to have 40% of its population living in cities by 2030 due to growing urbanization, similar to that of many other countries, calling for comprehensive development strategies. As per the Urbanization of the Human Population the processes underpinning the growth of cities in developed and developing countries were motivated by different drivers which led to the concept of Smart City [7].

According to Ministry of Urban Development (MoUD), India's current urban population is around 377 million, likely to be more than 500 million by 2050. According to McKinsey study by 2030, India will need to revive 468 cities with more than one lakh population to cope with fast urbanisation growth [9].

According to different academicians, smart city projects are designed to improve a city's economy, people, government, environment, mobility, and quality of life in six key areas. When a project incorporates more than one of these dimensions, it is seen more intelligent. They are frequently managed by several groups working together and use contemporary technology to create value on all fronts—economic, social, and ecological. All the definitions address specific components or facets of the Smart City concept. However, none of them singularly encompasses the entirety of what a Smart City should represent. Thus, in amalgamating these existing definitions, we can formulate a more comprehensive understanding. "A Smart City can be defined as a technologically advanced urban area that uses interconnected infrastructure, data analysis, and sustainable practices to enhance its overall functionality, improve the quality of life for its citizens, and promote economic growth in an environmentally friendly way." The study of these projects is also characterised by some more points which covers Innovation, Integration, Inclusion. The projects can be initiated either top-down (by government or institutions) or bottom-up (by citizens) and are driven by institutions that develop project objectives in line with the challenges and key performance indicators of the smart city vision.

2.2 Challenges in Developing a Smart City

As of January 2023, 68 Smart City projects have not fulfilled the Smart City Mission's physical goals, according to a study by a Parliamentary Committee. It's interesting to note that two Smart Cities—some by up to four times—have surpassed their project completion goals. The report also points out that these 32 successful Smart Cities may have overachieved in term of number of projects than necessary, inflating the overall finished projects. Therefore, the number of finished Smart City initiatives would be substantially fewer than first anticipated if these extra projects were eliminated.

The report highlights the constraints, including the COVID-19 pandemic and regional labor and land-related issues, and recognizes the desire of the Ministry in pursuing lagging Smart Cities, both financially as well as physically, to meet their marks within the new deadline of June 2023. The Committee is adamant that no further extension must be given until reasons for delays of projects are completely looked into and settled.

Total proposals submitted by 100 Smart Cities for projects under the Smart City Mission amount to Rs 2, 05, 018 crore. Of these, 7, 821 projects worth Rs 1, 81, 349 crore have commenced. So far, 5, 343 Smart City projects worth Rs 1, 00, 450 crore have been completed successfully. All these matrices indicate that many issues and factors are still hindering growth and implementation of large initiatives.

Smart cities face various challenges, as identified by researchers, encompassing project scale, diverse objectives, social, technological, and economic hurdles, climate implications, and more. Resolving these issues requires a shift from a technology-centric approach to smarter, established systems focused on optimizing resource utilization. Varghese Paul's exploration in the context of urbanizing India delves into alternative smart city concepts [12].

In the Indian scenario, [11] utilized the "Fuzzy Analytical Hierarchy Process" to pinpoint key obstacles hindering smart city development. They conducted a literature survey and consulted experts, highlighting significant impediments [11].

Major setbacks in smart city projects often stem from organizational and managerial difficulties, a point emphasized by [5]. Existing research predominantly focuses on technological aspects, neglecting organizational and policy considerations. Consequently, insights into effective project management and success factors remain limited, lacking an integrated smart framework across diverse contexts [5].

3. RESEARCH METHODOLOGY

The following literature review of the study has been conducted using the Preferred Reporting Items for Systematic Reviews and Meta-Analyses (PRISMA) technique. To locate, select, and include the publications for this study, a systematic literature review was conducted in adherence to every stage of the Preferred Reporting Items for Systematic Reviews and Meta-Analyses (PRISMA) procedure [10]. This paper analyses and discusses the problems in project management strategies for the effective deployment of smart cities. Prior scholars have conducted studies on implementation issues of smart cities; however, they are distributed in numerous scientific sessions and journal publications.

The method used in this research is a systematic review that attempts to identify, evaluate, and synthesize findings from research that are relevant and related to a particular study topic [8]. A systematic review can be described as a synthesis of primary research that covers the specific subjects and problems apart from promoting critical thinking. Systematic reviews synthesize findings from primary studies or research to make the information more comprehensive and balanced for the study. Therefore, the selection or identification of the relevant main research is a crucial step in the systematic review process [1].

Figure 16.1 depicts that there are three major stages of PRISMA, which are as follows: 1) Identification; 2) Screening 3) Included. It is imperative to collect primary research that are directly associated with this subject. In the third stage of the process, various of the exclusion criteria played an important role in the segregation of requisite data leading to exclusion, predominantly comprised of 1) Studies on Smart Cities in India 2) Research on Smart City Challenges and Barriers. 3) Studies on case studies of various smart cities 4) Research on Project Management in Smart Cities.

3.1 PRISMA Approach for Systematic Literature Review

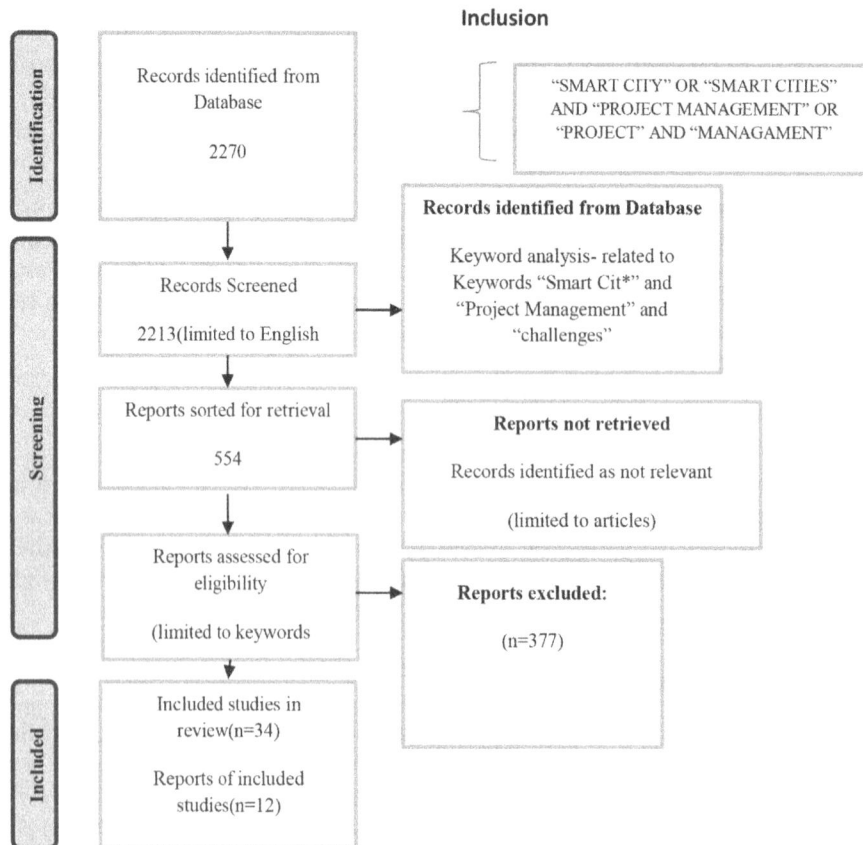

Figure 16.1 PRISMA approach for systematic literature review

Source: Authors

4. FINDINGS AND RESULTS

4.1 Yearly Distribution of the Retrieved Articles

On this graph, there is evidence of a steady increase from 2018 to 2023 in the number of publications, hence showing a trend where academic and scientific circles take on increasingly more interest in and pay closer attention to the management challenges of smart city initiatives. This tendency shows that as urbanization and integration of smart technologies into cities pick up speed, demand for scholarly investigation into how successful the management of such programs can be will also rise accordingly.

As shown in Figure 16.2, out of 676 papers published between 2018 to 2024, twelve (12) articles are included in this study. Thus, it could be argued that scholarly attention to the project management aspect of smart city had a continuous growth curve in the study of the subject. The visible rise in the studies could be seen after 2018 and continued till 2023.

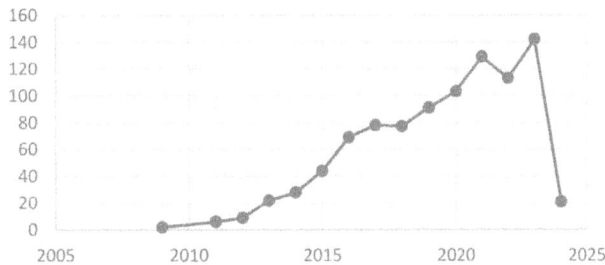

Figure 16.2   Documents published per year

Source: Authors

4.2 Country Wise Distribution of the Retrieved Articles

The graph shows an increase in the number of publications from 2018 to 2023, which demonstrates the academic and scientific interest in focusing on the challenges of managing smart city initiatives. The trend indicates that with increasing urbanization and integration of smart technologies into

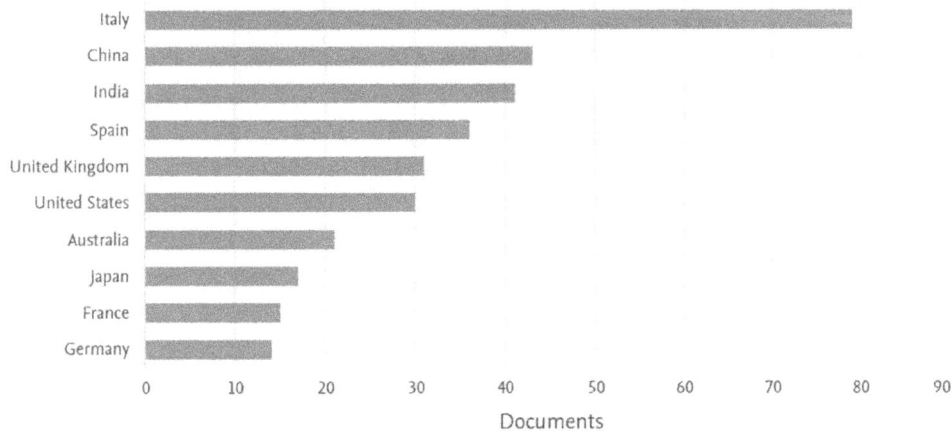

Figure 16.3   Country wise distribution of studies

Source: Authors

cities, the need for scholarly research in how these programs are managed effectively will also rise.

Analysis of study trends by country reveals a significant prevalence of documents in the Indian subcontinent, indicating a pronounced interest in the Smart City concept. This trend not only fosters scholarly inquiry but also contributes to the broader developmental landscape.

4.3 Key Word Re-occurrence Analysis

Figures 16.4, 16.5, and Table 16.1 present graphical and tabular visualizations of clusters of keywords in the smart cities and project management study issue identified through a systematic literature review. Visualizations of the interdisciplinary nature of smart city initiatives show clusters grouped around themes such as smart cities, project management, urban planning, decision-making processes, and sustainability. The images illustrate complex networks of themes that intersect in smart city initiatives: IoT, waste management, energy efficiency, and stakeholders. Table 16.1 details these linkages by showing the keywords used in the Scopus search along with the clusters found, revealing how these themes are related in the literature. This review underlines the need for integrated project management solutions capable of addressing the many facets of smart city redevelopment.

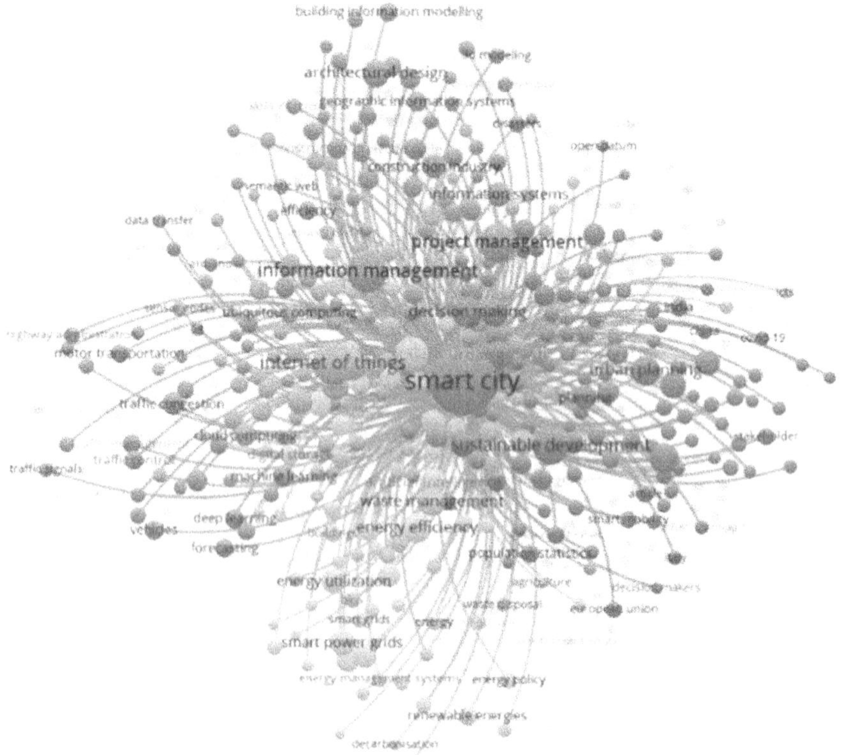

Figure 16.4 VOS analysis of scopus keywords without filter India

Source: Authors

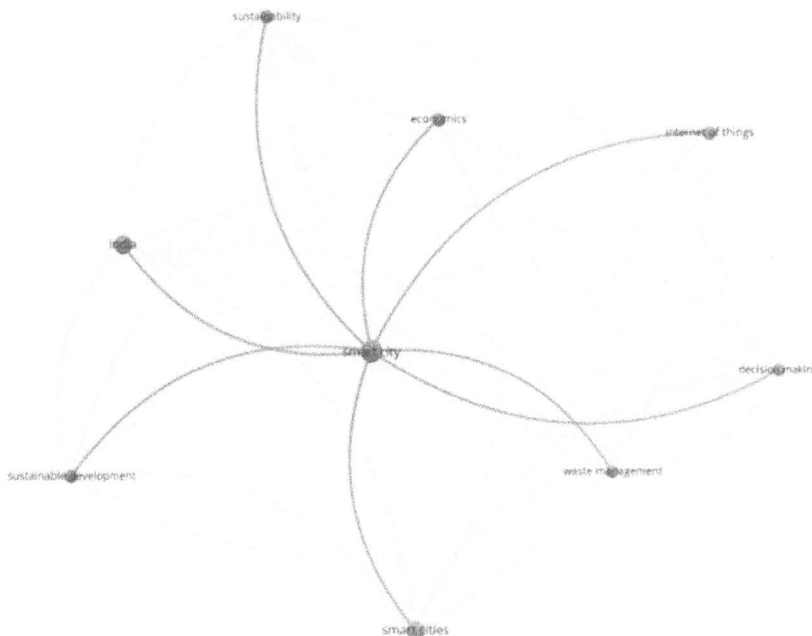

Figure 16.5   VOS cluster analysis of Scopus keywords with filter India

Source: Authors

4.4 Findings from PRISMA Analysis

All the existing literature studies related to the above-mentioned keywords with applied filter of India and without filters which covers global level studies. As we can see the above-mentioned table considering our three main keywords, in the Indian context there are still negligible or no studies done with respect to the challenges aspect of smart cities and project management whereas, when we compare it with the global level cluster figure it shows the relevant networks related to the same keywords which shows that there are some existing studies related to those parameters globally but not been incorporated much in Indian context studies. This is where we got to know about the research gap between these two important key words of the research project.

5. CONCLUSION

This systematic literature review, following the PRISMA methodology, delves into the complex challenges of managing smart city projects. The findings highlight that successful implementation of smart city initiatives requires a nuanced understanding and proactive approach to these intricacies. Integrating various technological systems while ensuring interoperability and data security poses significant hurdles. Equally important is the need for robust coordination among stakeholders, including government entities, private companies, and citizens, to align their expectations and objectives. The dynamic nature of urban environments demands agile project management frameworks that can adapt to evolving needs and incorporate emerging technologies. The review also points out that while there have been separate studies on the aspects of project management and the challenges faced by smart cities, there is a notable lack of research specifically addressing the

Table 16.1 Keywords used for scopus search

Key words used for Scopus search – Smart cities, Project Management			
Papers Uploaded for VOS Analysis with filter	No Filter-1597		Filter (India)-31
S. No.	Cluster Node	Cluster Ties keywords (Figure 2.5)	Cluster Ties keywords (Figure 2.6)
1	-	Decision Making, Information management, Sustainable development, Planning, Urban Planning, IOT, Information management, Project management, India, China, Covid-19, Stakeholders, Motor transportation, Waste management, Energy efficiency, Population statistics, Energy utilization, Waste management, Smart mobility, Construction industry, Architectural design, BIM, Traffic Signals, Congestion, Data transfer, Highway administration, Data transfer, Ecology, Open Datum, Efficiency, Semantic web, AI, Energy management systems	Sustainability, Smart cities/Smart city, Waste Management, Economics, IOT, Sustainable development
2	Project Management	Information management, Construction industry, Architectural design, 3d modelling, City construction, Decision making (smart city), Urban Planning, IOT (Automation, digital storage), Planning, Efficiency, AI	-
3	Smart Cities	Decision Making, Information management, Sustainable development, Planning, Urban Planning, IOT, Information management, Project management, India, China, Covid-19, Stakeholders, Motor transportation, Waste management, Energy efficiency, Population statistics, Energy utilization, Waste management, Smart mobility, Construction industry, Architectural design, BIM,	Sustainable development, Waste Management, Economics, IOT,
4	Decision Making	Urban Planning, Project management, Information management, Geographic information systems, Construction industry, Architectural design	Sustainability, Economics, IOT, Waste Management, Smart cities
5	Sustainable Development	Planning, Population statistics, Urban Planning, Sustainable cities, Energy utilization, Waste management, Smart city, Urban growth, Decision making, Solar grids	Sustainability, Smart cities/Smart city, Waste Management, Economics.

Source: Compiled by authors

challenges in project management within the smart city context. This gap is critical as it overlooks the interconnected nature of these challenges. Additionally, navigating the legal and regulatory frameworks, particularly concerning data privacy, security standards, and urban policies, is complex and requires careful attention. Socio-cultural factors, such as public resistance to change and the digital divide, must also be addressed to ensure inclusivity and project sustainability. Notably, the review identifies a pressing need for localized research in India, as existing global literature does not fully capture the unique challenges and opportunities of the Indian urban landscape. Future research should focus on creating solutions and frameworks tailored to the specific needs of Indian cities. In conclusion, this review underscores that smart city projects are not merely technological endeavors but complex socio-technical undertakings that demand a holistic approach. By acknowledging and proactively addressing these challenges, stakeholders can pave the way for the successful implementation of smart city initiatives, fostering more sustainable, efficient, and citizen-centric urban environments.

REFERENCES

[1] Adiyarta, K., Napitupulu, D., Syafrullah, M., Mahdiana, D., & Rusdah, R. (2020). Analysis of smart city indicators based on prisma : systematic review. In IOP Conference Series: Materials Science and Engineering, (Vol. 725, no. 1). https://doi.org/10.1088/1757-899X/725/1/012113.

[2] Bajpai, N. , & Biberman, J. (n.d.). Standard-Nutzungsbedingungen. http://hdl.handle.net/10419/249851.

[3] (2020). Challenges for smart cities in India our heritage challenges for smart cities in India. https://doi.org/10.13140/RG.2.2.24136.88329.

[4] Choudhary, S., Shrimali, H., & Shreemali, J. (2023). Stages and challenges in implementation of smart city project, Udaipur. *International Journal of Innovative Science and Research Technology*, 8(5), 2451–2456. https://doi.org/10.5281/zenodo.8017191.

[5] Chourabi, H., Nam, T., Walker, S., Gil-Garcia, J. R., Mellouli, S., Nahon, K., et al. (2012). Understanding smart cities: an integrative framework. In Proceedings of the Annual Hawaii International Conference on System Sciences, (pp. 2289–2297). https://doi.org/10.1109/HICSS.2012.615.

[6] Claudio Diogo Reis, L., Cristina Bernardini, F., Bacellar Leal Ferreira, S., & Cappelli, C. (2021). ICT governance in Brazilian smart cities: an integrative approach in the context of digital transformation. In ACM International Conference Proceeding Series, (pp. 302–316). https://doi.org/10.1145/3463677.3463682.

[7] Farrell, K. (2017). The rapid urban growth triad: a new conceptual framework for examining the urban transition in developing countries. *Sustainability (Switzerland)*, 9(8), 1407. https://doi.org/10.3390/su9081407.

[8] Kitchenham, B. (n.d.). Procedures for performing systematic reviews. https://www.researchgate.net/publication/228756057.

[9] Kumar, A. (n.d.). Preparing smart city development plan based on demographic and socio-economic analysis. https://doi.org/10.13140/RG.2.2.27514.18882.

[10] Mamani, T., Herrera, R. F., Muñoz-La Rivera, F., & Atencio, E. (2022). Variables that affect thermal comfort and its measuring instruments: a systematic review. *Sustainability (Switzerland)*, 14(3), 1773. MDPI. https://doi.org/10.3390/su14031773.

[11] Rana, N. P., Luthra, S., Mangla, S. K., Islam, R., Roderick, S., & Dwivedi, Y. K. (2019). Barriers to the development of smart cities in indian context. *Information Systems Frontiers*, 21(3), 503–525. https://doi.org/10.1007/s10796-018-9873-4.

[12] Varghese, P. (2016). Exploring other concepts of smart-cities within the urbanising Indian context. *Procedia Technology*, 24, 1858–1867. https://doi.org/10.1016/j.protcy.2016.05.238.

Advances in Construction, Real Estate, Infrastructure and Project Management – Anil Kashyap et al. (eds)
© *2026 Taylor & Francis Group, London, ISBN 978-1-041-13433-6*

17 Risk Analysis Using FMEA Technique for Infrastructure Projects

Avinash Bagul*

Assistant Professor,
School of Project Management, NICMAR University,
Pune, India

■ **ABSTRACT:** Risk management is an integral part of project planning activities. Failure Mode and Effects Analysis (FMEA) is the systematic process for identifying, evaluating, and mitigating failure modes for projects. In this work we have tried apply FMEA technique for infrastructure project to identify and manage potential risks that may impact in project duration, cost, or safety. As construction projects are complex and involve a wide range of stakeholders, materials, and activities, it is essential to proactively control potential failure modes. FMEA can help project teams reduce the chance of failures. Initially, different kind of risk associated with projects were identified through literature review and then FMEA methodology was applied identify prominent risk factors. Early identification of these risk factors can help construction managers minimize the instances of cost and time overruns and can aid in on time completion of the projects.

■ **KEYWORDS:** FMEA analysis, Infrastructure projects, Multi criteria decision making, Risk priority number, Risk analysis

1. INTRODUCTION

Execution of infrastructure projects on time requires project managers to be vigilant on project risk management. During execution of infrastructure projects there are numerous risks and uncertainties that may cause project failure, delays, and cost overruns. Therefore, it is essential to have project risk management plan in place. Contemporary trends in project risk management includes increased use of technology and automation such as AI and machine learning algorithms, blockchain and use of risk management software like Primavera risk analysis, Riskwatch etc. These tools help in automating risk identification, assessment, and risk mitigation [6]. Agile methodologies, initially

---

*Corresponding author: abagul@nicmar.ac.in

DOI: 10.1201/9781003669814-17

developed for software development, is nowdays used for project risk management. There is a growing trend to integrate project risk management with broader enterprise risk management frameworks. This ensures alignment of organizational goals and strategies [9]. Effective risk communication with stakeholders is emphasized in the today's project organizations which ensures that stakeholders are aware of potential risks and measures undertaken to mitigate them [18]. One of the important tactic for reducing the project risk is FMEA technique. In this study we have tried to apply this technique to identify and minimize risk in execution of the infrastructure projects. Several advantages of application of this technique in projects include risk mitigation, cost reduction, project quality assurance, improved project planning, enhanced safety, better communication and improved customer satisfaction.

FMEA is a structured and systematic process that is applied in the industry for identifying, evaluating, and mitigating failure modes. This process involves dissecting a system into parts/ functions/processes to establish how and why the failures might occur and their implications, as well as the likelihood of such failures. FMEA ranks different failure types using a system that comprises of numerical rating for occurrence, severity, and detection. From these assessments, FMEA computes a risk priority number (RPN) used by organizational entities to deal with the severest risks. Implementation teams may also formulate action plans aimed at mitigating risks related to reducing such likelihood. By proactively addressing possible problems, FMEA enhances quality, dependability, and customer satisfaction.

As risk is an integral part of any project, management of risk in execution of the projects is the sole responsibility of the project managers. Especially, for infrastructure projects, risk management is an essential element as these projects are complicated and frequently require huge capital investment. Successful execution of these projects ensures socioeconomic development of the country. Considering the proposed methodology, this paper has been organized as under: Section 1 provides detailed overview of the literature. Details on FMEA methodology are given in section 2. In the computation results section (section 3) methodology developed is verified using case study methodology for infrastructure project. The section 4 i.e. managerial implication and conclusion provides concluding remarks along with future research directions.

2. LITERATURE REVIEW

Project risk management is a strategic approach to risk attitude, risk communication and maintaining relationship with all the stakeholders to reduce project risk. Risk management contributes significantly to project success [5]. There exists direct linkage between the project output and the project related risk, [14]. Project risk is defined as an uncertain event which may have negative effect on project execution [10, 17]. Different techniques for managing project risk were extensively discussed by Cagliano et al. [4]. Project risk management techniques include fault tree analysis (FTA), Monte Carlo Simulation (MCS) and hazard operability study (HAZOP) and. FTA uses a top-down approach to identify the root causes of system failures. It involves constructing a fault tree diagram that shows the logical relationships between failures. The method is effective in identification of root causes of failures and is suitable for complex systems with interdependent failures. However, it requires detailed knowledge of the system and focuses on single failure events only. HAZOP is a qualitative technique involving systematic examination of process hazards and operability issues using guidewords to identify deviations from design intentions. The method is highly systematic and encourages team-based, multidisciplinary approach. It is effective in identification of process-related hazards. One of the limitation of the method is its qualitative nature which can lead to subjectivity. MCS uses statistical methods to model and analyse impact of risk by simulating a range of possible outcomes using probability distributions. The method provides a quantitative assessment of risk. It can handle complex, stochastic systems and offers a comprehensive view of potential outcomes and their probabilities. However, results obtained using MCS are highly dependent on the accuracy of input data and assumptions.

Project risk can be effectively managed through the application of the FMEA analysis technique. Ahmadi et al. [2] developed a risk management methodology using FMEA and multi criteria decision making (MCDA) techniques for Iranian Bijar-Zanjan highway construction project. The three stages of the methodology included detection of potential risks, applying a fuzzy FMEA for evaluation of hazards, and determining resolutions. Fuzzy AHP was applied to quantify cost, time, and quality. Each potential course of response to a risk occurrence was measured using SED index. "Raise of tar price" was identified as the risk event and strategies were suggested to mitigate this risk. Hayati and Abroshan [1] applied Fuzzy FMEA technique for risk assessment using for Tehran Subway Tunnelling Operations. The input parameters like frequency, occurrence, severity, and failure mode identification, were decided by specialists. In order to conduct a coherent and consistent analysis, fuzzy logic was employed in this work as a supplemental tool in conjunction with two computational methodologies. The outcomes demonstrated that in comparison to the conventional FMEA method, fuzzy FMEA methodology was found to be more adaptable and practical. [25] did project risk analysis for elevated metro rail projects using fuzzy failure mode and effect analysis. The risk considered were the risks associated with the pre, during and post implementation stages of the project execution. The main objective of the paper was to calculate risk severity by using fuzzy FMEA, followed by use of fuzzy EVM, detection tools, risks mitigation, and management strategies for all key operations of the elevated corridor metro rail project. Structured methodologies such as FMEA and EVM helped in identification of risk in the main activities of a project. The relationship among input characteristics in terms of risk probability, its impact as well as control/detection was based on fuzzy logic mapped in EVM and FMEA.

FMEA is a preventative risk analysis technique that aims to make the necessary modifications early on rather than after the occurrence of the event [26]. Its primary role is to evaluate the failure mode based on three dimensions: severity, occurrence, and detection. The risk is reduced to a manageable level by using this technique by identifying the significant failure's primary cause, risk's failure mode, and controlled using preventive measures. Salokolaei and Esmaili [23] developed a hybrid approach using AHP along with FMEA technique for risk assessment of the refinery construction projects. The technique was useful in making decisions on risk reduction, improving safety surrounding hazardous installations, preparing for emergencies, defining acceptable risk level, scheduling of industrial installation inspection and maintenance etc. Experts and professionals dealing with safety and environment have always considered the high risk associated with the gas and petrochemical industries due to their numerous and extensive hazards. This study sought to assess risk using multicriteria analytical hierarchy process together with FMEA. The results demonstrated that threats to oil and gas installations were hard and unique and ways were suggested to mitigate these threats. Sadidi et al. [21] used FMEA technique for assessment of tower crane safety indicators. A case of 150 tower cranes was conducted. In this study, thirty failure modes were analysed with highest RPNs observed at the base of sections, along the trolley body and at the jib holding split pins. Sakinah and Sutopo [22] used DMAIC and FMEA technique for Concrete Product Quality Improvement. Macura et al. [13] developed Risk Analysis Model for railway infrastructure projects in Serbia. Each risk incident was considered for frequency, severity, and ability to detect. The events were ranked using the Fuzzy Risk Prioritization Number (RPN). Lv et al. [12] used FMEA technique with Extended MULTIMOORA for operational risk evaluation for infrastructure projects.

Panenka and Nyobeu [15] developed a FMEA technique using an integration of assessment standards based on fuzzy logic for maintaining an aging infrastructure in Germany. Fuzzy measures were used to formulate important indicators such as risk number. Cheng and Lu [7] developed a novel approach for threat assessment based on FMEA technique and the fuzzy logic was applied to the construction of exhausted piles in the canals. Amini and Mojtaba [3] used FMEA technique for risk assessment of road tunnel project in Iran. The work front's insecurity was identified as the most likely threat for the project. Purwanggono and Margarette [19] developed Risk assessment methodology for an underpass infrastructure project by using fishbone diagram and FMEA method. The purpose of this study was to investigate the obstacles in the implementation of road infrastructure project. The fishbone diagram was used to identify the underlying causes of failure. The study results showed that creating standard operating procedure (SOP) guidelines is the most practical mitigation strategy to keep utilities from impeding project implementation. Patricio et al. [16] developed a risk mitigation model using FMEA technique with the objective of minimizing the high risk of worker accidents. Occupational safety hazards were identified by field research. It was determined that the suggested tool could detect risks and provide a broad and uniform diagnosis of the present risks facing the organization.

In recent years, several new approaches have been introduced along with FMEA techniques, particularly for infrastructure projects, like integration with advanced technologies, enhanced risk prioritization methods, integration with building information modelling, sustainability and resilience focus and automated tools and softwares. AI and machine learning algorithms are increasingly used to predict potential failure modes based on historical data, improving the accuracy and efficiency of FMEA [20]. Integration of IoT devices allows for real-time data collection and monitoring, enabling dynamic FMEA processes that can respond to real-time conditions [8]. Combining FMEA with MCDA techniques helps in better prioritizing risks by considering multiple criteria beyond traditional severity, occurrence, and detection metrics [27]. BIM-FMEA integration allows for enhanced visualization and analysis of potential failure modes, improving collaboration and decision-making [11]. Recent advancements in risk management emphasize assessing not just the immediate risks, but also long-term sustainability and resilience of infrastructure projects. This includes evaluating the impact of environmental factors and climate change on infrastructure performance [24]. Development of sophisticated software tools can automate various aspects of the FMEA process from data collection to risk prioritization and mitigation planning, enhancing its efficiency and accuracy Salah et al. [28].

Based on the above literature review, it can be concluded that risk assessment and its management for infrastructure projects has been demonstrated using case studies primarily from the foreign countries. However, their exists scope for implementation of this technique in the context of Indian Infrastructure Projects. The details of the methodology adopted for project risk mitigation is given in the following section.

3. METHODOLOGY

The stepwise methodology used to assess the risk for infrastructure projects using FMEA technique is given in the Figure 17.1.

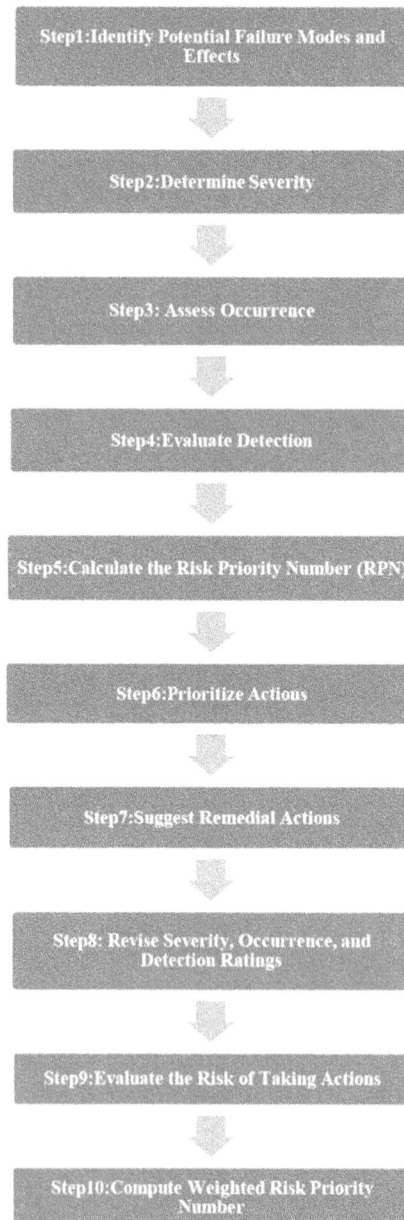

Figure 17.1 Risk management for infrastructure projects using FMEA technique

Source: Authors

Detailed description of these steps is given below.

Step 1: Identify Potential Failure Modes and Effects: Initially, list all possible way by which that the system, process, or product under evaluation can fail. Evaluate each failure mode's possible impact on system performance and client.

Step 2: Determine Severity: Find the severity of each of the failure mode by having discussion with the project managers.

Step 3: Assess Occurrence: Assess chance of occurrence of each failure mode. This phase involves determining the likelihood that the failure. To measure the occurrence, five point scale [1 (extremely unlikely) to 5 (very likely)] may be used.

Step 4: Evaluate Detection: Consider how well the current detection or control systems identify and lessen the impact of the failure modes. Give detection grade for each failure mode. This stage shows how well a failure can be identified before it impacts the system.

Step 5: Calculate the Risk Priority Number (RPN): Product of severity, occurrence, and detection scores for each failure mode gives Risk Priority Number.

Typically, RPN is calculated as:

$$\text{RPN} = \text{Severity} \times \text{Occurrence} \times \text{Detection} \tag{1}$$

All the failure modes are ranked using RPN values, obtained using equation (1). Failure modes with higher RPN values have more serious implications, they are more likely to occur, and are harder to detect. Therefore, they need to be given higher priority.

Step 6: Prioritize Actions: Sort the failure modes based on the RPN values. High RPN values signify a failure mode's substantial impact, high probability of occurrence, and limited efficacy of detection.

Step 7: Suggest Remedial Actions: Provide steps that should be undertaken to lessen the likelihood and severity of high-priority failure modes.

Step 8: Revise Severity, Occurrence, and Detection Ratings: Review severity, occurrence, and detection ratings for the impacted failure modes once corrective steps are taken.

Step 9: Evaluate the Risk of Taking Actions: Evaluate the risk involved in each suggested course of action.

Step 10: Compute Weighted Risk Priority Number: Multiply the amended RPN by the risk factor to get a weighted Risk Priority Number for each failure mode. A final priority to address failure modes is provided by the weighted RPN.

4. COMPUTATIONAL RESULTS

The above-mentioned methodology was verified with the help of data obtained from the missing link-infrastructure project in Pune, India. The interviews were conducted for 12 infrastructure managers to understand potential failure modes. Based on consensus, severity, occurrence and detection values for all the failure modes were arrived at and RPN values were calculated. Figure 17.2 below provides pareto chart for RPN values for the infrastructure project under consideration.

Pareto Chart of Risk Factor

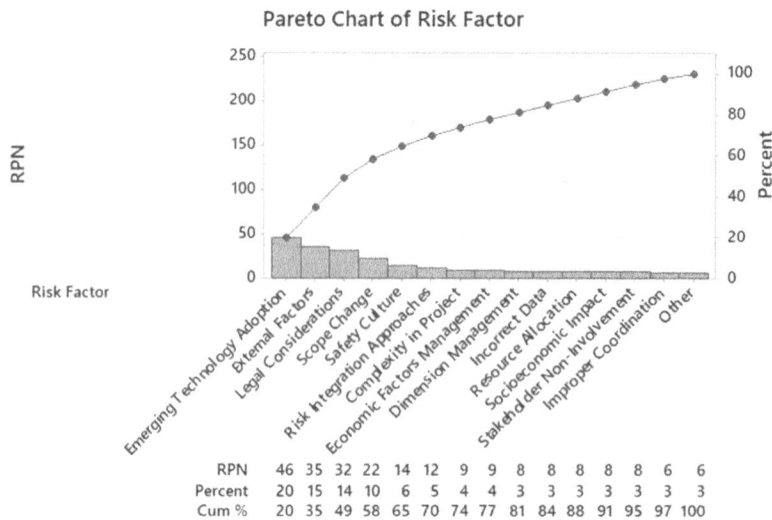

RPN	46	35	32	22	14	12	9	9	8	8	8	8	8	6	6
Percent	20	15	14	10	6	5	4	4	3	3	3	3	3	3
Cum %	20	35	49	58	65	70	74	77	81	84	88	91	95	97	100

Figure 17.2   Pareto chart for RPN values for infrastructure project

Source: Authors

5. Managerial Implications and Conclusion

Using above FMEA methodology and considering Pareto Analysis (Figure 17.2) it can be concluded that the prominent factors impacting risk for the infrastructure project under consideration were emerging technology adoption, external factors, legal considerations and scope change. Managers must give priority to these factors and develop risk mitigation strategies to minimize overall project risk.

Application FMEA methodology often encounters several challenges including subjectivity in assessment of severity, occurrence, and detection, time taken for implementation, unavailability of complete system data and the methodology may not adequately capture the interdependencies between different failure modes and their combined effects on the system. Subjectivity can be addressed by use of standardized criteria and scales along with training of team members. Focus on critical components and failure modes having highest impact on the system can reduce its implementation time. Development of robust data management practices can ensure completeness and accuracy of data which can enhance effectiveness of FMEA methodology implementation. The results obtained using FMEA methodology implementation may not be generalized on account of subjectivity. However, this methodology may prove to be beneficial in minimizing the risk associated with real estate projects as well.

Acknowledgment

I would like to thank NICMAR University, Pune for conducting this research.

References

[1] Abroshan, M. R. & Hayati, M. (2017). Risk assessment using fuzzy FMEA (case study: Tehran subway tunneling operations). *Indian Journal of Science and Technology*, 10(9), 1–9. doi: 10.17485/ijst/2017/v10i9/110157

[2] Ahmadi, M., Behzadian, K., Ardeshir, A., & Kapelan, Z. (2017). Comprehensive risk management using fuzzy FMEA and MCDA techniques in highway construction projects. *Journal of Civil Engineering and Management*, 23(2), 300–310. https://doi.org/10.3846/13923730.2015.1068847

[3] Amini, A., & Mojtaba, F. (2018). Risk assessment of Namaklan road tunnel using failure mode and effect analysis (FMEA). In *Tunnelling and climate change conference.*

[4] Cagliano, A. C., Grimaldi, S., & Rafele, C. (2015). Choosing project risk management techniques. A theoretical framework. *Journal of Risk Research*, 18(2), 232–248. https://doi.org/10.1080/13669877.201 4.896398

[5] Carvalho, M. M. D., & Rabechini Junior, R. (2015). Impact of risk management on project performance: the importance of soft skills. *International Journal of Production Research*, 53(2), 321–340. https://doi. org/10.1080/00207543.2014.919423

[6] Chapman, C., & Ward, S. (2011). How to manage project risk. John Wiley & Sons.

[7] Cheng, M., & Lu, Y. (2015). Developing a risk assessment method for complex pipe jacking construction projects. *Automation in Construction*, 58, 48–59. https://doi.org/10.1016/j.autcon.2015.07.011

[8] De La Torre Díez, I., López-Coronado, M., & Martin-Rodriguez, J. (2020). Integration of IoT with FMEA methodologies for improving risk management in smart cities. *Sustainability*, 12(12), 21–24. https://doi. org/10.3390/su12124970

[9] Fraser, J., & Simkins, B. (2016). Enterprise risk management: today's leading research and best practices for tomorrow's executives. John Wiley & Sons.

[10] Hillson, D. (2003). Effective opportunity management for projects: Exploiting positive risk. *CRC Press.*

[11] Kim, S.C. (2017). An analysis on risk factor of constructor with Fuzzy-FMEA. *Journal of the Korea Safety Management & Science*, 19(4), 43–52. http://dx.doi.org/10.12812/ksms.2017.19.4.43

[12] Lv, L., Li, H., Wang, L., Xia, Q., & Ji, L. (2019). Failure mode and effect analysis (FMEA) with extended MULTIMOORA method based on interval-valued intuitionistic fuzzy set: Application in operational risk evaluation for infrastructure. *Information*, 10(10), 313. doi:10.3390/info10100313

[13] Macura, D., Laketić, M., Pamučar, D., & Marinković, D. (2022). Risk analysis model with interval type-2 fuzzy FMEA—case study of railway infrastructure projects in the Republic of Serbia. *Acta Polytechnica Hungarica*, 19(3), 103–118.

[14] Mu, J., Peng, G., & MacLachlan, D. L. (2009). Effect of risk management strategy on NPD performance. *Technovation*, 29(3), 170–180. https://doi.org/10.1016/j.technovation.2008.07.006

[15] Panenka, A. N. D. R. E. A. S., & Nyobeu, F. (2018). Maintaining an aging infrastructure based on a fuzzy risk assessment methodology. *In 6th International Symposium on Reliability Engineering and Risk Management. Proceedings of the International Symposium, Singapore,* 31.

[16] Patricio, R. P., Catai, R. E., Michaud, C. R., & Nagalli, A. (2013). Model of risk management based in the FMEA technique–A case study in the construction of gabions. *Electronic Journal of Geotechnical Engineering,* 18, 4183–4199.

[17] Project Management Institute. 2008. "A guide to the project management body of knowledge (PMBOK Guide)", (5th ed.). Project Management Institute.

[18] Project Management Institute. (2017). A guide to the project management body of knowledge (PMBOK guide), (6th ed.). Project Management Institute.

[19] Purwanggono, B., & Margarette, A. (2017). Risk assessment of underpass infrastructure project based on IS0 31000 and ISO 21500 using fishbone diagram and RFMEA (project risk failure mode and effects analysis) method. In IOP Conference Series: Materials Science and Engineering, 277(1), 012039. doi:10.1088/1757-899X/277/1/012039

[20] Sader, S., Husti, I., & Daróczi, M. (2020). Enhancing failure mode and effects analysis using auto machine learning: A case study of the agricultural machinery industry. *Processes*, 8(2), 224. doi:10.3390/ pr8020224

[21] Sadidi, J., Gholamnia, R., Gharabagh, M. J., & Mosavianasl, Z. (2019). Evaluation of Safety Indexes of the Tower Crane with the FMEA Model (A Case Study: Tower Cranes Mounted in the City of Mashhad in 2016). *Pakistan Journal of Medical and Health Sciences*, 13, 607–612.

[22] Sakinah, S., & Sutopo, W. (2021). Analysis and design of precast concrete product quality improvement at PT. XYZ using the DMAIC and FMEA Method. In *Proceedings of the Second Asia Pacific International Conference on Industrial Engineering and Operations Management Surakarta, Indonesia*, 14–16.

[23] Salokolaei, D. D., & Esmaili, S. M. (2019). A hybrid approach based on AHP and FMEA approaches for risk assessment of refinery construction projects. *World*, 8(4), 35–41.

[24] Shakou, L. M., Wybo, J. L., Reniers, G., & Boustras, G. (2019). Developing an innovative framework for enhancing the resilience of critical infrastructure to climate change. *Safety Science*, 1(118), 364–378. https://doi.org/10.1016/j.ssci.2019.05.019

[25] Singha, M., & Debasis, S. (2017). Project risk analysis for elevated metro rail projects using fuzzy failure mode and effect analysis (FMEA). *International Journal of Engineering Technology Science and Research*, 4(11), 907–914.

[26] Wang, R., Feng, Y., & Yang, H. (2019). Construction project risk evaluation based on FMEA. In IOP Conference Series: Earth and Environmental Science IOP Publishing. 330(2), 22–41. doi:10.1088/1755-1315/330/2/022041

[27] Zhang, Z. X., Yang, L., Cao, Y. N., & Xu, Y. W. (2022). An improved FMEA method based on ANP with probabilistic linguistic term set. *International Journal of Fuzzy Systems*, 24(6), 2905–2930. https://doi.org/10.1007/s40815-022-01302-2

[28] Salah, B., Alnahhal, M., & Ali, M. (2023). Risk prioritization using a modified FMEA analysis in industry 4.0. *Journal of Engineering Research*, 11(4), 460–468.

Advances in Construction, Real Estate, Infrastructure and Project Management – Anil Kashyap et al. (eds)
© 2026 Taylor & Francis Group, London, ISBN 978-1-041-13433-6

18

Impact of ESG on Performance of Infrastructure Companies in India: An Exploratory Analysis

Swetha Kunchala[1],
Pragya Sinha[2],
Radhika Heda[3], Bharati Phadtare[4]
Advanced Project Management,
NICMAR University,
Pune, India

Harish Kumar Singla[5]
NICMAR Business School, NICMAR University,
Pune, India

■ **ABSTRACT:**

Purpose: The purpose of this study is to explore the relationship between ESG parameters and valuation of infrastructure companies in India.

Design/Methodology/Approach: The study is exploratory in nature and data has been collected through survey of professionals working in the industry and the questionnaire has been so designed that it captures the views of the respondents on all the three parameters of Environmental, Social and Governance aptly. The study is subjective in nature and is an assessment of the perceptions of the employees that whether integrating ESG into a companies' strategy is profitable and leads to increase in its performance or not. The dependent variable is the firm valuation whereas the independent variables are the various factors under the environment, social and governance pillars.

Findings: We expect our study to find a positive relationship between ESG and valuation of the infrastructure companies in India. We are hereby trying to determine the specific components of Environment, Social and Governance which are vital to create the value of the company. There is also an impact of ESG on long term sustainability.

Research limitations/implications: The quality and availability of data pertaining to ESG variables might present serious difficulties. A thorough analysis may be challenging to perform since infrastructure businesses may not publish ESG measures regularly.

---

[1]p2371008@student.nicmar.ac.in, [2]p2371040@student.nicmar.ac.in, [3]p2371046@student.nicmar.ac.in, [4]p2371069@student.nicmar.ac.in, [5]hsingla@nicmar.ac.in

DOI: 10.1201/9781003669814-18

Practical implications: This study will be useful for various stakeholders such as developers, government and infrastructure companies in order to make the policy framework which will lead to integration of ESG In valuation, this study will advise infrastructure companies to focus on the parameters which will increase their valuation.

Originality/value: The study is the original work of the authors and there is no conflict of interest.

■ **KEYWORDS:** ESG, India, Infrastructure, SUstainability, Valuation

1. INTRODUCTION

This research examines how organizations are evaluated by considering Environmental, Social, and Governance (ESG) factors, which hold importance in the infrastructure industry. It examines the influence of ESG factors on investment choices and valuation methods, and how society, the environment, and corporate activities are affected. The article additionally discusses the relationship between the ESG criteria and the 17 Sustainable Development Goals coined by the United Nations with the aim of promoting sustainable development and ethical corporate practices in the infrastructure sector.

Additionally, ESG factors influence the risk profile of a company which in turn can affect its valuation. Therefore, clear understanding of the subject is necessary to prepare for appropriate risk assessment and mitigation strategies. Furthermore, ESG disclosures are being made mandatory by regulatory bodies and studying their impact on valuation will aid in preparing the companies accordingly. The companies that have high ESG standards also build a rapport and trust with its investors as they perceive the company to be more reliable and less risky. It also fosters innovation and efficiency in areas like waste reduction, employing renewable sources of energy, social impact which in turn leads to cost saving and profitability.

India is a growing economy, and the government is allocating considerable budget for boosting the infrastructure of the country. Along with expanding the infrastructure of the nation, sustainability is becoming an indispensable parameter when it comes to formulating company strategy and meeting shareholders' expectations. Therefore, the companies are increasingly adopting ESG criteria in their decision making. However, research shows that not much studies have been carried out in India. Therefore, the paper is an attempt to capture the relation between ESG practices and the firm valuation scenario of infrastructure companies in India.

2. LITERATURE REVIEW

The literature on integrating Environmental, Social, and Governance (ESG) factors into investment decision-making and firm valuation is rich and multifaceted, as evidenced by several key studies. Schramade [15] introduces the "value-driver adjustment approach," advocating for the inclusion of ESG considerations in financial assessments to align investment decisions with broader societal and environmental goals. Buchanan et al. [4] and Sheikh [16] delve into how CSR initiatives positively influence firm value, highlighting that companies engaging in CSR practices are viewed more favourably by investors especially when coupled with strong corporate governance measures. Fatemi et al. [10] extends this by examining how varying ESG performance components affect firm value differently, emphasizing the nuanced impact of governance-related disclosures on moderating firm value. Broadstock et al. [3] explores the indirect link between CSR and innovation capacity, suggesting that socially responsible practices can enhance a company's ability to innovate, thus contributing to its long-term competitiveness.

Cornell [7] and Eliwa et al. [9] provide empirical evidence supporting the financial benefits of strong ESG practices, including lower borrowing costs and improved financial performance, reinforcing the idea that integrating ESG criteria into investment strategies can yield favourable financial outcomes while promoting sustainability. Lee et al. [13] further illustrate how robust ESG signals can enhance brand value across industries, suggesting that proactive ESG efforts contribute positively to corporate reputations and market positioning. Octaviani and Utama [14] examine how ESG factors, alongside corporate hedging strategies, influence stock price crash risk within the Indonesian energy sector, underscoring the risk management and value preservation benefits of comprehensive ESG frameworks.

Chao and Farrier [5] analyse Public-Private Partnerships (PPPs) in ESG projects, emphasizing how collaborative efforts between government and industry can drive social and environmental progress while creating value for stakeholders. Zhou et al. [22] highlight the mediating role of financial performance in the relationship between ESG activities and market value, emphasizing that improved ESG performance enhances company market value through enhanced operational capabilities. Aydoğmuş et al. [2] further underscore the positive relationship between ESG performance and firm value/profitability, using a robust dataset to demonstrate the financial returns associated with strong ESG performance metrics.

Koczar et al. [12] delve into practical methodologies for integrating ESG factors into corporate valuation processes, stressing the importance of accurate data and comprehensive analysis tools in leveraging ESG for valuation insights and risk management. Duncombe et al. [8] focus on ESG government risk and its impact on international IPO under-pricing, highlighting how regulatory environments and governance stability influence investor perceptions and initial market valuations. Xiao [18] explores the direct impact of ESG considerations on financial performance indicators such as profitability and risk management, revealing how ESG factors contribute to long-term business sustainability and investor confidence.

Cohen [6] discusses the broader economic implications of ESG risks, particularly how environmental and social risks can affect stock returns and investor decision-making across global markets. Jitmaneeroj [11] offers a detailed analysis of CSR components in the financial industry, using structural equation modelling to prioritize ESG indicators and sub-indicators for enhancing stock value. Xu [19] examines the evolving role of ESG ratings in corporate strategies, emphasizing their increasing importance in investor decisions and corporate governance practices.

Siwei & Chalermkiat [17] explore the mediation effect of debt costs on the relationship between ESG performance and firm value, highlighting how financial considerations interact with ESG practices to influence overall corporate valuation. Kim et al. [12] investigates the strategic impact of ESG initiatives on corporate reputation and firm value across different industries, underscoring the sector-specific benefits of robust ESG practices. Lastly Yahya [21] analyses investment patterns in BRICS countries, revealing how ESG performance influences asset allocation decisions and underscores the growing importance of sustainability in emerging market economies.

In summary, the literature review reveals a growing consensus on the positive correlation between robust ESG practices and enhanced firm value, supported by various empirical studies across different sectors and regions. These findings emphasize the multifaceted benefits of integrating ESG considerations into investment strategies and corporate governance frameworks, highlighting opportunities for both financial performance improvement and sustainable development.

3. RESEARCH METHODOLOGY

3.1 Questionnaire Design

In order to further evaluate the ESG parameters and the relation between the valuation of the company a questionnaire has been made keeping in mind relevant indicators which plays significant role in the way. We paid attention to the individual parameters of ESG as Environmental, Social and Governance while designing the questionnaire. The environmental factors that are taken into consideration are energy usage of the company, energy efficiency, measures to reduce energy consumption, treatment of surroundings, pollution and waste generated, energy conservation etc. The social factors like how the businesses maintain the relationship of the company with its employees in terms of health and safety, social work for the communities, safeguarding the interests of other stakeholders, decision making etc. are also included in the questionnaire. The factors such as the company's diverse leadership and accountability towards its stakeholders, transparent accounting methods, gender diversity, fair labour practices etc. come under the umbrella of Governance. This questionnaire is designed in a way that tries to cover all the aspects of the ongoing and current ESG factors along with the diverse group of people of different age, gender and their experiences. The approach taken for the data collection is based on the above-mentioned parameters which were leading to the objectives of this research.

We developed a questionnaire comprising of a multiple-choice question in which a scale was used which varies from 1 to 5 where answer 1, if strongly disagree with the statement, 2 if disagree, 3 if neutral, 4 if agree and 5 if strongly agree. The questionnaire was structured into three sections to ensure clarity and logical flow which can help the respondent to easily navigate.

3.2 Sample Profile

The questionnaire is designed so as to collect the demographic information of all the respondents such as age, gender and work experience in the industry. As each individual might have different perceptions about their organization's performance in the market based on their ESG initiatives, it becomes imperative to capture the responses from people of all age groups, gender and years of work experience. A total of 76 respondents responded to the survey. A brief profile of respondents is listed in Table 18.1.

Table 18.1 Demographic profile

Age	Nos	%
21 to 30 Years	68	89.5%
31 to 40 Years	4	5.3%
41 to 50 Years	4	5.3%
Gender		
Male	31	40.8%
Female	45	59.2%
Experience		
0 to 5 Years	70	92.1%
5 to 10 Years	6	7.9%

Source: Compiled by authors

3.3 Data Analysis Method

The data collected was analysed subjectively. The major purpose of the study is to capture the perception of the people working in the infrastructure industry towards integration of ESG into the companies' activities. The data was collected through circulating the survey forms to people working in the industry. All the responses were given a numerical value ranging from 1 to 5. After collection of data, cleaning and sorting of data was done to arrange them in tabular format. The analysis aims to highlight some key observations or trends in the response.

4. DATA ANALYSIS AND RESULTS

The data collected was arranged according to the questions and the number of responses in each category i.e. strongly disagree, disagree, neutral, agree and strongly agree (see Table 18.2).

Table 18.2 Response for each question

Questions	Strongly disagree	Disagree	Neutral	Agree	Strongly agree
My organization manages and minimizes water usage in its operations.	5.26%	10.53%	30.26%	39.47%	14.47%
My organization has implemented measures to reduce energy consumption.	3.95%	10.53%	27.63%	46.05%	11.84%
My organization has taken steps to reduce its carbon footprint	6.58%	14.47%	26.32%	35.53%	17.11%
My organization has taken initiatives to reduce, reuse, and recycle waste.	5.26%	17.11%	30.26%	34.21%	14.47%
My organization has measures in place to minimize air pollution from company's activities.	3.95%	9.21%	36.84%	36.84%	13.16%
My organization has measures in place to minimize water pollution from company's activities	3.95%	10.53%	36.84%	34.21%	14.47%
My organization has measures in place to minimize soil pollution from company's activities	7.89%	10.53%	31.58%	40.79%	9.21%
The organization has taken initiatives to utilize renewable energy sources	2.63%	10.53%	32.89%	39.47%	14.47%
The organization communicates its environmental performance and initiatives to stakeholders.	3.95%	9.21%	35.53%	36.84%	14.47%
The organization has policies and initiatives to support community engagement and development	7.89%	9.21%	25.00%	39.47%	18.42%
The organization takes measures to ensure fair labour practices throughout its supply chain.	3.95%	11.84%	26.32%	40.79%	15.79%
The organization takes initiatives or programs aimed at supporting underprivileged and marginalized communities.	5.26%	13.16%	38.16%	28.95%	14.47%
The organization prioritizes employee satisfaction and employee wellbeing.	5.26%	11.84%	32.89%	32.89%	17.11%
The organization engages with stakeholders to understand and address social concerns and priorities	3.95%	10.53%	31.58%	38.16%	15.79%
The organization take steps to ensure the health and safety of its employees and stakeholders.	2.63%	11.84%	32.89%	31.58%	19.74%
The organization manages and mitigates social risks, such as labour disputes or community resistance to operations	6.58%	7.89%	27.63%	42.11%	14.47%

Questions	Strongly disagree	Disagree	Neutral	Agree	Strongly agree
The organization make efforts to promote education, skills development, and empowerment within its workforce	1.32%	10.53%	27.63%	44.74%	15.79%
The company's board of directors is diverse in terms of its gender.	6.58%	10.53%	26.32%	43.42%	11.84%
The company has employees of different religious beliefs and practices	2.63%	11.84%	27.63%	38.16%	19.74%
The company has fair representation of individuals with physical impairments and disabilities	3.95%	13.16%	40.79%	26.32%	15.79%
The company has employees from different regions of the country.	5.26%	9.21%	26.32%	34.21%	25.00%
The company takes measures to ensure ethical conduct and compliance with laws and regulations.	3.95%	6.58%	28.95%	38.16%	22.37%
The company ensures transparency in its operations and decision making	1.32%	9.21%	31.58%	40.79%	17.11%
The company has integrated ESG related risks into its overall risk management framework.	3.95%	9.21%	35.53%	36.84%	14.47%
The company protects and promotes the rights of its shareholders.	3.95%	3.95%	25.00%	50.00%	14.47%
The company identifies and manages potential conflicts of interest.	6.58%	3.95%	30.26%	46.05%	13.16%
The company has policies and procedures to prevent bribery and corruption	1.32%	9.21%	28.95%	44.74%	15.79%
The company has gender diversity in its workforce.	9.21%	10.53%	35.53%	27.63%	15.79%
The company engages with local community by volunteering and community development projects	1.32%	11.84%	34.21%	34.21%	17.11%
The company ensures fair labour practices.	3.95%	9.21%	23.68%	50.00%	13.16%
The Brand Image of my organization has improved	5.26%	3.95%	35.53%	39.47%	15.79%
My Market Valuation of my company has improved substantially in recent years	2.63%	3.95%	34.21%	42.11%	17.11%
The Profitability position of my organization has improved in recent years	2.63%	7.89%	34.21%	32.89%	22.37%
The Order Book (New Upcoming Projects) position of my organization has improved in recent years.	2.63%	3.95%	31.58%	32.89%	27.63%
There is no relation between firm valuation and being environmentally friendly firm.	6.58%	13.16%	35.53%	31.58%	13.16%
There is no relation between firm valuation and being a good governing firm.	11.84%	11.84%	34.21%	22.37%	18.42%
There is no relation between firm valuation and being socially relevant firm.	9.21%	15.79%	28.95%	31.58%	14.47%

Source: Compiled by authors

5. DISCUSSION ON RESULTS

The table shows that out of the total responses recorded, the percentage of people who agree on their organization taking effective measures to reduce the impact on the environment from the company activities range from 34% to 41% with the mode of the data being around 36%. For the social parameters that were included in the questionnaire, the percentage of people who agree that their organization is taking social initiatives range from 28% to 42%. In the overall sense, there is a positive perception of the employees towards their organization taking socially driven decisions. This may be due to the fact that the people feel a sense of pride and trust in their organization and the personal values of the people align with the company's social initiatives. For the questions under the governance criteria, the percentage of employees who agree with their company being a good governing organization range from 26% to 50%. The results are supported by the findings of Alam et al. [1] and Yahya [20] where both of them reported that the companies can build resilience towards change in market conditions by fostering social responsibility and building positive relations with the local communities.

However, there is a mixed response from the employees when asked about the relation between the ESG initiatives their firm undertakes and the market valuation of the organization. This may be due to lack of knowledge of the subject. Nonetheless, 42% employees agree that the performance of their firm has increased in the recent years.

6. CONCLUSION

6.1 Summary

To conclude, employees perceive ESG initiatives of their firm positively especially when their personal values and beliefs are reflected in the decisions of the organization. It also leads to enhanced job satisfaction if there is transparency in the company's decision making. They believe that strong ESG practices enhance the reputation of the organization, financial performance, and its long-term viability. All of these contribute to positive valuation of the firm in the market. ESG performance of the company also has potential to mitigate risks, open new markets for the firm, thus creating a competitive advantage which leads to increase in overall valuation of the firm.

However, there may be concerns about the authenticity of the initiatives and the employees might consider them to just be marketing tactics. The employees might also be sceptical that the resource allocation for the ESG initiative will divert the focus from the core business operations or their own benefit. In such scenario, there is a negative view of the firm's ESG practices.

6.2 Implication for Stakeholders

The Environmental, Social and Governance initiatives have significant and varied implications for all the shareholders. It boosts employee morale and engagement as they feel part of a responsible and ethical organization. It can also provide opportunities for growth and learning in areas of sustainability and corporate social responsibility. ESG initiatives are critical in assessing long term risks and opportunities and may be used by investors for evaluating company performance and making investment choices. Companies that perform highly on the ESG parameters can be valued higher due to the perceived lower risks and greater long term growth potential. Suppliers are required to adhere to ESG standards which would improve their own practices leading to more sustainable supply chains. ESG initiatives are associate with long term value creation which would lead to increase in shareholders return over time. Shareholders benefit from the enhanced trust and reputation that comes with ESG practices.

There are also some challenges like balancing the resources between the business priorities and the ESG initiatives. Moreover, accurately measuring and reporting the impact of ESG initiatives requires a robust and transparent system. Sometimes meeting the diverse and conflicting expectations of the stakeholders might prove to be difficult.

6.3 Limitations and Scope for the Future

While this study has made valuable contributions to academia and practice, it is necessary to take on the note of the key limitations. This study has solely focused on the Indian companies which introduces the possibility of nations distinct characteristics. For the future studies it can include organisations from different sectors and countries to have a detailed understanding around the globe. The study has significant implications for all stakeholders, investors, policymakers, and managers which can be concluded to streamlining transparency in accordance with current stakeholders and their requirements. This literature believes that it can impact valuation of an organisation positively. However, there is limit to organisation's market value which can have an impact on ROE and ROI. This can also be a future scope for the further studies. The responses from the employees were majorly in the age group of 21–30. In other words, employees with very less years of experience might not be concerned with the ESG initiatives taken by the company. With the positive inclination towards the ESG & company's valuation imported from the survey, the initiative can help the organisation for better recognition and increase its value. The scope of the study is limited to the response from different organisations as a result, the findings cannot be generalized to small, medium, large-sized institutions.

REFERENCES

[1] Alam, A., Banna, H., Alam, A. W., Bhuiyan, M. B. U., & Mokhtar, N. B. (2024). Climate change and geopolitical conflicts: The role of ESG readiness. *Journal of Environmental Management*, 353. https://doi.org/10.1016/j.jenvman.2024.120284

[2] Aydoğmuş, M., Gülay, G., & Ergun, K. (2022). Impact of ESG performance on firm value and profitability. *Borsa Istanbul Review*, 22, S119–S127. https://doi.org/10.1016/j.bir.2022.11.006

[3] Broadstock, D. C., Matousek, R., Meyer, M., & Tzeremes, N. G. (2020). Does corporate social responsibility impact firms' innovation capacity? The indirect link between environmental & social governance implementation and innovation performance. *Journal of Business Research*, 119, 99–110. https://doi.org/10.1016/j.jbusres.2019.07.014

[4] Buchanan, B., Cao, C. X., & Chen, C. (2018). Corporate social responsibility, firm value, and influential institutional ownership. *Journal of Corporate Finance*, 52, 73–95. https://doi.org/10.1016/j.jcorpfin.2018.07.004

[5] Chao, A., & Farrier, J. (2022). Public-private partnerships for environmental, social, and governance projects: how private funding for infrastructure can produce mutual benefits for companies and the public. *International Conference on Sustainable Infrastructure 2021*.

[6] Cohen, G. (2023). The impact of ESG risks on corporate value. *Review of Quantitative Finance and Accounting*, 60(4), 1451–1468. https://doi.org/10.1007/s11156-023-01135-6

[7] Cornell, B. (2021). ESG preferences, risk and return. *European Financial Management*, 27(1), 12–19. https://doi.org/10.1111/eufm.12295.

[8] Duncombe, S., Park, M., Tarsalewska, M., & Trojanowski, G. (2023). ESG positioning in private infrastructure fundraising. *International Review of Financial Analysis*, 90. https://doi.org/10.1016/j.irfa.2023.102924

[9] Eliwa, Y., Aboud, A., & Saleh, A. (2021). ESG practices and the cost of debt: evidence from EU countries. *Critical Perspectives on Accounting*, 79. https://doi.org/10.1016/j.cpa.2019.102097

[10] Fatemi, A., Glaum, M., & Kaiser, S. (2018). ESG performance and firm value: the moderating role of disclosure. *Global Finance Journal*, 38, 45–64. https://doi.org/10.1016/j.gfj.2017.03.001

[11] Jitmaneeroj, B. (2023). Prioritizing CSR components for value enhancement: Evidence from the financial industry in developed and emerging markets. *Heliyon*, 9(5). https://doi.org/10.1016/j.heliyon.2023.e16044

[12] Kim, Y., Jang, H., & Seok, J. (2023). Uncovering the Relationship between ESG Practices and Firm Value: The Role of Reputation and Industry Sensitivity. Asia Marketing Journal, 25(4), 207–218. https://doi.org/10.53728/2765-6500.1620

[13] Koczar, J., Zakhmatov, D., & Vagizova, V. (2023). Tools for considering ESG factors in business valuation. *Procedia Computer Science*, 225, 4245–4253. https://doi.org/10.1016/j.procs.2023.10.421

[14] Lee, M. T., Raschke, R. L., & Krishen, A. S. (2022). Signaling green! firm ESG signals in an interconnected environment that promote brand valuation. *Journal of Business Research*, 138, 1–11. https://doi.org/10.1016/j.jbusres.2021.08.061

[15] Octaviani, F. A., & Utama, C. A. (2022). Impact of Corporate Hedging and ESG on Stock Price Crash Risk: Evidence from Indonesian Energy Firms. *Indian Journal of Corporate Governance*, 15(2), 149–169. https://doi.org/10.1177/09746862221129341

[16] Schramade, W. (2016). Integrating ESG into valuation models and investment decisions: the value-driver adjustment approach. *Journal of Sustainable Finance and Investment*, 6(2), 95–111. https://doi.org/10.1080/20430795.2016.1176425

[17] Sheikh, S. (2018). Is corporate social responsibility a value-increasing investment? Evidence from antitakeover provisions. *Global Finance Journal*, 38, 1–12. https://doi.org/10.1016/j.gfj.2017.08.002

[18] Siwei, D., & Chalermkiat, W. (2023). An analysis on the relationship between ESG information disclosure and enterprise value: A case of listed companies in the energy industry in China. *Cogent Business and Management*, 10(3). https://doi.org/10.1080/23311975.2023.2207685

[19] Xiao, K. (2023). How does Environment, Social and Governance Affect the Financial Performance of Enterprises? *SHS Web of Conferences*, 163, 04015. https://doi.org/10.1051/shsconf/202316304015

[20] Xu, Y. (2023). The impact of ESG performance on company valuation: a case study based on Industrial Bank. In *Business, Economics and Management FEIM*.

[21] Yahya, H. (2023). The role of ESG performance in firms' resilience during the COVID-19 pandemic: Evidence from Nordic firms. *Global Finance Journal*, 58. https://doi.org/10.1016/j.gfj.2023.100905

[22] Yu, Z., Farooq, U., Alam, M. M., & Dai, J. (2024). How does environmental, social, and governance (ESG) performance determine investment mix? New empirical evidence from BRICS. *Borsa Istanbul Review*. https://doi.org/10.1016/j.bir.2024.02.007

[23] Zhou, G., Liu, L., & Luo, S. (2022). Sustainable development, ESG performance and company market value: Mediating effect of financial performance. *Business Strategy and the Environment*, 31(7), 3371–3387. https://doi.org/10.1002/bse.3089

Advances in Construction, Real Estate, Infrastructure and Project Management – Anil Kashyap et al. (eds)
© 2026 Taylor & Francis Group, London, ISBN 978-1-041-13433-6

19 Relationship of Infrastructure Development with Real Estate Demand

Poulomee Ghosh*

Assistant Professor,
SREFM, NICMAR University Pune

Prajjval Tripathi and Utkarsh Dwivedi

MBA ACM Student,
NICMAR University Pune

■ **ABSTRACT:** The Indian government invests significant money in infrastructure building, amounting to more than 3% of the GDP. Infrastructure building directly and indirectly has an effect on economic growth and the real estate demand is grossly affected by infrastructure at all levels. The primary objective of this paper is to establish a theoretical framework of how infrastructure affects real estate demand through a literature review and then assess how access to urban infrastructure and city sizes affect real estate demand through primary surveys. About 40 research papers that relate infrastructure with real estate were reviewed to establish the theoretical framework. Data of the 134 responses received from multiple cities were analysed to determine the priorities of various infrastructure segments and impacts. The key findings of the paper are that infrastructure creation has a macro-level effect of boosting the economy leading to an increase in jobs, income and thus real estate demand. At the micro or neighbourhood level, infrastructure improves liveability but may have negative externalities of pollution, congestion and increased cost of living. Proximity to public transport is the top priority when deciding home locations followed by commercial development, education and healthcare.

■ **KEYWORDS:** Infrastructure development and access, Real estate demand, Macro and micro levels, City sizes

1. INTRODUCTION

The government has spent a whopping 23 lakh crore rupees on infrastructure over the last three years (FY22-FY24). The infrastructure focus is apparent when one looks at the capital spending

*Corresponding author: pghosh@nicmar.ac.in

DOI: 10.1201/9781003669814-19

to GDP ratio, which has almost doubled from 1.6 per cent of GDP in 2018–19 to 3.2 per cent of GDP in 2023–24 [9]. The government's increased infrastructure-led spending is not unwarranted given India's infrastructure gap and the economic potential fuelled by its growing urbanisation and demographic dividend. About 31.16% of the country's population i.e. 461 million people live in Indian cities and 42% of this urban population is living in cities of sizes smaller than 0.3 million. The urban population has a growth rate of 2.3% [1]. While the growing urban population does demand infrastructure support in terms of basic services of water, energy, transport public transport etc., there is also a huge potential of economic infrastructure to boost the economic activities in cities. Cities of India are expected to generate 75% of the country's income by 2031. To realise the ambition of Viksit Bharat 2047, 70% of the required infrastructure is yet to be built for which $ 840 billion needs to be invested in the coming years [14]. The increase in the proportion of infrastructure investment thus is in tune with the requirements.

While national infrastructure projects and programs such as highway building through Bharatmala, or ports through Sagarmala or logistics integration through Gati Shakti programs are major investment arenas, it is the urban transformation that is seemingly critical to the country's growth. One of the crucial segments of infrastructure is urban infrastructure given the importance of cities in India's growth trajectory. Urban infrastructure comprises water, sewer, waste management, storm water, energy, urban streets, public transport, street lighting, health care, education, housing and public parks [13]. Several programs such as PMAY 2 housing for all, AMRUT, Swatchh Bharat Abhiyan, Smart Cities Mission, CITIIES 2, Metro Rail, Waterfront Development, Streets improvement etc. are being implemented to augment the city's infrastructure and fill the gap. It is these programs related to urban infrastructure that are closest to urban dwellers and play a significant role in determining the livability and the cities and their neighbourhoods. As the access to these infrastructures are an absolute necessity for urban dwellers, their presence has a significant influence on the urban growth and development of the land surrounding it. It influences the land values and the property market and then feeds back into further infrastructure requirements.

At a macro level when the economy is growing, people have more disposable income to spend, which can lead to increased demand for real estate. Economic growth can create jobs, increase incomes, and boost consumer confidence, increase productivity, facilitate trade and commerce. Transportation networks, airports, metrorail, and highways act as catalysts to real estate demand [21]. The growth and absorption of real estate properties reflect how well the economy is performing and real estate and construction sectors in themselves contribute majorly to the economy [22]. At a micro or neighbourhood level, infrastructure improves livability making the location attractive thereby increasing real estate demand. The real estate demand in return requires further infrastructure upgrades [5].

However, infrastructure investments will lead to economic growth only through strategic vision and good governance failing with the debt incurred in failed projects could lead to economic turmoil [12]. Infrastructure for basic services such as water or public transport may also have some negative impacts on society during construction and later if not operated effectively. This paper first aims to establish a theoretical framework between the linkages of infrastructure and real estate demand based on a literature review. Secondly, it explores the preferences and priorities of various infrastructure access for choosing a property through a primary survey.

2. LITERATURE REVIEW

The primary purpose of the literature review is to assimilate how infrastructure development positively or negatively affects livability and residential real estate and interpret a theoretical framework relating to infrastructure development and real estate. On the same lines the literature review is divided into three parts, first on the negative effects of infrastructure development, second on the positive effects of infrastructure development and third on past theoretical framework and an interpretation. Based on the most common infrastructure considered in the literature, the parameters were shortlisted and used for the second part of the study.

2.1 Positive Effects of Infrastructure on Real Estate

Several studies have measured how infrastructure and public services relate to housing prices and have found a positive result. Water, electricity, good roads, waste disposal systems, drainage systems, medical centers, and recreational facilities affect real estate values [2]. Annual rent increases when there is an increase in electricity supply, water supply and improvement on access road [3]. Infrastructure investment in High-Speed Rail in China has resulted in a significant boost to housing prices, especially in small and medium-sized cities [24]. The study by Mei, Zhao, Lin, & Gao [16] found that green space within 135 meters of a flat raised the cost of a home by 3.4–4.6%. Xiao, Wang, & Fang [25] found that the percentage of urban green spaces had a positive effect on residential property prices in Shanghai. Yang, Chau, Szeto, Cui, & Wang [26] found that infrastructure development has a significant positive impact on real estate prices, the impact is strongest in areas with good transportation infrastructure, such as roads and railways. Mesthrige & Maqsood [17] established that there is a 6.5% of property value premium after the construction announcement; and a higher up to 6.7% after the operation of the metro line in HongKong. In emerging economies, as the announcement of infrastructure projects is marred with uncertainties, property prices remain unaffected in the initial days but increase once the project is completed [19]. Cervero [4] observed that new light rail stations in the San Francisco Bay Area lead to significant property value increases within a quarter-mile radius. It can thus be said that infrastructure projects if successful eventually lead to increased property prices and real estate demand.

2.2 Negative Effects of Infrastructure on Real Estate and Liveability

Few studies studying the relationship between infrastructure investment and housing prices have found negative or varying results. Gan, Ren, Xiang, Wu, & Cai [8] in a study in China discovered that the effect of public services on housing prices is related to urbanization levels, various regions and disposable income. The relationship is non-linear following an inverted U shape meaning while public services do play a role in property price increase, after a certain point as affluent residents occupy an area, the effect of additional public services seems to reduce property prices. Efthymiou & Antoniou [6] found that the planned metro line in Thessaloniki, Greece, affects property prices and rents and there is a negative impact on purchase prices due to construction externalities. Another study found that conflict over infrastructure projects in suburban zones of Paris led to decreased property values due to uncertainty and negative externalities [23]. Zhu [28] in a study about a cancelled transport project in UK found that the property prices around the location where a proposal of a railway station got cancelled say an increase in property value due to reduced chances of negative externalities. Similarly, the announcement of new infrastructure project with potential negative environmental impact has a reduced willingness to pay as observed in Queensland Australia [18].

Seo, Salon, Kuby, & Golub [20] in their study found that positive externality (i.e., accessibility) of highways and light rail accrues at exits and stations, whereas nodes and links of highways and light rail have negative effects. Zhang, Li, Lownes, & Zhang, [27] observed that most properties in the Stamford area show appreciation towards rail service and depreciation to bus service. Researchers have also observed that the considerations and effects of infrastructure on real estate vary with city sizes. Small cities, with their relatively smaller number of amenities and features, tend to have fewer numbers of variables that operate to determine house prices [15]. Thus, it can be said that infrastructure projects do negatively affect real estate, especially in the initial years of the project and if it does not bring any direct benefit to the residents of the location.

2.3 Theoretical Frameworks in Literature and Interpretation

The theoretical models essentially trace how infrastructure projects affect real estate and why. Good urban infrastructure leads to better accessibility and connectivity that attracts business and increases land prices. It also boosts productivity and profitability which in return creates further real estate demand [7]. Hong & Hong-Ping [10] in their framework showed that transport infrastructure improves accessibility which increases services, enforces culture and attracts investments. This then leads to the growth of commercial real estate. At the same time improved accessibility leads to development, migration and increased occupancy of residential spaces thereby boosting real estate. Zhu, Zhu, Liu, Zhang, & Yuan [29] discussed that an urban rail transit system interacts with economic, social and environmental points such as malls, schools and gardens and boosts property prices in the immediate periphery thereby creating economic impact. Jin, et al. [11] argued that there are trade-offs to an infrastructure project, particularly public transport. While public transport increases accessibility and reduces travel time and cost, it also degrades the micro-environment and causes noise pollution. Public transit leads to land value increase causing housing prices to also increase. These trade-offs are considered by citizens in their residential location choices.

Figure 19.1 represents a theoretical framework relating urban infrastructure projects and real estate demand. It is based on the positive and negative effects of infrastructure projects on real estate and the theoretical framework suggested in the past literature. At a macro-level, government infrastructure initiatives and policies lead to economic growth by increasing productivity and reducing the cost of production as well as job creation, which leads to job creation and increased per capita income. This in return boosts the demand for homes and office spaces thereby increasing real estate demand. At a micro level, infrastructure projects bring about public transport accessibility, improved infrastructural services of water, sewer and waste management, and better social infrastructure of parks, schools, and clinics. This also makes the location attractive and brings about economic activities thereby leading to job opportunities. But at the same time, there are some negative externalities of noise, air pollution, congestion and loss of neighbourhood character particularly during the infrastructure construction period. The trade-offs of these positive and negative effects decide the quality of life or livability of that location and if the advantages are more, the real estate demand of the location increases.

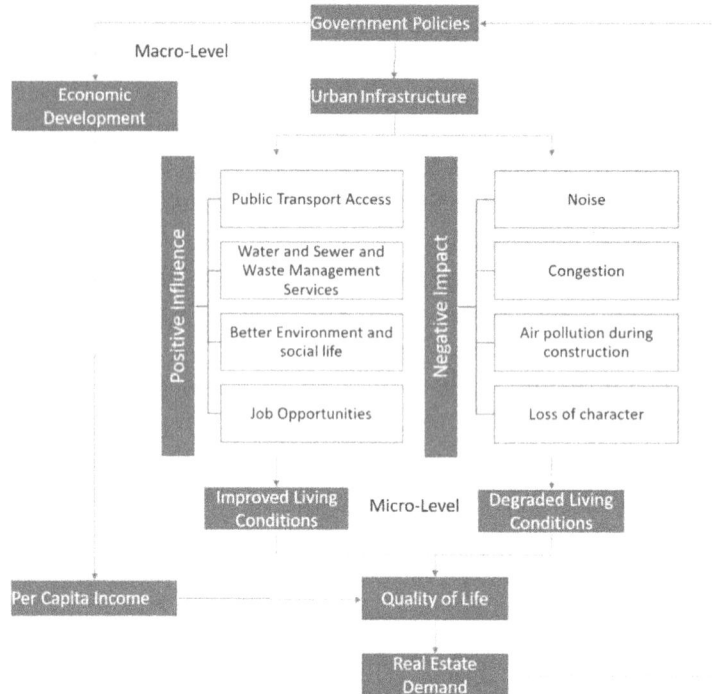

Figure 19.1 Relationship of urban infrastructure and real estate demand

Source: Authors

Based on the above framework and the kind of infrastructure considered in the past literature the parameters considered for further study are proximity to public transport; availability of schools and educational institutes; access to healthcare facilities; nearness to shopping centers; quality of road and transportation network; adequacy of water and sanitation facilities; green spaces; safety and security.

3. METHODOLOGY

The second part of the study involves exploring the preferences and priorities of various infrastructure access for choosing a property. For the same, a questionnaire survey was conducted across multiple cities in India including metropolitan and non-metropolitan cities.

3.1 Questionnaire Design and Survey

The questionnaire had four major parts. The first part captured the respondent's profile including name, city, locality, age, education, and occupation. The second part captured the housing preferences of the respondents including type of housing, ownership, location, rental change etc. The third part captured infrastructure impacts including common concerns such as environmental degradation and rising property prices, highlighting areas for sustainable development. One aspect of this part was to enumerate and rate important factors considered for property decisions where the parameters identified in the literature were rated on a 5-point Likert scale ranging from not important to very important. The last part of the questionnaire survey was open-ended for the responders to flag specific concerns, observations and opinions.

Survey conducted between July 2023 to February 2024 through online circulation of Google forms through various professional networks. Responses were collected from cities of 16 major states of India.

3.2 Responses

A total of 134 responses were recorded from across India. Of the 134, 60% of the responders are from the age group of 25 to 34. About 25 % in the age group 35 to 59 and the remaining 15 % from the age group 18 to 24 and above 60. 45% of the responders resided in stand-alone apartments, 24% in townships, 13% in residential societies, and the remaining 18% in detached homes. 45% i.e. 60 responses are from metro cities and 55% i.e. 74 responses are from non-metropolitan cities. Table 19.1 indicates the proportion of responses from various states.

Table 19.1 Responses from various states

State	Percentage of respondents
Madhya Pradesh	30%
Maharashtra	27%
Karnataka	8%
Uttar Pradesh	6%
Delhi	5%
West Bengal	5%
Andhra Pradesh	4%
Gujarat	3%
Haryana	2%
Telangana	2%
Assam	2%
Chhattisgarh	2%
Rajasthan	2%
Bihar	1%
Himachal Pradesh	1%
Tamil Nadu	1%

Source: Compiled by authors

60% of the respondents reside in self-owned properties and 40% are tenants. In non-metro cities proportion of self-owned properties reported is high i.e. 81% compared to 42% in metropolitan cities.

3.3 Analysis Technique

In the first part of the analysis cross-tabulation of data sets was made and observations were made. Some of the cross-tabulations were for metropolitan and non-metropolitan cities versus their housing rents, priorities for location selection, housing sizes etc. In the second part, the Likert scale data related to the importance of infrastructure parameters for location selection was analyzed using the Relative Importance Index (RII). The following equation is used for RII:

$$RII = \frac{\sum_{i=1}^{5} Wi \times ni}{A \times N} \tag{1}$$

RII = Relative Importance Index for each parameter

i = Severity levels in the Likert Scale

Wi = Weightage for each severity level (range from 1 to 5)

ni = Number of responses (frequencies) for each severity level

A = Maximum weightage of a parameter (i.e., 5 in this case)

N = Total number of respondents

Based on the observations from cross-tabulation and RII analysis, a conclusion is drawn.

4. RESULTS AND DISCUSSION

The results and discussion are explained in two parts, one, showing the observation from cross-tabulation and, two, results from RII.

4.1 Cross Tabulation Results

Cities of different sizes and growth rates have different behaviour of real estate and infrastructure demand. The larger the population and growth, the higher the demand for residential properties and thus a thriving market with increasing property prices. Higher property prices cumulated with in-migration trends lead to higher rental demands in larger or metropolitan cities. A similar trend is observed in this study. Figure 19.2 shows the distribution of renters across rent brackets for metropolitan and non-metropolitan cities.

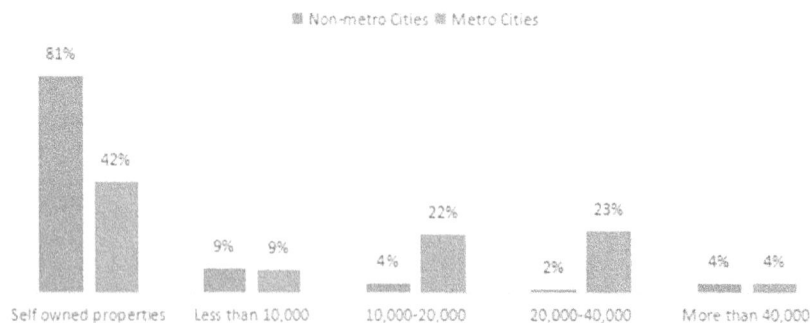

Figure 19.2 Metropolitan cities v/s non-metropolitan cities rent distribution

Source: Authors

It can be observed that the proportion of renters is almost double in metropolitan cities and non-metropolitan cities. The proportion of metropolitan respondents in the higher rent bracket of 20 to 40 thousand per month is highest compared to a very small proportion of non-metropolitan respondents in the same bracket. This shows higher rents in metropolitan cities. Figure 19.3 shows the increase in rent compared to previous year.

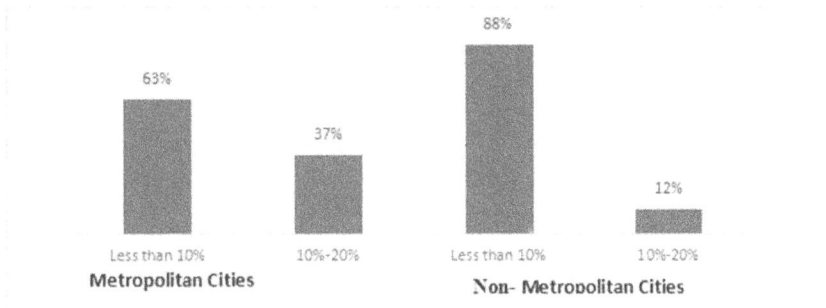

Figure 19.3   Rent change in metropolitan cities v/s nonmetropolitan cities

Source: Authors

The majority of respondents from both metropolitan and non-metropolitan cities have reported a less than 10% increase in rent compared to the previous year. However, 25% of additional respondents in metropolitan areas experienced more than a 10% increase compared to the past year. This indicates a stable to moderately growing rental market in both kinds of cities.

The study also sought to determine which respondents valued more: having a higher quality of life which may be away from employment zones or living near their place of employment despite externalities. This directly relates to the importance of transportation infrastructure particularly public transport and the ease of access it may provide. It further reflects on the city structure and distribution of residential and work zones. In this study 58% of the respondents reported that they would prioritise a higher standard of living with larger houses even if it means a longer commute to work and 42% indicated that they would prioritise proximity to work even if it meant expensive homes and congested areas. Figure 19.4 shows the proportion of respondents from metropolitan and non-metropolitan cities and their preference of proximity to work or standard of living.

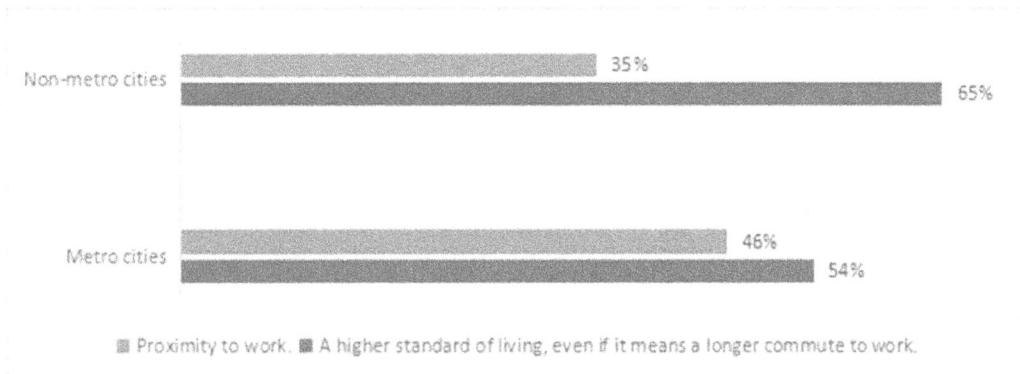

Figure 19.4   People's preferences by city size

Source: Authors

The preference for a better standard of living is consistent across metropolitan and non-metropolitan cities even if the proportion of respondents preferring proximity to work is much higher in metropolitan cities. This significant inclination emphasizes the evolving priorities of individuals in urban settings, emphasizing a desire for a well-rounded and enhanced lifestyle beyond immediate work accessibility. This insight carries substantial implications for city infrastructure, suggesting a need to focus on creating communities with robust amenities, green spaces, and an overall high quality of life. It also shows that commute is greater haste and commute time is longer in larger metropolitan cities and thus a proportion of citizens prioritising proximity to work for housing location decisions in these cities is larger. These cities need robust public transport while maintaining the wholistic well-being through infrastructure and services. However, the trend shifts in other cities, indicating that for a significant portion of individuals in these areas, the hassle of longer travel times is outweighed by the desire for an enhanced standard of living. This shift suggests that in non-metropolitan cities, where commuting might be less arduous and living spaces potentially more affordable, people don't prioritise travel times for the benefits associated with an improved lifestyle.

Figure 19.5 indicates the same priorities of the standard of living versus proximity to work against sizes of homes.

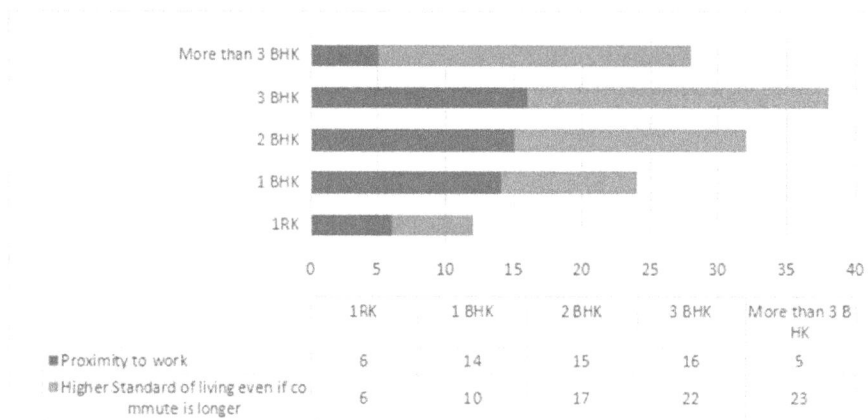

	1RK	1 BHK	2 BHK	3 BHK	More than 3 BHK
■ Proximity to work	6	14	15	16	5
▥ Higher Standard of living even if commute is longer	6	10	17	22	23

Figure 19.5 People's preferences as per size of residence

Source: Authors

It can be observed that people living in smaller homes which could mean the younger population or lower/middle-income group prioritise proximity to work much more than people living in larger homes which can be said to be the higher income group. Preference to the standard of living over proximity to work seems to have a relationship with home sizes and thus income. This shift may also suggest that people may be ready to put up with longer commutes in exchange for larger homes, given that properties in commercial districts are typically more expensive. The observed trade-off between commute convenience and living space provides insight into the decision-making process considering infrastructure in location selection for homes.

While the benefits of infrastructure vary with different sectors, the adverse effects particularly at the time of construction have certain commonalities. Figure 19.6 indicates the proportion of respondents who selected the most adverse impact they experienced from infrastructure projects around them.

The majority of respondents reported increasing traffic congestion as the main negative impact. This implies that the development or alteration of urban infrastructure, to improve accessibility and connection, can unintentionally result in increased traffic, which could affect daily commutes and general mobility. Concerns about environmental deterioration followed closely, indicating knowledge of the possible detrimental ecological effects connected to urban infrastructure initiatives. This

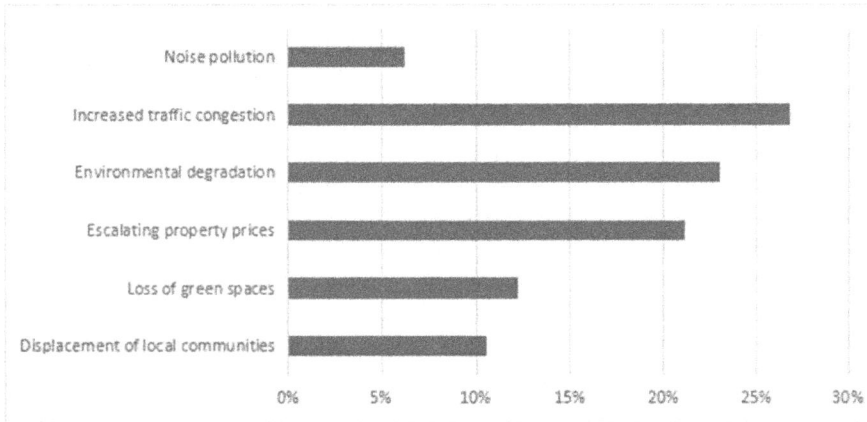

Figure 19.6 Adverse effects of urban infrastructure

Source: Authors

awareness points to the necessity of environmentally sustainable solutions to urban infrastructure. Furthermore, participants reported rising real estate prices as a concern. While increasing property values is considered an improvement in the real estate market, from a resident's perspective this could lead to affordability of housing challenges.

In summary, the results of this part of the study highlight the diverse concerns that people have about their home selection and the implications of infrastructure. Given the focus on problems like traffic jams, the environment, and real estate costs, comprehensive and sustainable urban planning strategies are required to solve these challenges and promote local growth.

4.2 Infrastructure Prioritisation Results

The last part of the study explores the essential urban infrastructure amenities people consider while deciding property purchase. A list of eight criteria was provided to the respondents, who were then

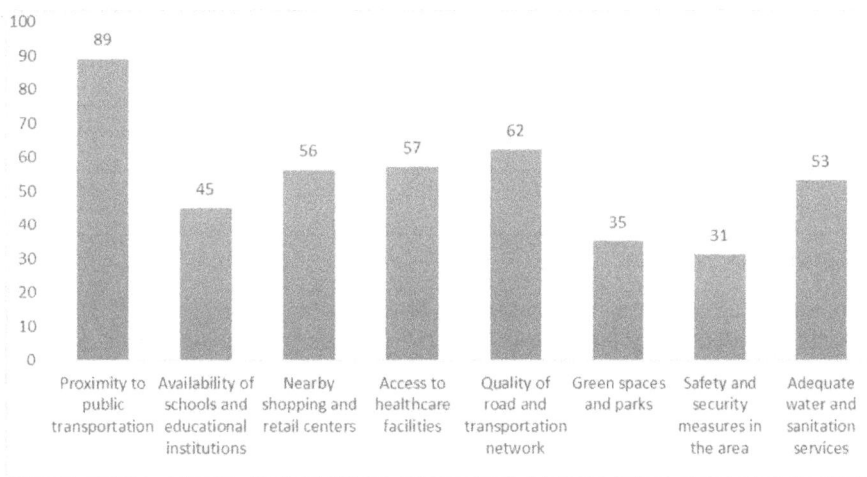

Figure 19.7 Urban infrastructure facilities governing the decision to buy a property

Source: Authors

asked to choose their top three choices. According to survey results, the counts for each factor are shown in Figure 19.7. This sheds light on the elements of urban infrastructure that have a major impact on the perceived value and desirability of a home, giving a useful overview of the important factors people take into account when making property-buying decisions.

It can be observed that proximity to public transportation is the most preferred factor when individuals are making decisions about purchasing a property. This highlights the pivotal role that easy access to public transit plays in influencing housing choices. The convenience and efficiency offered by proximity to public transportation not only improves daily commutes but also contributes to the overall accessibility and connectivity of a property, making it a top priority for potential buyers. Other significant factors are the quality of roads and transportation networks, access to healthcare, and nearby shopping and retail options. These preferences collectively suggest that individuals value surrounding urban infrastructure and amenities along with the housing unit. Quality road networks and efficient transportation contribute to the overall convenience of living in a particular area, while access to healthcare and nearby shopping centres further adds to the liveability and desirability of a property. In essence, the survey findings emphasize the paramount importance of transportation-related factors in property-buying decisions.

Further in giving importance to infrastructure in property decisions and demand, the essential elements found to be influencing real estate in the literature including airports and logistic parks, public transport, commercial development, and facilities for healthcare and education were considered. Respondents to the survey were asked to rank the impact of these variables on the demand for real estate on a 5-point Likert rating scale, where 5 represents the greatest impact and 1 is the least impact. RII analysis was carried out to rank these factors. Table 19.2 indicates the results of RII analysis.

Table 19.2   Result of RII analysis

Factors considered	Importance					Total	A*N	RII	Rank
	Least Important	Slightly Important	Neutral	Fairly Important	Very Important				
	1	2	3	4	5				
Commercial development	0	0	18	39	75	585	660	0.886364	2
Public transport	0	4	11	35	82	591	660	0.895455	1
Education and healthcare	0	0	20	47	65	573	660	0.868182	3
Airports and logistic parks	5	14	20	43	50	515	660	0.780303	4

Source: Compiled by authors

A strong consensus regarding the perceived impact of various factors on real estate demand has been revealed by the Relative Importance Index (RII). The previously observed importance respondents assigned to public transport are found to have the most positive influence on real estate demand. The participants' collective recognition of the importance of accessible and efficient transport systems in influencing property values is reflected in this acknowledgement, which emphasises their importance. Commercial development comes in second place. Thriving commercial areas can stimulate demand for real estate and that the attractiveness of a location is dependent on business activities and economic vigour. It also supports fulfilment of daily needs. Healthcare and Education Facilities have a significant impact, however, they rank marginally lower than Public Transportation and Commercial Development. The recognition of these amenities as significant variables highlights the comprehensive approach to considerations that home buyers of real estate contemplate. The focus on healthcare and education infrastructure highlights how important a well-rounded living environment is to potential real estate growth and the value people associate with the future of children and health. Airport and Logistic Parks received lowest rank, indicating that respondents thought they had less of an impact on demand for real estate. Airports and Logistic Parks are not directly relevant to the quality of life at the micro-level and respondents didn't see them as essential as the other parameters even though they might be playing essential roles at the macro-level.

The understanding gained from this study is crucial for stakeholders in the real estate industry, guiding them in making informed decisions based on the priorities and preferences revealed by the surveyed participants. The insights into the dynamics of infrastructure and real estate relationships at the micro and macro level are relent in infrastructure policy formulation and real estate market assessment.

5. CONCLUSION

The main finding of this study is the intricate relationship of infrastructure with real estate. Infrastructure creation boosts the economy, creates jobs, increases income and thus increases real estate demand. Better access to infrastructure services improves liveability of neighbourhoods. But infrastructure projects hampers certain aspects like pollution, congestion and increases cost of living, congestion etc. The size of cities affects real estate ownership, type, demand and rental value. Ownerships are higher in non-metro cities. Rent increases are higher in metropolitan cities. The larger the size of the city, more is the trend of multifamily/multi-story living. Residents from cities of different sizes value different factors while choosing homes. Comparatively more citizens value residing in proximity to work in metro cities. However, the preference for better standard of living over proximity to workplace is dominant across cities. Proximity to public transport is the top priority when deciding home locations followed by commercial development, education and healthcare. Increased traffic congestion, environmental degradation and property price escalation are the major concerns related to infrastructure projects.

REFERENCES

[1] URBANET. (2018, July 31). *Urbanisation in India – Infographics.* Retrieved from www.urbanet.info: https://www.urbanet.info/urbanisation-in-india-infographics/

[2] Ajibola, M., Awodiran, O., & Salu-Kosoko, O. (2013). Effects of infrastructure on property values in unity estate, Lagos, Nigeria. *International Journal of Economy, Management and Social Sciences,* 2(5), 195–201.

[3] C.Onyejiaka, J., & Okpala, C. A. (2020). The contributions of urban infrastructure to residential real estate investment value in Awka, Anambra state. *British Journal of Environmental Sciences,* 8(3), 13–32.

[4] Cervero, R. (2020). The transit metropolis: A 21st century perspective. In E. Deakin, *Transportation, Land Use, and Environmental Planning* (pp. 131–149). ELSEVIER. doi: https://doi.org/10.1016/B978-0-12-815167-9.00007-4

[5] Chen, H., Zhang, Y., Zhang, N., Zhou, M., & Ding, H. (2022). Analysis on the spatial effect of infrastructure development on the real estate price in the Yangtze River Delta. *Sustainability,* 14(13), 7569. doi: https://doi.org/10.3390/su14137569

[6] Efthymiou, D., & Antoniou, C. (2015). Measuring the effects of transportation infrastructure location on real estate prices and rents: investigating the current impact of a planned metro line. *EURO Journal on Transportation and Logistics,* 3(3), 179–204. doi: https://doi.org/10.1007/s13676-013-0030-4

[7] Faster Capital. (2024). *Urban infrastructure development and its effects on economic productivity.* https://fastercapital.com/content/Urbanization-and-CGE-Models--Understanding-Urban-Economic-Dynamics.html#Urban-Infrastructure-Development-and-its-Effects-on-Economic-Productivity.html

[8] Gan, L., Ren, H., Xiang, W., Wu, K., & Cai, W. (2021). Nonlinear influence of public services on urban housing prices: A case study of China. *Land,* 10(10). doi: https://doi.org/10.3390/land10101007

[9] Gupta, S. (2024). *Budget 2024: Need for Infra Push 2.0 to make India's roads & railways one of world's best.* https://economictimes.indiatimes.com/news/economy/infrastructure/budget-2024-need-for-infra-push-2-0-to-make-indias-roads-railways-one-of-worlds-best/articleshow/107282175.cms?from=mdr

[10] Hong, T., & Hong-Ping, W. (2014). Impacts of the urban rail transit on the real estate values. *Information Technology Journal,* 13(5), 960.

[11] Jin, T., Cheng, L., Liu, Z., Cao, J., Huang, H., & Witlox, F. (2022). Nonlinear public transit accessibility effects on housing prices: Heterogeneity across price segments. *Transport Policy,* 117, 48–59. doi: https://doi.org/10.1016/j.tranpol.2022.01.004

[12] Kelly, S. (2015). Estimating economic loss from cascading infrastructure failure: a perspective on modelling interdependency. *Infrastructure Complexity,* 2, 7. https://doi.org/10.1186/s40551-015-0010-y

[13] Knight Frank. (2020). *India Urban Infrastructure Report 2020.* Mumbai: knightfrank.co.in/research. chrome-extension://kdpelmjpfafjppnhbloffcjpeomlnpah/https://content.knightfrank.com/research/1901/documents/en/india-urban-infrastructure-report-2020-indian-real-estate-residential-office-6914.pdf

[14] Kouamé, A. T. (2024). *Gearing up for India's Rapid Urban Transformation.* World Bank Group: https://www.worldbank.org/en/news/opinion/2024/01/30/gearing-up-for-india-s-rapid-urban-transformation#:~:text=Since%20nearly%2070%20percent%20of, percent%20of%20GDP%20per%20annum.

[15] Lipscomb, C. (2003). Small cities matter, too: the impacts of an airport and local infrastructure on housing prices in a small urban city. *Review of Urban & Regional Development Studies,* 15(3), 255–273. doi: 10.1111/j.1467-940X.2003.00076.x

[16] Mei, Y., Zhao, X., Lin, L., & Gao, L. (2018). Capitalization of urban green vegetation in a housing market with poor environmental quality: evidence from Beijing. *Journal of Urban Planning and Development,* 144(3), 05018011. doi: https://doi.org/10.1061/(ASCE)UP.1943-5444.0000458

[17] Mesthrige, J. W., & Maqsood, T. (2022). Transport infrastructure, accessibility and residential property values: evidence from Hong Kong. *Built Environment Project and Asset Management,* 12(2), 163–179. doi: https://doi.org/10.1108/BEPAM-01-2021-0019

[18] Neelawala, P., Briggs, M., Robinson, T., & Wilson, C. (2015). The impact of project announcements on property values: an empirical analysis. *Australasian Journal of Environmental Management,* 22(3), 340–354. doi: http://dx.doi.org/10.1080/14486563.2015.1028110

[19] Panchapagesan, V., Venkataraman, M., & Ghosh, C. (2021). On the Impact of Infrastructure Improvement On real estate property values: Evidence from a quasi-natural experiment in an emerging market. *IIM Bangalore Research Paper,* 646. doi: https://dx.doi.org/10.2139/ssrn.3890048

[20] Seo, K., Salon, D., Kuby, M., & Golub, A. (2019). Hedonic modeling of commercial property values: distance decay from the links and nodes of rail and highway infrastructure. *Transportation,* 46, 859–882. doi: https://doi.org/10.1007/s11116-018-9861-z

[21] Sisodiya, S. (2023). *The interplay of infrastructure development and real estate in India: A symbiotic relationship.* https://timesofindia.indiatimes.com/blogs/voices/the-interplay-of-infrastructure-development-and-real-estate-in-india-a-symbiotic-relationship/

[22] Tirumala, R. D. (2022). Chapter 9: The confluence of real estate and infrastructure: a research agenda. In P. T. Miao, *A Research Agenda for Real Estate* (pp. 165–180). elgaronline. https://www.elgaronline.com/edcollchap/edcoll/9781839103926/9781839103926.00017.xml

[23] Torre, A., Pham, V., & Simon, A. (2015). The ex-ante impact of conflict over infrastructure settings on residential property values: The case of Paris's suburban zones. *Urban Studies*, 2404–2424. doi: https://doi.org/10.1177/0042098014546499

[24] Wang, R., Ye, L., & Chen, L. (2019). The impact of high-speed rail on housing prices: Evidence from China's prefecture-level cities. *Sustainability, 11*(13), 3681. doi: https://doi.org/10.3390/su11133681

[25] Xiao, Y., Wang, D., & Fang, J. (2019). Exploring the disparities in park access through mobile phone data: Evidence from Shanghai, China. *Landscape and Urban Planning*, 181, 80–91. doi: https://doi.org/10.1016/j.landurbplan.2018.09.013

[26] Yang, L., Chau, K. W., Szeto, W. Y., Cui, X., & Wang, X. (2020). Accessibility to transit, by transit, and property prices: Spatially varying relationships. *Transportation Research Part D: Transport and Environment*, 85, 102387. doi: https://doi.org/10.1016/j.trd.2020.102387

[27] Zhang, B., Li, W., Lownes, N., & Zhang, C. (2021). Estimating the impacts of proximity to public transportation on residential property values: An empirical analysis for Hartford and Stamford areas, Connecticut. *ISPRS International Journal of Geo-Information*, 10(2), 44. doi: https://doi.org/10.3390/ijgi10020044

[28] Zhu, W. (2024). The effects of infrastructure projects on house prices and rents: evidence from the HS2 extension cancellation in the UK. *Journal of Sustainable Business and Economics, 7*(2), 76–94. doi: http://dx.doi.org/10.26549/jsbe.v7i2.19155

[29] Zhu, Z., Zhu, Y., Liu, R., Zhang, L., & Yuan, J. (2022). Examining the effect of urban rail transit on property prices from the perspective of sustainable development: evidence from Xuzhou, China. *Buildings*, 12, 1760. doi: https://doi.org/10.3390/buildings12101760

Advances in Construction, Real Estate, Infrastructure and Project Management – Anil Kashyap et al. (eds)
© 2026 Taylor & Francis Group, London, ISBN 978-1-041-13433-6

20 | Exploring Integration of Building Information Modelling (BIM) and Last Planner System (LPS)—A Systematic Literature Review (SLR)

Vetrivel N.[1],
Pavithraa G.[2], Charen K.[3],
Manoj Varshan[4], and Deepak M. D.[5]
NICMAR University, Pune

■ **ABSTRACT:** In the current era of industrial and enterprise evolution, construction projects are getting more complex and demanding as they suffer from lack of technological advancements and workflow optimization. To overcome this situation, there has been a growing interest in the adaptation of the latest technological approaches and providing better and more reliable results. One such approach is the integration of BIM (Building Information Modelling) and LPS (Last Planner System) among construction projects. The research method adopted in the study was based on PRISMA approach. Based on the analysis, 44 papers were scrutinized for further study. These papers were further evaluated using the Theory Context Characteristics Methodology (TCCM) approach to systematically determine and understand the future research directions. Through the SLR, the study helps to manifest the existing practices, barriers, and potentials in the Integrating LPS and BIM in the construction industry. The integration of BIM and LPS presents a promising opportunity to revolutionize construction project management by combining the visualization and data capabilities of BIM with the collaborative and iterative planning approach of LPS.

■ **KEYWORDS:** Last planner system, Building information modelling, Systematic literature review, PRISMA, TCCM

1. INTRODUCTION

1.1 Background

Modern construction projects suffer from a lack of technology breakthroughs and workflow efficiency, making them more complex and demanding. While productivity in other industries has increased since the 1960s, worker productivity in the construction sector has decreased. Seventy

[1]vetrivel171999@gmail.com, [2]pavithraaguna98@gmail.com, [3]charenkumaran@gmail.com, [4]manojvarshan27@gmail.com, [5]deepakmd.md@gmail.com

DOI: 10.1201/9781003669814-20

percent of projects are over budget and delayed right now. The integration of Building Information Modelling (BIM) and the Last Planner System (LPS) has been the focus of recent research as a solution to these problems. It has been demonstrated that Lean and BIM approaches work well together. Research indicates that BIM can function as a tool for reducing waste, strengthening lean building methods, and enhancing project results. Construction project management could undergo a revolution with the combination of LPS's collaborative and iterative planning style with BIM's data visualization and capabilities. This collaboration can improve quality, safety, cost control, and schedule adherence by streamlining several project phases.

It is possible to overcome the constraints of 2D CAD technology and handle intrinsic construction process concerns by using lean and BIM simultaneously, as demonstrated by recent case studies. All things considered, combining BIM with LPS has the potential to significantly improve construction project performance.

1.2 Aim and Objectives

The aim is to perform a systematic literature review to identify major benefits, challenges, and future research directions.

Objectives:

1. Identify the relevant research papers, and journals on integrating the Last Planner System and BIM using PRISMA.
2. Analyze the theory, contextual background, characteristics, and methodologies used in the studies.
3. Identify the successful strategies and challenges in implementing them and propose future research areas on advancing the integration of LPS and BIM in the construction industry.

1.3 Scope

The study will encompass articles and conference papers sourced from online databases such as Scopus, Web of Science, and ResearchGate, all written in English. The literature review will concentrate on recent publications from the past 15 years to understand the development and progress in the integration of Building Information Modelling (BIM) and Late Planner Systems (LPS). The primary focus will be on the construction industry in India, with additional coverage of select papers from other countries to evaluate the efficiency of this integration and identify existing barriers. While the study will not propose a framework for integration, it will suggest areas for further research.

2. Literature Review

2.1 Lean Construction

The lean construction philosophy maximizes production flexibility and transparency by managing projects through the reduction of cycle times and essential processes. The Last Planner System (LPS), one of Lean's useful concepts, aids in achieving these objectives. LPS encourages flexible and proactive management throughout the building process, in contrast to conventional project management approaches like the critical path method, which detects deviation after the fact.

Through cooperative planning engaging all stakeholders, LPS promotes realistic workflows that provide flexibility in response to changes and unanticipated events. This system uses both short-

and long-term planning to guarantee consistent processes and dependable outcomes. Nevertheless, inadequate human resources for efficient application, reluctance to change, and a insufficient training are obstacles to the acceptance of LPS in the construction sector.

2.2 Building Information Modelling (BIM)

In the construction industry, BIM (object parametric modelling) is used to generate digital models of structures. It has nine dimensions: planning (4D), costing (5D), sustainability (6D), facility management (7D), safety (8D), and lean construction (9D). Lean Project Delivery System (LPDS) stages are supported by BIM, which improves supply chain management and lowers human error in quantity take-offs.

Using an object-based file format and data architecture, Industry Foundation Classes (IFC), created by building SMART, offer an informative framework to address data management in the Architecture, Engineering, and Construction (AEC) sector.

There are four stages of BIM maturity:

Low Cooperation: 2D CAD drawings are used. Uses a Common Data Environment (CDE) to manage project data in a somewhat collaborative manner.

Complete Collaboration: Uses 3D modelling and incorporates dimensions for managing costs (5D) and time (4D).

Complete Integration: Incorporates dimensions such as 6D, providing a common model in a cloud-based setting that is available to all project members.

The five main Level of Detail (LoD) levels are 100, 200, 300, 400, and 500. The LoD describes the data required in the 3D model at different project stages. In new construction, the 3D model is created from the ground up using a point-to-BIM procedure; in older buildings lacking documentation, different techniques are employed. Maintaining model ownership, encouraging teamwork, and bringing academic knowledge up to date with industrial improvements are among the organizational and legal concerns.

2.3 Last Planner System

A proactive method for managing and completing projects with the least amount of waste possible while maintaining sustainability is lean construction. It centres on Procedure and Flow,

Organizing as opposed to responding, establishing frameworks to accomplish objectives, overseeing the supply chain for the project, and ongoing development.

The LPS, a collaborative planning method involving top management, is a crucial element in lean construction. Project planning is divided into five stages by LPS:

Master planning: creates a project vision by defining goals, schedules, deliverables, and scope.

Phase planning: divides the overall plan into manageable chunks and uses pull planning to pinpoint the essential tasks for every benchmark.

Look-ahead planning: Proactively schedules tasks over the next three to six weeks, taking labour and material limits into account.

Weekly Planning: Creates thorough action plans for the future week, which are then evaluated several times to eliminate roadblocks during lookahead planning.

Daily Huddles: Hold daily review sessions to set goals, track advancement, and quickly resolve problems.

Percent Plan Complete (PPC), which monitors the accomplishment of scheduled work and pinpoints planning errors or the underlying reasons for unfinished tasks, serves as a proxy for LPS dependability and efficacy. When it comes to meeting project deadlines, a greater PPC is preferred.

3. RESEARCH METHOD

3.1 SLR using PRISMA

The detailed methodology for the study uses SLR (Systematic Literature Review) as a research tool to select the right research papers that will help frame the objectives for the research. To acquire a better understanding of the effective integration of BIM and LPS, research papers from Scopus, Research Gate, and IGLC papers were referred to. PRISMA (Preferred Reporting Items for Systematic Reviews and Meta-Analyses), consisting of three-level advanced steps was followed to arrive at an unbiased, relevant set of research papers excluding the outliers.

STEP 1: Identification: General keywords such as LEAN and BIM or Building Information Modelling or Building Information Modelling and Construction were used. This resulted in a total of 508 documents. Though the result consisted of the majority of documents on the Engineering subject area, there were a lot of outliers too such as Environmental science, social science, etc.

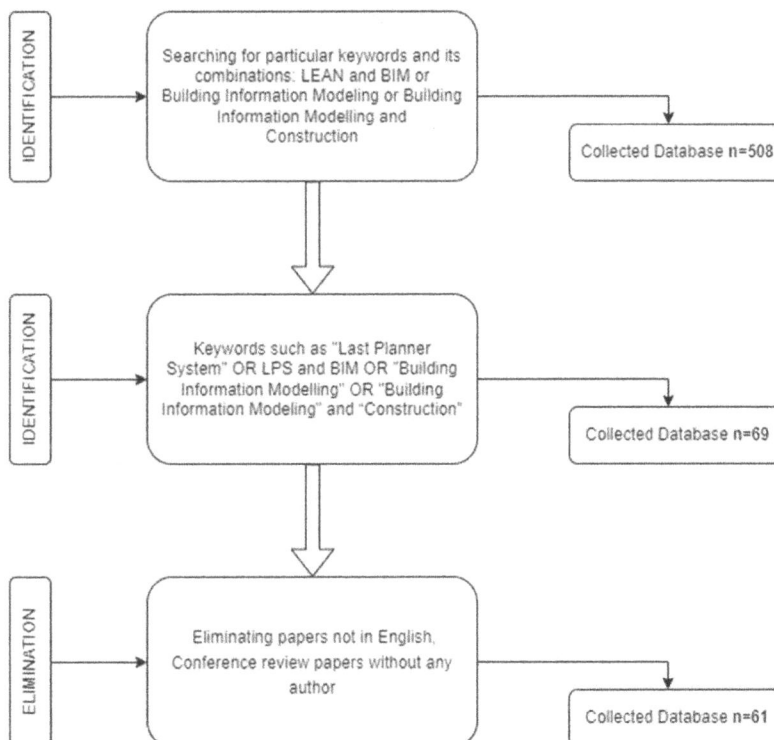

Figure 20.1   PRISMA process flowchart

Source: Author's compilation

STEP 2: Screening: To further narrow it down, more keywords like Last Planner System and Construction were added. The Scopus research search is shown in Figure 20.2. After doing this, 69 papers remained among which, 49.3% – Conference papers, 34.1% – Articles, 10.1% – Conference reviews, 5.8% – Reviews.

STEP 3: Eliminating: These results were narrowed down by eliminating a few irrelevant papers (i.e., papers not in English and conference review papers without any author). 8 such documents were eliminated and are left with 61 papers in total.

4. RESULTS AND DISCUSSION

Based on the detailed method as stated in the earlier section, the results are discussed as follows:

The first section of the result discusses the bibliometric profile, which includes publication, authorship, and citation. The research environment, including its theories, context, characteristics, and methodology (TCCM components), is then described.

4.1 Publication Trends Over Time

Analysis of the data shows the distribution of published documents over a selected timeline. The graph shows a clear development on areas of BIM-LPS over two non-overlapping time period. The first-one being 2016 and the second year being 2020 Figure 20.2.

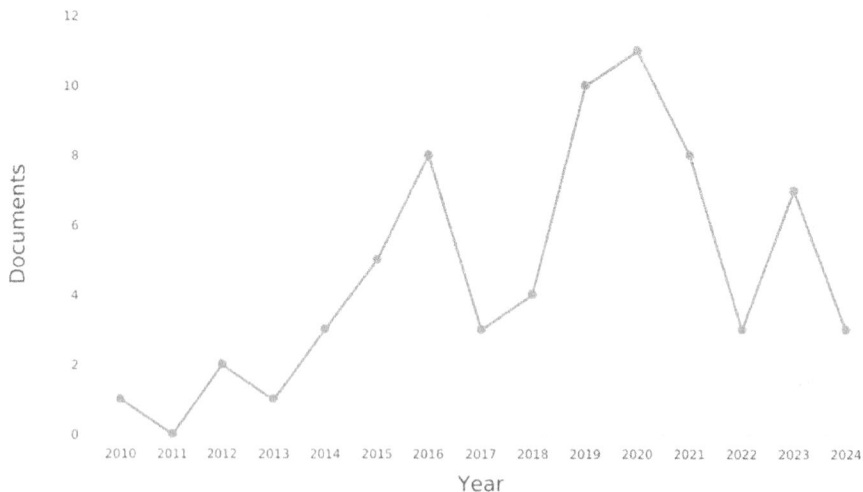

Figure 20.2   Publication over the years

Source: Scopus

4.2 Publication by Country

Developed countries such as Brazil, Finland, and United States have higher research publications on this topic when compared to developing countries such as India, China, France, and Canada Figure 20.3. This may be due to many reasons but one of the major reasons is resistance to change among various stakeholders.

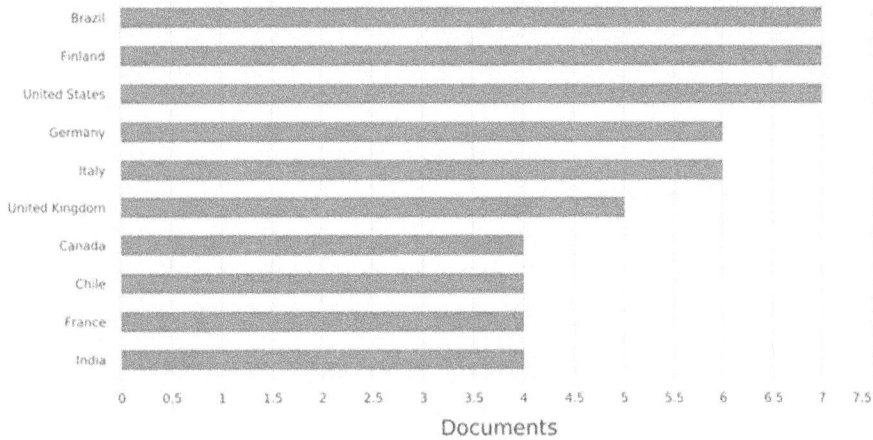

Figure 20.3   Publication by country

Source: Scopus

4.3 Keyword Analysis

Keywords serve as a framework for the research topic that eventually leads to the gap and objectives. To determine the co-occurrence of keywords, author citations, and document citations, a bibliographic analysis was conducted using VOS-Viewer. It was used to understand the interconnection between the keywords and overlay visualization as shown in Figure 20.4. This image helps in understanding the development in these areas over the time period.

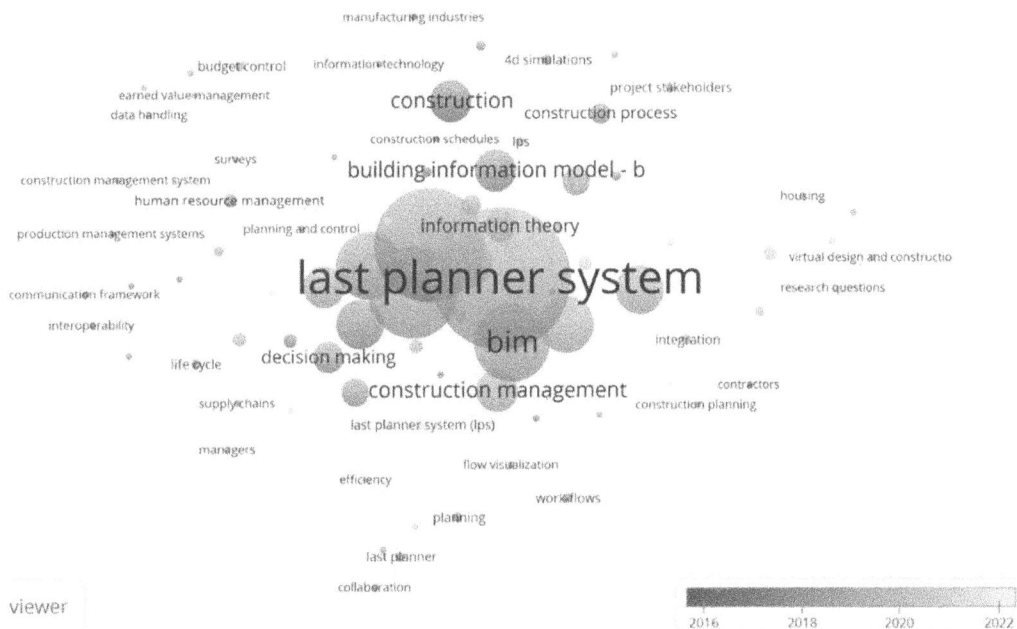

Figure 20.4   Mapping of keyword overlay visualization

Source: Vos Viewer

The size of the circle also directly correlates with the frequency of occurrence of the keyword. This clearly indicates a strong correlation between areas like "Lean construction", "BIM", "Last planner system", "Integrated Project Delivery", "4D BIM" and many others as shown in Figure 20.6. So, it can be understood that these are the current topic of interest and has better corelation with each other. It is also understood that the latest research trend is more towards "Construction 4.0, digital engineering, organization behaviour, agile management, pre-construction, construction life cycle" with integrating BIM and Last Planner System.

5. TCCM FRAMEWORK FOR INTEGRATING LPS AND BIM

The analysis was done using the TCCM framework, where T is Theory, C is Context, C is Characteristics and M stands for Methodology. This is used to systematically identify the progress of the study topic and to better understand future research directions.

5.1 Theory

The overall papers can be categorized into the following three theories. i) Integration theory ii) Complementary theory iii) Adaption theory.

Integration Theory

In this theoretical approach, two concepts or methodologies are normally integrated. In this case, it's predominantly BIM and Last Planner System. These theories are about how combining concepts can have a greater effect when compared to one concept. This theory includes concepts such as: SYNERGY: It will emphasize the strength of both concepts to combine into a much more efficient model or framework. EFFICIENCY: Building Information Modelling is focused on Digital representation and management of project data, when integrated with LPS which focuses on collaborative planning and scheduling, it helps in optimizing the project performance and outcomes.

Complementary Theory

According to this theory, one concept could help the other concept to overcome its flaws or limitations. It can be done even by just the use of lean tools to help to improve limitations in a BIM technique. These theories are done with the help of identifying and utilizing both complementary tools and techniques.

Adaption Theory

This theory focuses on the process of adopting and diffusing innovative concepts or technologies. It examines factors influencing the adoption process, including organizational culture, technological readiness, stakeholder collaboration, and perceived benefits. This theory helps us understand the challenges and facilitators of integrating new concepts into construction practices.

5.2 Context

The papers were analyzed to understand the most frequently studied countries and regions. It can be understood from the figure that the publications of papers or the research on integrating BIM and LPS is high in developed countries as compared to developing countries. It can be regarded so because of the early adaptation of these tools and techniques in those countries. Also, in some countries, it is mandated by the Government to implement BIM methodology in their projects.

There are various reasons why the research papers are minimum in developing countries and one major reason is that, when people are not ready to accept or adapt to new change researchers are not into developing and expanding the scope of that area. Now, as the adaptation of BIM is increasing in India, various research is undertaken to get the maximum output upon combining those with various principles such as Lean (LPS).

5.3 Characteristics

By examining the relationships found in the studies, this section looks at how integration affects growth. It emphasizes the primary ways that integration affects growth and the frameworks that are employed to account for these consequences. It has been determined that BIM directly contributed to the attainment of the lean objectives by supporting proactive management, control, communication, transparency, visualization, and rapid planning, among other aspects. However, concurrently, lean procedures facilitate the adoption and implementation of BIM, which also prevent the BIM paradigm from being twisted and encourage the model's continuation and application after the design phase [1].

Similarly, to fully investigate the use and quantitative analysis of BIM-LC technology and techniques in ACE projects in China, a case study of a project in Dezhou, China was conducted [2]. In this project, the Lean Construction Management teams were functioning similarly to an EPC. To manage all aspects of the lifecycle, including design, procurement, transportation, and construction, the owner engaged an engineering procurement construction contractor. The key results emphasize the potential benefits this provides to project participants, as proven by its ability to reduce implementation issues, improve construction efficiency, reduce carbon emissions, and minimize waste during construction. EPC contractors can therefore benefit from using the BIM-LC approach [2].

5.4 Methodology

The literature reviewed included both qualitative and quantitative assessments of the data collected. The chosen studies used a variety of methodological techniques, where most of the research used qualitative techniques for data collecting regarding efficiency and barriers in implementing new technologies/ methodologies. It can be concluded from the research findings that the most often used methodological approach was case study analysis, which was followed by interviews, Literature reviews, etc. Some of the quantitative methods were performance metrics and cost-benefit analysis, and documental analysis.

6. EXISTING BIM AND LEAN FRAMEWORKS

At this phase, a detailed analysis was done on the TCCM report, and further papers were filtered out that were more relevant to INTEGRATED BIM AND LPS SYSTEM. The papers were studied in detail to gain an understanding of the most recent developments in lean construction and BIM.

Research has explored the potential of BIM to represent the procedural aspects of construction projects through the lens of Lean principles, termed KanBIM [3]. The study also highlights the integration of BIM with relevant technologies to implement a "PULL FLOW" approach, derived from Lean methodology. This integration aims to minimize variability in construction processes. The central goal of this framework is to design and assess a BIM-driven system that enhances production planning and facilitates daily production management at construction sites [5].

VisiLean represents another framework designed to integrate the Last Planner workflow with BIM. Its primary purpose is to enhance collaboration among production teams and site personnel, particularly the Last Planners, while supporting lean production management on construction sites. This framework emphasizes two key components of the production system: (i) representation of the production management process and (ii) visualization and representation of the product. Additionally, it highlights the need to address specific requirements in future research, including (i) improving communication among site operatives and (ii) ensuring accurate information delivery to enable more informed decision-making [4].

7. CONCLUSION

The lag in construction productivity as compared to other industries is a backlog for the construction sector in the world. This minimal productivity is because of the labour-intensive nature of this sector. So, to enhance the productivity of the construction field all around the world there are various tools and techniques are being adopted. The adoption of these tools and techniques is far slower in developing countries as compared to the developed nations which adapted to these years ahead of us. Building information modelling is one such tool whose application to the construction industry is yet to reach its height. Lean principles once developed for the manufacturing industry have now had a major impact on the construction industry.

From the analysis done through the VOS viewer, it is identified that the adoption of these principles and techniques has gained value only over the past decade. The use of BIM in the construction sector improves the visualization of work beforehand and helps to reduce delays, whereas the Lean concept aids in the optimization of the construction workflow. Integrating both concepts to optimize construction productivity and improve efficiency in the emerging trend. According to the literature, using Lean without BIM has some benefits, and vice versa. BIM and Lean are more effective when applied jointly, rather than individually. Lean and BIM improve workflow by eliminating out-of-sequence work, reducing interference among trade crews, and boosting job continuity. There are numerous barriers to applying BIM and LPS in the construction business. Inadequate IT implementation for the Last Planner System (LPS) can result in insufficiently developed BIM integration skills, such as data management, construction sequencing, quantity take-offs, coordination, and supply chain management. This highlights a significant gap in the availability of a robust conceptual framework for effectively integrating BIM and LPS in construction processes.

Also, there are various frameworks that are being modelled or formulated to integrate BIM and Last planner system efficiently based on various case studies discussed and in various contexts all over the world, but the limitations are that the applicability of those frameworks is only within the context of that research. Pertaining only to that limited scope and minimum sample size, these tools and frameworks could not be a generalized solution that can be applicable to different projects. So, a possible future research approach could be looking into coming up with a framework or a tool that could be applicable for more than just a particular project i.e., a more generalized framework or a tool where BIM and Last Planner System are efficiently integrated.

REFERENCES

[1] Andújar-Montoya, M. D., Galiano-Garrigós, A., Echarri-Iribarren, V., & Rizo-Maestre, C. (2020). BIM-LEAN as a methodology to save execution costs in building construction-An experience under the spanish framework. *Applied Sciences (Switzerland)*, 10(6), 1913. https://doi.org/10.3390/app10061913

[2] Hei, S., Zhang, H., Luo, S., Zhang, R., Zhou, C., Cong, M., & Ye, H. (2024). Implementing BIM and lean construction methods for the improved performance of a construction project at the disassembly and reuse stage: a case study in Dezhou, China. *Sustainability (Switzerland)*, 16(2), 656. https://doi.org/10.3390/su16020656

[3] Sacks, R., & Barak, R. (2008). Impact of three-dimensional parametric modeling of buildings on productivity in structural engineering practice. *Automation in Construction*, 17(4), 439–449. https://doi.org/10.1016/J.AUTCON.2007.08.003

[4] Koskela, L. J. (2011). Visilean: Designing a production management system with lean and BIM. https://www.researchgate.net/publication/289858189

[5] Sacks, R., Barak, R., Belaciano, B., Gurevich, U., & Pikas, E. (2012). KanBIM workflow management system: prototype implementation and field testing. In *Lean Construction Journal*. www.leanconstructionjournal.orgwww.leanconstructionjournal.org

Advances in Construction, Real Estate, Infrastructure and Project Management – Anil Kashyap et al. (eds)
© 2026 Taylor & Francis Group, London, ISBN 978-1-041-13433-6

21

Assessing Disparities and Opportunities: A Comparative Review of Carbon Markets in Developing and Developed Nations

Rukhsar Sayed[1]
Research Scholar,
NICMAR University, Pune

Smita Patil[2]
Associate Professor,
NICMAR University, Pune

■ **ABSTRACT:** In the worldwide initiatives to mitigate the effects of climate change and minimize greenhouse gas emissions (GHGs) the world is looking for instruments that can transform the climate crisis. Many countries are looking at Carbon Markets as a fragment of a solution since they remain an integral part of international climate change policies. The carbon market operates contrarily for different economies around the world. In this paper, a comparative review has been highlighted to analyze the variables on which carbon markets depend, the trading methods employed, carbon pricing mechanism, advantages, and challenges faced by a developing and developed nation for Carbon Trading Practices (CTP). In addition, the study also discusses research goals, areas of concern, and knowledge gaps related to the application of carbon credits to improve awareness of their significance in mitigating climate change. The paper also studies the development of carbon markets and emissions trading to provide insight into current events. Policies governing carbon markets would be more likely to support the transition to a low-carbon economy if they were more prepared to take previous mistakes into account when creating new policies.

■ **KEYWORDS:** Carbon markets, Carbon credits, Emission trading system, Carbon trading practices, Developing nations, Developed nations

1. INTRODUCTION

Climate change is a global environmental problem that has been associated with the increasing concentration of greenhouse gases (GHGs) [10]. With the rise in climate change, there is a growing

[1]rukhsar.ph2@pune.nicmar.ac.in, [2]spatil@nicmar.ac.in

DOI: 10.1201/9781003669814-21

demand for energy consumption as the requirement for refrigeration, cooling and air conditioning increases hence the production of electricity enhances to meet requirement, leading to higher GHG emissions. To promote energy conservation and emission reduction, the development of carbon emission trading market has been adopted as one of the mitigation methods. In order to encourage and enforce global GHG emission reduction, the United Nations Framework Convention on Climate Change (UNFCCC) has been deploying a variety of Emission Trading System (ETS) techniques [6]. The annual value of the global carbon markets exceeds $175 billion, impacting companies across all economic sectors [3]. The carbon market generally refers to a framework that provides a trading system where carbon emission or carbon equivalent gases are bought and sold with the unit of transaction being carbon credits or carbon emission allowances. Carbon markets are created by public and private entities to prevent climate crisis by supporting the actions of GHG emission mitigation. The creation of carbon markets has encouraged various sectors and industries to reduce their emission by limiting, monitoring and modifying their operational approaches, also this provides a platform for innovators and researchers to develop environmentally friendly methods for industrial operations. Both carbon-based greenhouse gas emissions (such as CO_2 and methane) and non-carbon-based emissions (such as nitrous oxide) can be included in carbon markets [2]. The notion of carbon trading was channelled with numerous environmental implementations and policies. That led to the signing of the Kyoto Protocol in 1997 which was the first vehicle for emissions trading in GHGs. While very few transactions have occurred directly between nations as a result of the Kyoto Protocol, the European Union and many other authorities have since sought emissions trading as a means of reducing their greenhouse gas emissions. By a considerable margin, carbon markets now represent the largest class of environmental or emissions trading markets globally in terms of both volume and market value [7]. The expansion of carbon markets over the past two decades reflects a growing global commitment to addressing climate change through market-based mechanisms. Even while there are still obstacles to overcome, the ongoing development and integration of these markets bode well for more extensive and efficient cuts in greenhouse gas emissions.

Carbon credits are traded on the carbon market; a carbon credit is a tradeable security. The owner is allowed to discharge one ton of greenhouse gas equivalent (CO_2e) or carbon dioxide (CO_2) [9]. Auctions are used for the allocation of these carbon credits. The sole factors influencing the cost of carbon permits are the supply and demand for permits, which are determined by the cap and the marginal cost of reducing carbon [4]. International initiatives like the Paris Agreement and the Kyoto Protocol promoted the reduction of emissions by means of carbon taxes, carbon trading, and green development. [8]. A cap-and-trade system has been implemented under the Kyoto Protocol to limit the greenhouse gas emissions of industrialized nations [1]. Carbon credit markets fall into three groupings: mandatory vs voluntary, allocation vs offset, and international vs regional markets [5] which is illustrated in Figure 21.1.

Wang et al. studies that the carbon credit system can effectively quantify and reward reductions in carbon emissions from individual lifestyle choices [38]. The potential savings from reduced GHG emissions could translate to significant carbon credits, offering financial returns through carbon trading markets [37]. He and Song identifies from their study that the potential savings from reduced GHG emissions could translate to significant carbon credits, offering financial returns through carbon trading markets.

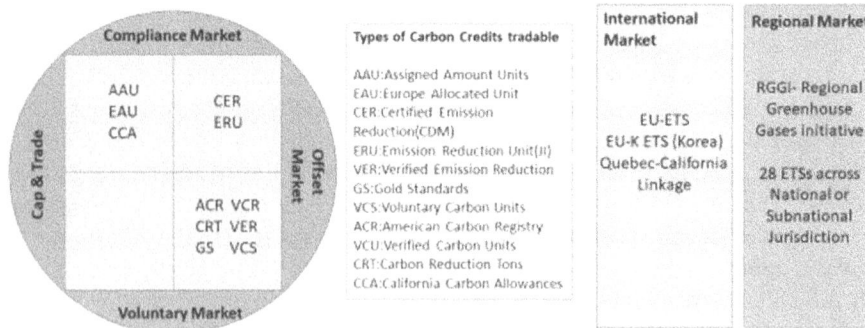

Figure 21.1 Different carbon markets

Source: Authors

The aim of comparing carbon markets in developing and developed nations is to assess their effectiveness, equity, and economic impacts. This comparison highlights best practices, identifies challenges unique to each context, and informs policy adjustments to enhance global cooperation in mitigating climate change.

2. METHODOLOGY

First, literature might be found using databases such as EBSCO, IEEE Explore, ProQuest, Web of Science, Google Scholar, and Scopus. You might also look at books and specialist magazines. When searching the literature for this study, the Scopus database was taken into account. The database found several relevant journals that were examined to obtain access to current studies carried out by researchers in the carbon market for the construction industry domain, as well as study-specific journals, using the analysis result section.

"Carbon Market" OR "Emission Trading System" OR "Cap and Trade" OR "Carbon Price" OR "Carbon Credit" were the search terms used to look through the SCOPUS databases. This keyword search string produced 935 results on May 17, 2024, and they were further filtered to find potentially relevant material. In addition to SCOPUS databases, a few other resources were preferred for understanding the concept of carbon markets and emission trading systems. In addition to reports from the World Bank, ICAP, NCM, Ministry of Power, Ministry of New and Renewable Energy, Ministry of Petroleum and Natural Gas, and "Shodhganga," an Indian digital repository for Ph.D. theses and dissertations, a few ISO standards for quantifying, measuring, reporting, and verifying the GHG emission and carbon accounting were also reviewed. The initial screening of literature using filters in SCOPUS is illustrated in Figure 21.2, out of which 93 literatures were studies for the review of carbon market in developed and developing countries.

The main focus area remains on the developing and developed nation's (i) policy frameworks, (ii) market mechanism and market participation, (iii) financial mechanism and instruments, (iv) economic impact, (v) disparities in outcome, and (vi) opportunities for improvement of carbon market. A comparative analysis is preformed of the above-mentioned highlights, followed by challenges and opportunities, policy recommendations, and conclusions.

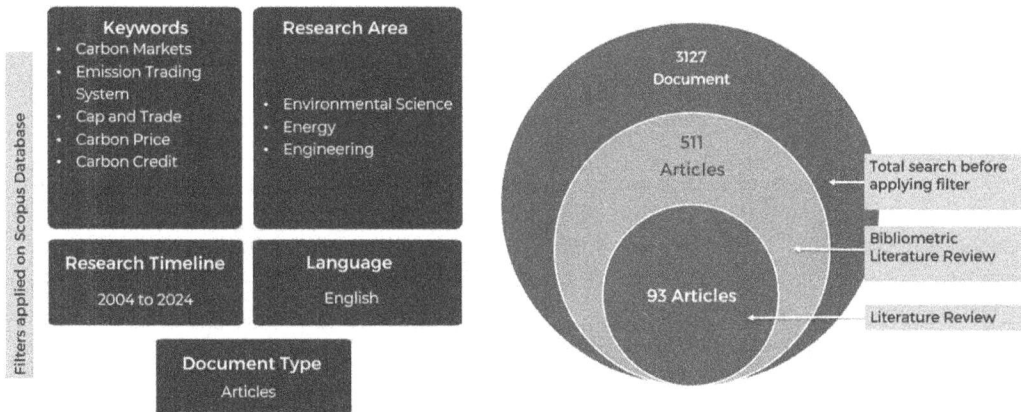

Figure 21.2   Literature screening and selection

Source: Authors

The research questions identified by the thorough study of the literature are as follows:

1. What are the key differences in carbon markets between developing and developed nations?
2. How do these disparities impact the effectiveness of carbon markets?
3. What opportunities exist for improving carbon markets in both contexts?

Secondly, while selecting sources for the topic the following criteria were prioritized:

1. Relevance: Make sure that the sources discuss carbon markets in detail, with an emphasis on developing and developed country comparisons, including cap-and-trade programs, carbon taxes, and offset methods.
2. Credibility: Pick reliable writers and organizations, such scholarly publications, governmental agencies, and global organizations (like the World Bank and UNFCCC). For intellectual rigor, peer-reviewed papers are preferred.
3. Recency: To represent recent changes in the carbon markets, use sources that offer the most recent information and analysis. However, consider earlier sources to put things in perspective.
4. Comprehensiveness: Choose resources that provide in-depth study on case studies, market dynamics, regulatory frameworks, and economic repercussions.
5. Geographic and Sectoral Coverage: To offer a comprehensive picture of the carbon markets in both emerging and developed countries, make sure that a variety of locations and industries are included. To understand the carbon markets mechanism in different geographical location various countries having different demographics are selected. The sectoral coverage is inclusion of different sectors into various carbon trading schemes by the stakeholder depending on the economy of the country or region.
6. Holistic Viewpoints: Incorporate a range of opinions, scathing analyses, triumphs, and setbacks, representing different stakeholders such as the public sector, private sector, and civil society.

3. COMPARATIVE ANALYSIS

3.1 Policy Frameworks

For developed nations, the cap-and-trade mechanism is usually adopted where the limit of emissions is gradually decreased over time. Allowances are distributed through a combination of free allocation and auctioning. For the EU ETS, the allocation of carbon credits is based on an auctioning system, whereas for China's national ETS, carbon credits are allocated freely using the benchmarking method [11, 12]. Robust MRV (Monitoring, Reporting, and Verification) systems ensure transparency and compliance, providing a fair chance of trade among participants [13]. Many differences exist within various developed nations depending on the scope and coverage of the carbon market. For instance, specific sectors are covered in the Regional Greenhouse Gas Initiative (RGGI), while broader sectors are included in California's Cap-and-Trade Program. California's linkage with Quebec facilitates market liquidity and flexibility [14]. Revenues collected from carbon market auctions are utilized differently across regions. For example, RGGI revenues focus on energy efficiency investments [15].

The policy framework for developing nations functions differently. Policies are gradually implemented in phases, with each phase refined based on observations from governmental bodies, regulatory agencies, and international organizations [12]. Initially, allocation of allowances is free, transitioning to more auctions over time as more sectors participate in the market. Robust MRV systems are implemented to maintain accuracy, transparency, and compliance [13]. Market stability depends on measures to prevent price volatility and ensure smooth market operations [12]. The policy framework followed in different ETS is demonstrated in Figure 21.3.

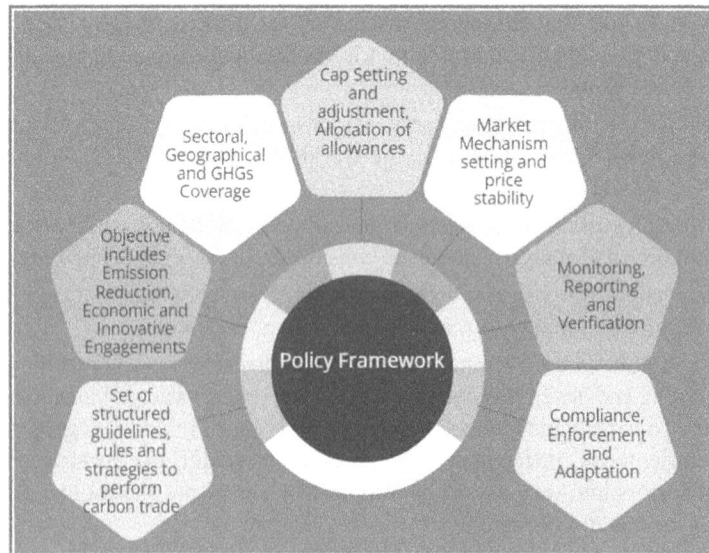

Figure 21.3 Policy framework of carbon market

Source: Authors

3.2 Market Mechanism and Market Participation

The market mechanisms of carbon markets vary between developing and developed nations. Industrialized countries commonly use complex cap-and-trade systems, such as the California Cap-and-Trade Program and the EU Emissions Trading System (EU ETS), which facilitate allowance trading and set emission limitations [11, 14]. These systems are often complemented by robust price stability measures and auctions to enhance market liquidity and efficiency [13]. In contrast, developing countries often adopt experimental or nascent cap-and-trade systems, like China's national ETS, which initially relies on the free distribution of allowances before progressively introducing auctioning mechanisms [12]. Offset programs, such as India's Perform, Achieve, and Trade (PAT) scheme, are also prevalent in developing markets, focusing on promoting energy efficiency in key industries [16, 17]. International financial and technical assistance plays a crucial role in enhancing the efficacy and scalability of both developed and developing carbon market systems [13]. The mechanisms of different emissions trading systems have been studied and are exemplified in Figure 21.4. The market participation comparative analysis in developed nations and developing nations involved in the carbon market is shown in Table 21.1.

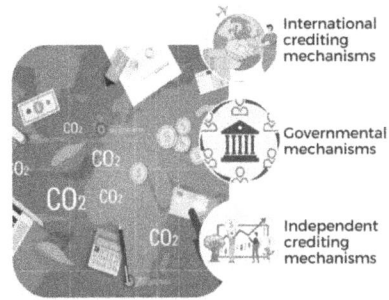

Figure 21.4 Different market mechanism for ETS

Source: Authors

Table 21.1 Market participants of carbon market

Stakeholders	Developed Nation	Developing Nation
Regulate Entities	Power plants, Industrial Manufacturers, Aviation	Power Sector, Industrial Sectors
International Organization	-	World Bank, UNFCC
Financial Institutions	Banks, Investment Firms, Carbon Exchange	Banks and Carbon Funds
Government Bodies	Regulatory Agencies	National Ministry and Local Authority
Environmental Groups	NGOs	Banks and Carbon Funds
Market Traders	Brokers and Traders	Businesses
Research and Academia	-	Universities and Institute

Source: Compiled by authors

3.3 Financial Mechanisms and Instruments

The various phases of the carbon market's evolution, as well as the unique requirements and environments of industrialized and developing countries, are reflected in monetary arrangements and techniques. While developed nations leverage sophisticated financial tools and market mechanisms, developing nations often rely on international support and phased approaches to build their carbon market [18]. These mechanisms provide market stability and allow participants to hedge against carbon price fluctuations. Revenues from these mechanisms are typically reinvested in climate initiatives, renewable energy, and energy efficiency projects to support sustainable development and

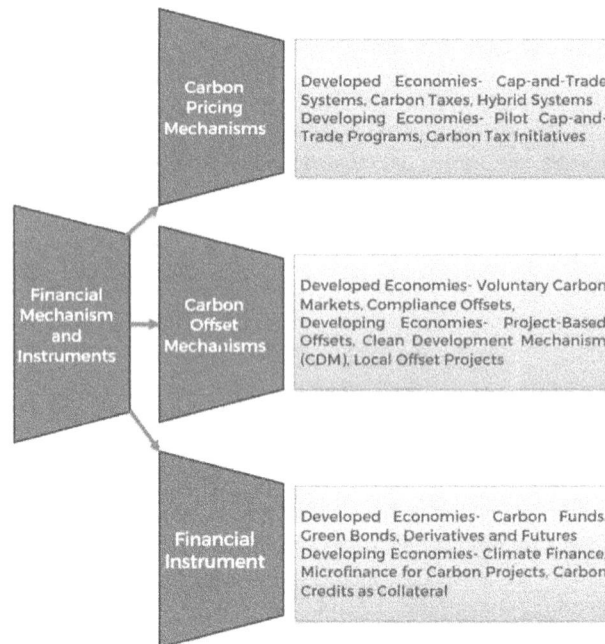

Figure 21.5 Financial mechanisms and instruments used for carbon market

Source: Authors

emission reduction goals. Figure 21.5 provides a summary of financial mechanisms and instruments for the carbon market used in developed and developing nations.

3.4 Economic Impact

The economic impact of the carbon market on developed nations is determined by setup costs, which is initially a high compliance cost for companies to enter the credit markets [7]. Revenue generation is significant from the auctioning allowances which is reinvested in climate projects and can be used to support vulnerable industries and communities to promote economic equity [43]. When a carbon market is developed high liquidity and availability of financial instruments are there, that facilitates the active trading of credits and allowances and it also manages the risk of high price volatility in the market. Carbon markets help in the incentivization of the companies that innovate and invest in low-carbon technologies, thereby providing a shift in the market that provides new green jobs and industries contributing to the economic growth. The competition is enhanced internationally and it can also affect international trade across the nations and lead to contribution to sustainable economic growth and public health benefits [44]. Unlike developed nations developing nations have the cost of compliance which is an initial and ongoing cost that is mitigated by subsidies and international support. The revenue generation is limited in the early stages but has potential growth, and used for sustainable development. Market liquidity is a challenge in the developing carbon market as the participation and liquidity are low, and require investment in infrastructure. Although, technological innovation provides incentives to companies for cleaner technology and energy-efficiency, however, a lot of technological innovation depends on international technology transfer from developed nations. Also, another aspect is carbon markets provide social and environmental benefits by improving public health and the environment [20]. Comparing the scale and functioning of carbon markets in developed and developing nations – the California carbon market and ETS in China show different dynamics. Market Value and Revenue: Carbon market in California which started in 2012 has an overall market size of $ 25 billion and has produced auction revenues of $20 billion [19]. All of these funds are flowed back into renewable energy, energy efficiency, and electrification initiatives. On the other hand, China's ETS, launched in early 2021, sold about 2.4 billion CNY ($1.2 billion) in its initial year of operation since it is a new system and depends mainly on free allowance instead of auctions [21]. Cost of Allowances: California presently has allowance prices between $30 to $50 per tonne due to high market demand and well-developed system of measures on the prices stabilization. Lower allowance prices ($8–$12 per ton) of China also suggest the growth of this system and attempts at causing least impact in industries [22, 23]. Savings and Costs: Cost reductions for both regulatory compliance savings are between $20–$40 per unit in the two markets with California enjoying the most efficient [24]. It is assumed that administrative costs amount to $3–$5 billions in California market and $1–$3 billions in [25]. The highlight shows how California is all about trying to get as much money as possible toward climate change efforts and how China is slowly transitioning towards more market development.

3.5 Disparities in Outcome

Developed nations typically achieve significant emissions reductions through mature carbon markets and robust regulatory frameworks. For instance, the EU ETS has achieved significant emissions reductions [44]. Between 2005 and 2020, emissions from sectors covered by the EU ETS fell by about 35%. In 2019, the power sector emissions in the EU decreased by 15% compared to the previous year [26]. Industrial emissions under the EU ETS also saw a decline, albeit at a slower pace, with a 2% reduction in 2019 [27]. The carbon market has not significantly hampered economic growth, demonstrating that emissions can be reduced while maintaining economic stability [28].

The effectiveness of carbon markets in developing nations varies widely. Some markets, the EU ETS's long-standing operation and mature regulatory framework have resulted in steady and substantial emissions reductions, particularly in the power sector [44]. Conversely, China's ETS, although the largest, is still developing and exhibits inconsistent results due to economic and infrastructural challenges [29]. Developed nations like those in the EU benefit from robust regulatory mechanisms and enforcement capabilities, ensuring compliance and effectiveness [36]. In contrast, developing nations often struggle with regulatory enforcement, data reliability, and economic constraints that hinder the full potential of their carbon markets [30]. The economic and infrastructural maturity of developed nations supports the integration and effectiveness of carbon market [33]. Developing nations, while making progress, often face significant infrastructural and economic barriers that impede the optimal functioning of their ETS [32]. Figure 21.6 shows the gist of disparities that occurs in a Carbon Market.

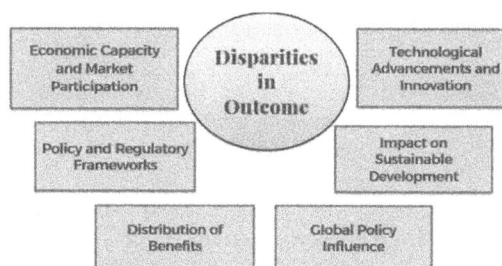

Figure 21.6 Disparities in outcome

Source: Authors

3.6 Opportunities for Improvement

Developed nations have already established carbon trading markets, but there is scope for improvement in several domains such as enhancing the market stability, further integration of other global markets and standardizing the regulations facilitating smoother functioning, increasing cap stringency by tighter emission caps and sectoral expansion to broaden the market's impact, and better utilization of the auction revenue generated to be directed to the innovative technology and infrastructure development for long term sustainability [36].

However, the improvement differs for the developing nation in that includes Capacity building by providing technical training programs for government officials, industry participants and institutions to gain expertise in the carbon market operations [39]. Robust institutions have to be set up to look after the market operations and ensure transparency and compliance. International support is required to build market infrastructure, develop the market, and technology transfer to improve low-carbon and energy-efficient technologies [32]. Market expansion has to be performed by pilot projects in various sectors, regional collaboration needs to be exhilarated, and a more market integrated approach needs to be inculcated for best practice [34]. Improving the MRV system will enhance the trust and credibility of the system, also a third-party verification system can be introduced to improve transparency and accountability [9].

4. CONCLUSION

There are notable differences in emissions reductions, market development, and economic implications between the carbon markets in industrialized and developing countries. Developed countries have economic development and successful carbon reductions because of their sophisticated regulatory frameworks, technological innovation, and stable markets. On the other hand, developing countries struggle with issues including unstable markets, high compliance costs, and restricted access to technology. Nonetheless, there exist significant prospects for enhancement via the development of capacity, global assistance, integration of markets, and improved financial institutions. Reducing these gaps and taking advantage of these chances can promote more efficient and just international carbon markets. The brief comparison of the carbon market for developed and developing nations is shown in Table 21.2.

Table 21.2 Summary of comparative analysis

Factors	Developed Nation	Developing Nation
Policy Framework	Comprehensive and stringent regulations; integrated with broader climate goals (e.g., EU ETS, California Cap-and-Trade)	Emerging and evolving regulations; focus on capacity building and gradual implementation (e.g., China's ETS, India's PAT Scheme)
Market Mechanism	Emerging and evolving regulations; focus on capacity building and gradual implementation (e.g., China's ETS, India's PAT Scheme)	Pilot and phased cap-and-trade systems; reliance on offset mechanisms; gradually expanding sectoral coverage (e.g., China's ETS, Mexico's Carbon Market Pilot)
Financial Mechanism and Instruments	Sophisticated financial instruments, including futures, options, and derivatives; significant use of auctioning for allowance allocation	Limited financial instruments; initial reliance on free allocation with plans to introduce auctioning; need for international financial support

Disparities in Outcome	Generally effective in reducing emissions; high market liquidity; significant investment in clean technologies	Mixed outcomes; variability in market stability and effectiveness; challenges in achieving significant emissions reductions due to economic and developmental priorities
Opportunities for Improvement	Enhancing market stability mechanisms; further integration with other global carbon markets; increasing the stringency of caps	Market expansion in pilot ETS for sectoral coverage, strengthening MRV, enhancing stakeholder participation

Source: Compiled by authors

REFERENCES

[1] Chaudhry, D. (2008). A brief study of the voluntary carbon markets, recent and future trends with special focus on India. https://ssrn.com/abstract=1334905/http://dx.doi.org/10.2139/ssrn.1334905

[2] Agriculture and Forestry Offsets in Carbon Markets: Background and Selected Issues. (2021). https://crsreports.congress.gov

[3] Calel, R. (2013). Carbon markets: a historical overview. *Wiley Interdisciplinary Reviews: Climate Change,* 4(2), 107–119. https://doi.org/10.1002/wcc.208

[4] Cramton, P., & Kerr, S. (2002). Tradeable carbon permit auctions: how and why to auction not grandfather. *Energy Policy,* 30(4), 333–345

[5] Gren, I. M., & Zeleke, A. A. (2016). Policy design for forest carbon sequestration: A review of the literature. *Forest Policy and Economics,* 70, 128–136. https://doi.org/10.1016/j.forpol.2016.06.008

[6] Lam, P. T. I., Chan, E. H. W., Yu, A. T. W., Cam, W. C. N., & Yu, J. S. (2014). Mitigating climate change in the building sector: Integrating the unique characteristics of built facilities with emissions trading schemes. *Facilities,* 32(7–8), 342–364. https://doi.org/10.1108/F-04-2013-0035

[7] Newell, R. G., Pizer, W. A., & Raimi, D. (2014). Carbon markets: Past, present, and future. *Annual Review of Resource Economics,* 6(1), 191–215. https://doi.org/10.1146/annurev-resource-100913-012655

[8] Raina, N., Zavalloni, M., & Viaggi, D. (2024). Incentive mechanisms of carbon farming contracts: A systematic mapping study. *Journal of Environmental Management,* 352. Academic Press. https://doi.org/10.1016/j.jenvman.2024.120126

[9] Woo, J., Fatima, R., Kibert, C. J., Newman, R. E., Tian, Y., & Srinivasan, R. S. (2021). Applying blockchain technology for building energy performance measurement, reporting, and verification (MRV) and the carbon credit market: A review of the literature. *Building and Environment,* 205. https://doi.org/10.1016/j.buildenv.2021.108199

[10] Yang, Y., & Cheng, L. (2017). Operational efficiency evaluation and system design improvements for carbon emissions trading pilots in China. *Carbon Management,* 8(5–6), 399–415. https://doi.org/10.1080/17583004.2017.1387033

[11] European Commission (2023). EU Emissions Trading System (EU ETS). https://ec.europa.eu

[12] ICAP (2023). Emissions Trading Worldwide: Status Report 2023. International Carbon Action Partnership. https://icapcarbonaction.com

[13] World Bank (2022). State and Trends of Carbon Pricing 2022. https://worldbank.org

[14] CARB (2023). California Cap-and-Trade Program: overview. California Air Resources Board. https://arb.ca.gov

[15] RGGI (2022). The regional greenhouse gas initiative: investment of proceeds. https://rggi.org

[16] BEE (2023). Perform, achieve, and trade (PAT) scheme. *Bureau of Energy Efficiency,* India. https://beeindia.gov.in

[17] Hasselknippe, H. (2003). Systems for carbon trading: An overview. *Climatic Change,* 65(2), 107–133. https://doi.org/10.1016/j.clipol.2003.09.014

[18] Finon, D. (2019). Emissions trading schemes: Design choices and experiences in the EU and elsewhere. *Energy Policy*, 128, 159–170. https://doi.org/10.1016/j.enpol.2019.01.019

[19] Newell, R. G., Pizer, W. A., & Raimi, D. (2013). Carbon markets 15 years after Kyoto: Lessons learned, new challenges. *Journal of Economic Perspectives*, 27(1), 123–146. https://doi.org/10.1257/jep.27.1.123

[20] Popp, D. (2011). International technology transfer, climate change, and the clean development mechanism. *Review of Environmental Economics and Policy*, 5(1), 131–152. https://doi.org/10.1093/reep/req018

[21] Li, J., Wang, Y., & Zhang, Y. (2023). China's emissions trading system: performance and prospects. *Energy Policy*, 165, 112927. https://doi.org/10.1016/j.enpol.2022.112927

[22] Borenstein, S., Bushnell, J., & Wolak, F. A. (2021). California's cap-and-trade market through 2030: A preliminary supply/demand analysis. *Energy Institute at Haas Working Paper*, 314R. https://haas.berkeley.edu/wp-content/uploads/WP314R.pdf

[23] Huang, Y., Yang, L., & Zhang, Z. (2012). Carbon pricing in China: Progress and challenges. Energy Policy, 51, 134–141. https://doi.org/10.1016/j.enpol.2012.06.045

[24] Wu, X., Tian, Z., & Guo, J. (2022). A review of the theoretical research and practical progress of carbon neutrality. *Sustainable Operations and Computers*, 3, 54–66. https://doi.org/10.1016/j.susoc.2021.10.001

[25] Zhang, G., & Bi, S. (2023). Inhibition or promotion: the impact of carbon emission trading on market structure: evidence from China. *Frontiers in Energy Research*, 11, 1–19. https://doi.org/10.3389/fenrg.2023.1238416

[26] Bayer, P., & Aklin, M. (2020). The European union emissions trading system reduced CO_2 emissions despite low prices. *Proceedings of the National Academy of Sciences*, 117(16), 8804–8812. https://doi.org/10.1073/pnas.1918128117

[27] European Commission. (2023). The EU Emissions Trading System in 2020: Trends and projections. Retrieved from https://climate.ec.europa.eu/news-your-voice/news/eu-carbon-market-continues-deliver-emission-reductions-2023-10-31_en

[28] European Commission. (2023). Carbon Market Report: Emissions from EU ETS stationary installations fall by over 9%. Retrieved from https://climate.ec.europa.eu/news-your-voice/news/record-reduction-2023-ets-emissions-due-largely-boost-renewable-energy-2024-04-03_en

[29] Tong, D., Zhang, Q., Zheng, Y., Caldeira, K., Shearer, C., Hong, C., et al. (2019). Committed emissions from existing energy infrastructure jeopardize 1.5 °C climate target. Nature, 572(7769), 373–377. https://doi.org/10.1038/s41586-019-1364-3

[30] Organisation for Economic Co-operation and Development (OECD). (2023). Carbon pricing: A development and trade reality check. Retrieved from https://www.oecd.org/environment/carbon-pricing-a-development-and-trade-reality-check.htm

[31] Tuerk, A., Mehling, M., Flachsland, C., & Sterk, W. (2009). Linking carbon markets: concepts, case studies and pathways. *Climate Policy*, 9(4), 341–357. https://doi.org/10.3763/cpol.2009.0621

[32] CEPS, M. E., & Egenhofer, C. (2018). Tools to boost investment in low-carbon technologies five possible ways to create low-carbon markets in the EU. 1–11. https://www.ceps.eu/ceps-publications/tools-boost-investment-low-carbon-technologies/.

[33] Kollenberg, S., & Taschini, L. (2015). The European Union emissions trading system and the market stability reserve: optimal dynamic supply adjustment. *SSRN Electronic Journal*. https://doi.org/10.2139/ssrn.2610213.

[34] Lee, W. J., & Mwebaza, R. (2022). New strategy for innovative RD&D in view of stakeholder interaction during climate technology transfer. *Sustainability*, 14(14). https://doi.org/10.3390/su14148363.

[35] Musier, R., & Adib, P. (2009). *Emerging Carbon Markets and Fundamentals of Tradable Permits. Generating Electricity in a Carbon-Constrained World*. 1st ed. Elsevier Inc. https://doi.org/10.1016/B978-1-85617-655-2.00003-1.

[36] OECD/IEA, and IRENA. (2017). Perspectives for the Energy transition: Investment needs for a low-carbon energy system. 5–16. https://www.irena.org/DocumentDownloads/Publications/Perspectives_for_the_Energy_Transition_2017.pdf.

[37] Sun, L., Cao, X., Alharthi, M., Zhang, J., Taghizadeh-Hesary, F., & Mohsin, M. (2020). Carbon emission transfer strategies in supply chain with lag time of emission reduction technologies and low-carbon preference of consumers. *Journal of Cleaner Production,* 264. https://doi.org/10.1016/j.jclepro.2020.121664.

[38] Wang, T. Gong, L., Lei, T. & Luo, Q. (2023). Carbon credit system as an incentive to green travel: a literature review. https://doi.org/10.1061/9780784484869.017

[39] Ping, P., & Qin, Z. (2011). Using Carbon trading to seize the commanding heights of economic development. *Energy Procedia,* 5, 74–78. https://doi.org/10.1016/j.egypro.2011.03.014.

[40] Gong, Z., & Bi, S. (2023). Inhibition or promotion: the impact of carbon emission trading on market structure: evidence from China. *Frontiers in Energy Research,* 11, 1–19. https://doi.org/10.3389/fenrg.2023.1238416.

[41] World Bank Group. (2018). *The potential for climate auctions as a mechanism for NDC implementation.,* https://openknowledge.worldbank.org/server/api/core/bitstreams/a2b91ee2-af74-55d4-b785-3ad43c9407b0/content

[42] United Nations Division for Sustainable Development. (2012). *The transition to a green economy: Benefits, challenges and risks from a sustainable development perspective.* https://www.unep.org/resources/report/transition-green-economy-benefits-challenges-and-risks-sustainable-development

[43] Bayer, P., & Aklin, M. (2020). The European Union Emissions Trading System reduced CO_2 emissions despite low prices. *Proceedings of the National Academy of Sciences of the United States of America,* 117(16), 8804–8812. https://doi.org/10.1073/pnas.1918128117

Advances in Construction, Real Estate, Infrastructure and Project Management – Anil Kashyap et al. (eds)
© 2026 Taylor & Francis Group, London, ISBN 978-1-041-13433-6

22 Identification and Analysis of Critical Success Factors in TQM in Indian Real Estate Construction Sector

Nilesh Agarchand Patil[1]

Assistant Professor,
School of Construction, NICMAR University,
Pune

■ **ABSTRACT:** Total Quality Management (TQM), is essential in the Indian construction sector to improve project efficiency, save costs, and guarantee customer satisfaction. Construction companies can meet schedules, minimize risks, and achieve higher quality standards by putting TQM principles, such as continuous improvement, employee involvement, and customer satisfaction. Better project outcomes and a competitive edge are achievable through TQM's promotion of a culture that fosters innovation and accountability. To satisfy the changing needs of stakeholders, maximize resources, and handle difficulties in India's rapidly expanding construction sector, TQM provides a strategic framework that promotes sustainable growth. This article used scholarly research papers to understand TQM principles in real estate operations. It then explored practical TQM implementation in the real estate sector using questionnaires to gather insights from industry practitioners. By bridging the gap between theory and practice, this research enriched its findings using quantifiable data. The findings from this study focuses on how well the Indian construction industry implements TQM by using these eight prominent critical success factors (CSFs), which are employee empowerment and involvement, education and training, top management commitment, supplier quality management, information and analysis, process management, strategic quality management and customer satisfaction.

■ **KEYWORDS:** Indian, Construction, TQM, Critical success factors

1. INTRODUCTION

After World War II, the necessity to compete in a global market where superior quality, reduced prices, and quick development are critical to market leadership gave rise to the idea of Total Quality Management (TQM). TQM is now seen as a basic necessity for any business in order to compete, let alone dominate, its industry. It is a method of planning, establishing,

*Corresponding author: npatil@nicmar.ac.in

DOI: 10.1201/9781003669814-22

and understanding each activity of the process while eliminating unnecessary steps routinely followed in organizations. TQM is a philosophy that makes quality values the driving force behind leadership, design, planning, and improvement activities. The methods of TQM were prevalent until the end of the twentieth century, after which the Integrated Quality Management System (IQM) emerged at the beginning of the twenty-first century, integrating all relevant systems for competitive advantages [1].

Encouraging all members of an organisation, from entry-level workers to senior executives, to focus on improving quality and attaining customer satisfaction is the cornerstone of Total Quality Management (TQM). "A management philosophy and company practices that aim to harness the human and material resources of an organisation in the most effective way to achieve the objectives of the organisation" is how the British Standards Institution defines Total Quality Management [2]. According to ISO 8402, TQM is "a management approach of an organization centered on quality, based on the participation of all its members, and aiming at long-term success through customer satisfaction and benefits to all members of the organization and society" (International Organization for Standardization [6]. ASQ also defines TQM as a management approach centered on quality, based on organization-wide participation, and aimed at long-term success through customer satisfaction.

In the real estate industry, a crucial part of the global economy, TQM significantly affects the quality of life for individuals and communities by shaping the built environment. Therefore, maintaining high levels of productivity and performance is of paramount importance. TQM has become a strategic approach to tackle these challenges by promoting a culture of continuous improvement and customer-centric processes. India's real estate sector, an essential part of the economy, faces challenges such as regulatory complexities, quality issues, and competition. To win bids, Indian construction companies must be highly competitive and produce high-quality work within budget and timeframe [4]. To stay competitive, businesses need to implement highly effective quality policies that prioritize TQM, which is especially crucial for the Indian construction sector looking to expand its global market share [7]. By embracing the TQM philosophy and exceeding quality requirements, construction companies can make evolutionary changes to their organizations and stand out from competitors [5].

TQM is a well-established approach to enhancing organizational performance and quality. It acknowledges quality as a strategic objective by minimizing rework and maximizing profitability to achieve customer satisfaction [3]. Through continuous quality improvement, TQM engages all employees in the endeavour to boost customer happiness and attain outstanding performance [8]. According to the American Society of Quality (ASQ), TQM is a quality-centered management strategy that emphasises employee involvement across the entire organisation and strives for sustained success via client satisfaction.

The Construction Industry Institute (CII) describes that the TQM is a collection of improvement-centered procedures and methods used in a changed management environment, it is best described as a journey rather than a destination [9]. According to the idea of continuous improvement, this environment must exist for the duration of the business, and the techniques will be applied consistently. The improvement process never ends; therefore, no true destination is ever reached [8]. This paper seeks to comprehensively understand TQM implementation within the Indian Real Estate Sector by examining multiple critical success factors (CSFs) in TQM through a range of research papers and performing quantitative analysis.

2. RESEARCH METHODOLOGY

The Research Methodology plays a pivotal role in any project work, as it enables a comprehensive understanding of the subject matter by analysing numerous research papers from both Indian and Global authors. The methodology for this study focuses on identification of Critical Success Factors (CSFs) of total quality management with the perspective of Indian real estate sector. Objective of this research is to analyse TQM in the Indian Real Estate sector by identifying its critical success factors. Through our research, we aim to highlight the organizations implementing TQM and not implementing TQM in the Real Estate sector. For the research data is collected by following ways. Figure 22.1 shown below has further classified the research steps in detail.

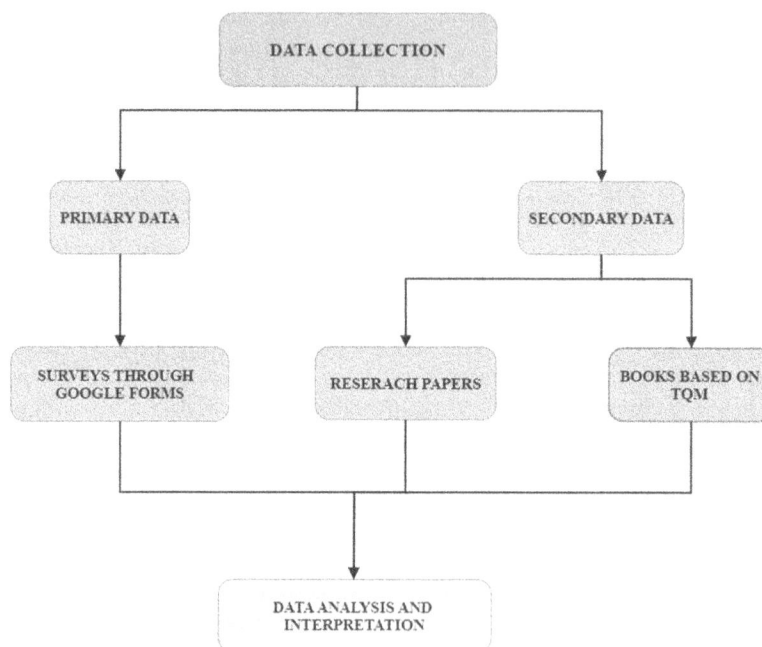

Figure 22.1 Research methodology

Source: Authors

2.1 Secondary and Primary Data

The secondary data is existing data previously collected by someone else and used by the author in this research and analysis. This data is collected through reading research papers, and articles published at National and Global level. The research articles are selected on the basis of two criteria: reputable publications and recently published articles.

The primary data is collected through the online survey. For this, the questionnaire has been circulated using Google forms and collected the information regarding TQM from the construction industry experts. The questionnaire is circulated with four categories of experts from quality management background, which includes managers from private sector, engineers from public sectors, consultants and researchers.

Further this data can be classified into Qualitative and Quantitative:

2.2 Qualitative and Quantitative Data

Qualitative data is information used to define characteristics, attributes, or qualities that is not reliant on numerical values. It would be beneficial to discuss the positive changes brought about by TQM and the potential disadvantages of not implementing it.

Quantitative data refers to information that can be measured and expressed numerically. This can include statistics regarding the CSFs implemented in TQM practices.

A well-planned research design serves as a navigational tool for conducting a study, facilitating the selection of appropriate information gathering methods, sample sizes, and metrics to be evaluated. The research design for this project involves data collection methods and analysis and interpretation of data using CSFs in the context of Indian Real Estate scenario.

3. DATA COLLECTION

Data Collection includes all the methods and procedures that are considered and implemented for gathering data for analysis and interpretation. The data collection process was multifaceted, drawing on various sources to capture a comprehensive understanding of TQM practices in real estate. Primarily, the research journey was guided by insights from scholarly research papers, offering a foundational understanding of TQM principles and their adaptation within the unique context of real estate operations. Leveraging the insights and frameworks established in existing literature, this study embarked on a quest to delve deeper into the practices of TQM within the real estate sector. Furthermore, recognizing the expertise and insights offered by industry professionals, this research also integrated the use of questionnaires. These questionnaires were meticulously designed to extract firsthand experiences, perceptions, and challenges faced by professionals directly involved in implementing TQM practices within real estate organizations. By engaging with individuals possessing rich domain knowledge and practical insights, this study endeavours to bridge the gap between theoretical frameworks and on-the-ground realities, thereby enriching the depth and applicability of its findings. The amalgamation of research papers and insights gathered from industry experts serves as the foundation of this project. The process of collecting data involves a number of specific steps, which are clearly depicted in

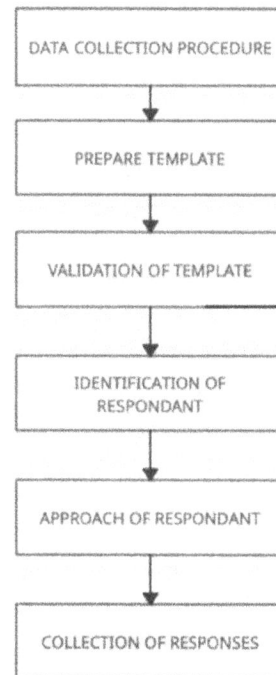

Figure 22.2 Data collection procedure

Source: Authors

Figure 22.2. These steps are carefully designed to ensure accurate and reliable data collection. The key steps include:

3.1 Prepare Template

This step involves designing a tool to collect data. This instrument can be a survey, questionnaire, interview guide, or any other format suitable for gathering the information you need for your research project. The study has identified 8 CSFs and prepared the questionnaire through Google form for the same.

3.2 Validation of Template

Once you have your data collection instrument one needs to test it to ensure it yields the information you require. This process is known as validation. The author consulted with experts in the industry who validated the questionnaire through pilot study.

3.3 Identification of Respondent

This step involves determining who will participate in this study. Author needed to establish the target population and then develop a sampling strategy to select the participants from that population.

3.4 Approach of Respondent

This refers to the method to get in touch with the expert participants and invite them to take part in your study. The author approached several participants through Email, Telephone, LinkedIn, and other social sites.

3.5 Collection of Responses

This is the stage where you actually gather the data from your participants. This will depend on the type of data collection instrument you are using. The author sends the Google forms to the participants after communicating with them. The study has received 101 responses from several participants.

4. DATA ANALYSIS AND FINDINGS

The four categories of respondents have been selected for this study, which include government, private, consultants, and researchers. A total of 101 responses have been collected. Initially, the demographic analysis of respondents was conducted, as shown in Figure 22.3. The three parameters considered for respondent selection are qualification, experience, and the number of projects handled. Almost more than 90% of respondents have highest qualification is graduation.

From above Figure 22.3, it has been observed that more than 55% of the respondents have a minimum experience of 5 years in quality management work. Also, the maximum 5 projects have been handled by more than 53% of respondents from the real estate sector of India.

Additionally, the study gained important insights on the application of Total Quality Management in the Indian construction industry by using a thorough technique of descriptive statistics. Descriptive statistics were used to summarize the survey responses and provide a clear snapshot of the TQM practices across the surveyed organizations. By calculating measures such as mean and standard deviation, the study aimed to understand the central tendency and variability in respondents' ratings of TQM implementation. The Data Analysis comprises of statistical analysis of 8 Critical Success Factors (CSFs) based on the responses of the respondents. The eight CSFs include top management commitment, process management, supplier quality management, employee empowerment and involvement, education and training, information and analysis, customer satisfaction, and strategic quality management. The mean and standard deviation are used to analyse the 101 responses. The

Highest Qualification of Respondent
101 responses

- PhD
- Post Graduate (PG)
- Post Graduate Diploma (PGD)
- Under Graduate (UG)
- Diploma

Experience of the Respondent
101 responses

- 1-5 years
- 6-10 years
- 11-15 years
- 16-20 years
- more than 20 years

Number of the Projects handled by Respondent
101 responses

- 1-5 Projects
- 6-10 Projects
- 11-15 Projects
- 16-20 Projects
- More than 20 Projects

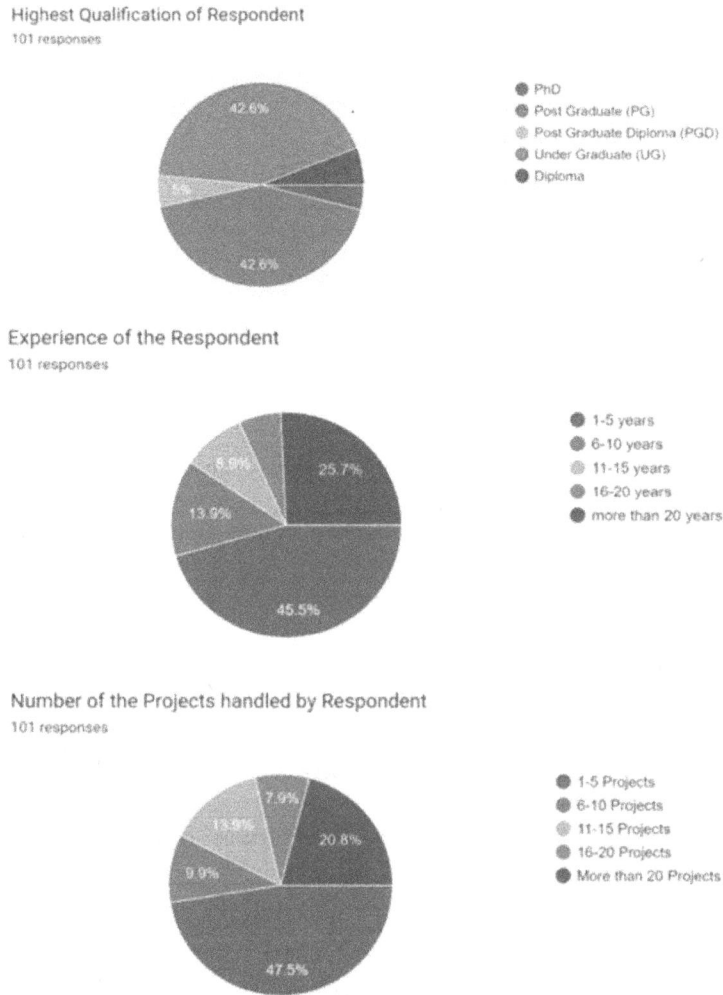

Figure 22.3 Respondents demographic information

Source: Authors

mean value is calculated for the respondent on a 1–5-likert scale. Table 22.1 summarizes the analysis of all eight CSFs, and all CSFs are briefly explained below:

4.1 CSF 1: Process Management

In real estate, process management involves effectively handling various stages of property development, from planning and acquisition to construction, marketing, and sales. This includes streamlining procedures to ensure efficient project execution, adhering to regulations, and optimizing workflows to enhance overall productivity.

4.2 CSF 2: Education and Training

In real estate, education and training are vital for professionals to stay updated on market trends, regulations, and best practices. It includes continuous learning about property valuation, legal

Table 22.1 CSFs of TQM for real estate construction in india

Sr. No.	CSFs with parameters	Mean Value	Standard Deviation
1	**Process Management**		
1.1	Resource Optimization: Efficiently allocate resources to maximize output while minimizing waste and costs.	3.41	1.079
1.2	Workflow Automation: Implement automated systems to streamline processes and enhance operational efficiency.	3.18	1.033
1.3	Quality Control Measures: Institute rigorous quality checks at every stage to maintain high standards & minimize errors.	3.52	1.188
1.4	Continuous Process Improvement: Foster a culture of ongoing refinement and enhancement in operational workflows for sustained efficiency gains.	3.28	1.069
2	**Education and Training**		
2.1	Personalized Learning Paths: Develop tailored training programs to meet individual employee learning needs.	3.19	1.239
2.2	Hands-On Workshops: Conduct practical, hands-on workshops to reinforce theoretical knowledge with real-world application.	3.41	1.234
2.3	Mentorship Programs: Establish mentoring initiatives to provide support for skill development and career growth.	3.41	1.079
2.4	E-Learning Platforms: Utilize digital platforms to deliver flexible and accessible learning modules for remote or self-paced training.	3.04	1.363
3	**Customer Satisfaction**		
3.1	Feedback Mechanisms: Implement robust feedback systems to gather and act upon customer opinions and suggestions.	3.36	1.205
3.2	Timely Issue Resolution: Ensure swift resolution of customer concerns to maintain satisfaction and trust.	3.61	1.104
3.3	Personalized Services: Tailor services to individual customer needs and preferences for a more personalized experience.	3.40	1.167
3.4	Post-Service Follow-Up: Conduct follow-ups post-service to ensure sustained customer satisfaction and address any lingering issues.	3.39	1.157
4	**Top Management Commitment**		
4.1	Resource Allocation Prioritization: Allocate resources in alignment with quality objectives to emphasize their importance.	3.36	1.205
4.2	Visible Leadership Support: Demonstrate visible and active support from top management to reinforce quality commitment.	3.53	1.045
4.3	Setting Clear Objectives: Establish clear and measurable quality objectives to guide the organization.	3.58	1.051
4.4	Employee Engagement: Encourage a sense of ownership and dedication to quality objectives by involving staff members in decision-making processes.	3.55	1.144
5	**Supplier Quality Management**		
5.1	Performance Metrics Monitoring: Regularly monitor supplier performance against predefined quality metrics and standards.	3.40	1.123

Sr. No.	CSFs with parameters	Mean Value	Standard Deviation
5.2	Collaborative Improvement Initiatives: Collaborate with suppliers to jointly identify and address areas for quality improvement.	3.35	1.187
5.3	Risk Mitigation Strategies: Prepare a risk mitigation plan to reduce the risks brought on by variations in supplier quality.	3.45	1.044
5.4	Supplier Development Programs: Provide suppliers with assistance and training to improve their quality standards and competencies.	3.11	1.232
6	**Employee Empowerment and Involvement**		
6.1	Recognition Programs: Plan and establish efforts to honour and reward staff members for their contributions to quality-related projects.	3.36	1.171
6.2	Cross-Functional Teams: Promote cooperation amongst various teams or departments to address quality-related problems as a group.	3.44	1.153
6.3	Empowerment through Decision-Making: Empower employees to make decisions that impact quality within their respective roles.	3.65	1.090
6.4	Training for Quality Advocacy: Provide training to equip employees to become advocates for quality within the organization.	3.42	1.125
7	**Information and Analysis**		
7.1	Data-Driven Decision-Making: Make use of data analytics to inform choices and tactics for improving quality.	3.30	1.205
7.2	Real-Time Monitoring Systems: Implement systems for real-time monitoring of quality-related metrics and key performance indicators.	3.35	1.161
7.3	Root Cause Analysis: Conduct thorough analyses to identify underlying causes of quality issues for effective resolution.	3.46	1.127
7.4	Accessible Reporting Mechanisms: Ensure easy access to comprehensive reports for stakeholders to facilitate informed actions.	3.45	1.212
8	**Strategic Quality Management**		
8.1	Risk Assessment and Mitigation: To find possible quality concerns and create mitigation plans, do thorough risk assessments.	3.50	1.083
8.2	Alignment with Organizational Goals: Ensure that quality management strategies align seamlessly with broader organizational objectives.	3.49	1.110
8.3	Continuous Improvement Philosophy: Encourage a culture of ongoing quality management practice improvement across the entire company.	3.45	1.253
8.4	Scenario Planning for Quality: Develop contingency plans to address potential deviations from quality standards and mitigate their impact.	3.41	1.142

Source: Compiled by authors

aspects, negotiation skills, a customer service, ensuring that professionals are well-equipped to serve clients effectively.

4.3 CSF 3: Customer satisfaction

In real estate, customer satisfaction revolves around meeting and exceeding client expectations. This encompasses providing transparent information, excellent service, timely responses, and post-purchase support, all aimed at ensuring a positive experience for buyers, sellers, and renters.

4.4 CSF 4: Top Management Commitment

Real estate requires strong leadership committed to quality, ethical practices, and long-term vision. This involves setting clear quality objectives, allocating resources effectively, and ensuring that the organization's values align with delivering high-quality properties and services.

4.5 CSF 5: Supplier Quality Management

In real estate, managing suppliers involves partnerships with construction firms, material suppliers, and service providers. Ensuring high-quality materials and services are procured, maintaining adherence to timelines, and fostering collaborative relationships are crucial for successful property development.

4.6 CSF 6: Employee Empowerment and Involvement

Real estate success relies on empowered employees who contribute to quality property development and customer service. Creating a culture where employees feel valued, involved in decision-making, & encouraged to suggest improvements fosters a more engaged and committed workforce.

4.7 CSF 7: Information and Analysis

Real estate relies heavily on data analysis for market trends, property valuation, and investment decisions. Access to accurate and timely information, coupled with analytics that aid in forecasting market movements, supports informed decisions in buying, selling, and property development.

4.8 CSF 8: Strategic Quality Management

Strategic quality management in real estate involves aligning quality objectives with organizational goals, mitigating risks associated with market fluctuations, and constantly improving processes to adapt to changing market demands. It includes scenario planning, risk assessment, and a continuous focus on improvement.

From the above analysis, the results suggest that the mean is between 3 and 4 for most of the critical success factors. This means that most of the processes are implemented in the organization properly. The standard deviation falls between 1 and 1.2, which shows a buffer for the mean or how much the mean can deviate from its value. This standard deviation between 1 and 1.2 is normal for this case. The standard deviation balances the results.

5. CONCLUSIONS AND FUTURE DIRECTIONS

According to a recent study's findings, a sizable portion of businesses in the Indian construction sector are actively integrating Total Quality Management (TQM) concepts into their operations. The findings reveal that the average rating for TQM implementation is moderate to high, indicating that most of the surveyed organizations are actively embracing TQM principles. The study shows that consulting firms are more likely to use TQM within a structured framework, while contractor companies tend to use TQM without considering a structured framework. The results show a low standard deviation, which suggests that there is consistency in responses, and the implementation of TQM is not highly varied among the surveyed organizations. This suggests there is a certain level of uniformity in the perception of TQM effectiveness.

The study also finds that TQM is followed in organizational processes within the Indian construction sector, as evidenced by the good mean rating. This is in line with TQM principles that aim to enhance quality and efficiency in all aspects of operations. Organizations that embrace TQM are likely to identify critical success factors and address areas for enhancement, contributing to sustained success. Based on the positive mean rating, it can be inferred that organizations in the Indian construction sector are considering and adopting TQM practices. However, the study highlights specific areas where lower ratings were observed, which organizations should focus on to further strengthen the effectiveness of TQM implementation.

There are certain restrictions on the survey that could compromise the precision and applicability of its findings. First of all, because the sample size is restricted to Pune, the results might not fairly reflect the variety of perspectives and experiences seen throughout the Indian construction sector. Secondly, the respondent count was limited to over 101 responses, which reduces the statistical power of the analysis. To improve the project results, it is recommended to expand the sample size to cover the entire country.

A well-structured framework is needed to encourage contractor companies to adopt Total Quality Management (TQM) principles. This approach should be consistent and effective, like that of consulting firms, and can ensure that TQM principles are implemented efficiently across different industry organizations. Emphasizing a culture of continuous improvement is crucial in organizations. This involves identifying areas for improvement regularly and actively working towards implementing changes to enhance quality and efficiency. Employees must have access to training and development programs in order to improve their comprehension of TQM approaches and principles. This will empower them to contribute actively to the success of TQM initiatives within their organization.

REFERENCES

[1] Abdul-Aziz, A. R. (2002). The realities of applying total quality management in the construction industry. *Structural Survey*, 20(2), 88–96.

[2] British Standards Institution. (1992). *Total quality management* (BS 78501:1992).

[3] Chin-Keng, T., & Abdul-Rahman, H. (2011). Study of quality management in construction projects. *Chinese Business Review*, 10(7), 542–552.

[4] Erande, S. S. and Pimplikar, S. S. (2016). total quality management in Indian construction industry. *International Research Journal of Engineering and Technology*, 3(6), 685–691

[5] Faihan, A., Rushami, Z. Y., & Rabiul, I. (2013). Relationship Between Total Quality Management Practices and Contractors Competitiveness. *American Journal of Applied Sciences*, 10(3), 247–252.

[6] International Organization for Standardization (ISO 8402:1994). (1994). *Quality management and quality assurance*.

[7] Patyal, V. S., Ambekar, S., & Prakash, A. (2020). Organizational culture and total quality management practices in Indian construction industry. *International Journal of Productivity and Performance Management*, 69(5), 895–913.

[8] Rumane, A. R. (2024). *Total Quality Management: Applications and Concepts for Construction Projects*. CRC Press.

[9] Thomas, H., & Oswald, J. L. (1992). Guidelines for Implementing Total Quality Management in Engineering and Construction Industry. *Construction Industry Institute (CII)*, Doc 74.

Advances in Construction, Real Estate, Infrastructure and Project Management – Anil Kashyap et al. (eds)
© 2026 Taylor & Francis Group, London, ISBN 978-1-041-13433-6

23 Feasibility Study of Prefabricated Container Houses in India

Shashank B. S.[1] and Ashish Rastogi[2]

Assistant Professor,
NICMAR University
Pune

■ **ABSTRACT:** There is cause for concern about the severe housing shortage that the socioeconomically disadvantaged population in India experiences. Based on contemporary research, there is a growing scarcity that is escalating in an inevitable manner, impacting both recognised and informal stakeholders within the housing industry. The traditional methods of constructing buildings have limited opportunities for alternative building techniques that could yield substantial cost and time savings. The main objective of this research is to evaluate the feasibility of using renovated shipping container homes as an alternative to traditional brick-and-mortar dwellings in India's post-disaster housing initiative. Comparative research was conducted on a real-life scenario of a post-disaster housing project in India. A proposed design was put forth, incorporating recycled shipping containers as structural elements, in order to examine the disparities in cost and time between a comparable programme and location. An analysis was conducted to compare the traditional building with a container-based design and a conventional precast design. The examination of the structure provides insights into the feasibility and structural integrity of the proposed solution. The preliminary cost estimation indicates a cost reduction of 1.9 lakh rupees per unit, resulting in a cumulative cost savings of 3.66 crores for the entire project, which consists of 192 units. The container building offers substantial time savings, with a completion period of around 3.5 months, in contrast to the 8-month duration required for the conventional building. The findings indicate that containers present a viable and environmentally friendly choice for prefabricated buildings, with the potential for expedited completion and reduced expenses compared to traditional alternatives. Several additional benefits include the modular aspect, the possibility to relocate, and the ease of transport.

■ **KEYWORDS:** Prefabricated houses, Shipping containers, Post-disaster mitigation, Comparative study

[1]shashankbs@nicmar.ac.in, [2]arastogi@nicmar.ac.in

DOI: 10.1201/9781003669814-23

1. INTRODUCTION

Globally, there are about 1.7 crore structurally sound shipping containers that are idle after the active life of 8–10 years in freight transport [13]. However, their technical life can last for 20–25 years more if maintained properly [12]. It is not economical to transport the empty containers back to their origin, but keeping them in the destination ports takes up a lot of storage space. Moreover, they cannot be disposed of in landfills; recycling requires huge amounts of energy; and the melting process results in the emission of greenhouse gases. With the help of recent technological innovation, these shipping containers can be reused efficiently for building purposes. There are many cases of successful usage of shipping containers around the globe for classrooms, homes, emergency shelters, offices, hotels, etc.

Shipping containers are manufactured in two major commercial sizes, designated as "20'-HC-6.0m and 40'-HC-12.0m". They measure 2.4 metres in width and 2.7 metres in height, with HC denoting a high cube. Containers are made up of joint corrugated panel members for the top and bottom walls, double upright sidewalls, and rear walls that are joined at the edges. They have built-in beams, a plywood flooring system, and a sturdy construction. The majority of the preparatory tasks for conversion into habitable residences can be completed in factories, such as disinfection and cleaning, opening cutting, strengthening and joining, surface preparation and painting, and so on. Interior finishing and services can be completed in-place. It takes around 8000 KWh to recycle a 3.63-tonne shipping container using a Basic Oxygen Furnace (BOF) or Electric Arc Furnace (EAF). However, transforming it into a residence requires only 5% of the energy, or 400 KWh. The biggest barrier to widespread container use is the stigma associated with these unsightly metal boxes abandoned in shipyards. However, if reused correctly, it reduces a significant carbon footprint without causing indoor air temperatures to rise. The building made mostly of shipping container material is no less better to the traditional construction in terms of embodied energy and thermal performance in hot regions [14]. The coastline length of India amounts to 7500 km, and there is a total of 213 major and minor seaports along these coasts. Enormous amounts of trade are happening in these ports, and there are plenty of empty cargo containers sitting idle in the dockyards. India also has several cargo container manufacturers. These facts suggest that construction using containers is a good option to explore in India [12]. Architectural projects seldom use containers in India. Few firms, like Studio Alternatives (Pune) and Container Solutions India (Bangalore), are experimenting with containers in projects like farmhouses, office spaces, vocational training schools, etc. But there are no widespread practices in India that use containers. The container costs vary from INR 56, 000 to 65, 000 for the 20'x8' standard used container and INR 96, 000 to 125000 for the 40'x8' container [12]. There are places like California (US) that have clearly defined regulations and norms for the usage of containers in construction, but in India, the National Building Code does not address this method of construction.

Access to housing is a constitutional right in India. Still, the urban housing shortage is estimated to be 18.78 million. And the national slum population amounts to 6.5 crore. Huge quantities of waste are generated in construction—530 million metric tonnes (MT) of C&D waste in 2013. These figures call for urgent, sustainable intervention in the Indian housing sector. Objectives of the study is to identify the challenges and opportunities posed by the selected rehabilitation project in the context of intervention using containers and to check the feasibility of the proposed solution through comparative analysis with the existing project based on cost, time, and sustainability.

2. LITERATURE REVIEW

2.1 Design

Specifics about the solution for successful design completion, and the scaled model of the structure or site constructed from prescribed materials, in this case foam board and corrugated paper. These models were assessed on their accuracy, workmanship, and architectural vision, and are a terrific way to understand the 3-dimensional space of any building layout [2].

Exploring container design by looking into other solutions for reusing these steel boxes, which are usually left unused or expensively recycled in a complex manufacturing process that not only costs money but also consumes energy. Explains how the container handles the load and compares its expected lifespan. After analysing six cases, it is concluded that, despite their very systematic and regular shape, shipping containers can be reused to create not only functional spaces but also various configurations of architectural programmes or functions ranging from a very simple residential unit to a small office building extending to a shopping mall or a cultural centre that can accommodate a larger number of users [15].

2.2 Thermal and Indoor Environment

A comparative study between a container building (called a retainer) and a national building of the same size and made of conventional materials, both situated in Chennai. Comparison in terms of embodied energy and indoor thermal performance. Both buildings were evaluated for thermal performance. The average indoor temperature of the greenhouse is about 0.38 °C less than that of the conventional building. It is not a significant difference, but it proves that the thermal performance of Green container is no way inferior to that of conventional buildings [14]. Comparing the indoor environment of ordinary classrooms and container classrooms in several schools in Austria. Data loggers and sensors were installed in each school to measure the temperature, and relative humidity sensors were installed in the four classrooms of each school. Container schools had high temperatures in the winter, resulting in higher heating costs. During the summer months, container schools had a relatively lower temperature than ordinary schools. The CO_2 concentration in the container schools was high due to the higher occupancy density. Overall, container classrooms had inferior environmental quality compared to ordinary classrooms [9–11].

2.3 Structure

By using finite element analysis programme to do numerical analysis using LUSAS analysis software. The structural solution for a 20' and 40' HC container home for a single family was presented as a case study. The openings and additional supports were modelled. With the available steel structure code procedures, it is possible to evaluate a container building, but it is difficult. Assumptions and various simplifications had to be made. Components that require additional strengthening were listed. The paper concluded that even though the process is difficult, the feasibility of such a system for construction should be recognised [3, 4].

Structural guidelines for using ISO containers for non-shipping applications such as construction that will help in forming the design requirements. The models of containers were made using HyperMesh, SolidWorks, and Abaqus software. The basic model as well as seven other modified models of containers were analysed for different loading scenarios. Plots of maximum force applied and displacement at yielding of different modifications were presented as graphs. A significant reduction in lateral resistance was observed when walls in the loading direction were removed. The original model was the strongest [1].

2.4 Life Cycle Analysis

The energy was calculated during the structures' development and operating phases during a 15-year period. The aim was determined using the Life Cycle Energy Assessment (LCEA) and Life Cycle Cost Assessment (LCCA) techniques. A prefabricated house has a 29.7% lower life cycle cost and 25.7% lower life cycle energy than a container house. It is preferable to build larger residences with a higher number of inhabitants. The investigation revealed that housing with greater floor areas is more energy and cost-efficient [5, 13, 16].

2.5 Versatility of Container Architecture

The long-term psychological well-being of earthquake survivors can be influenced by the type of temporary housing they were provided with. An analysis of individuals living in converted container homes, wooden dachas, and a control group unaffected by the earthquake revealed consistent findings [3, 8, 17].

3. RESEARCH METHODOLOGY AND DATA COLLECTION

3.1 Methodology Adopted

The project site is analysed in detail, and design considerations are derived. The design is executed with the help of data collected from real-life case studies as well as from standards and site-specific constraints.

3.2 Data Collection on Standards: ISO and CSC Standards

The specifications, serviceability, structural strengths, and applications of shipping containers are outlined in various documents provided by the International Organisation for Standardisation (ISO) and the International Convention for Safe Containers (CSC). Table 23.1 provides a comprehensive list of the ISO container standards that are adhered to by the majority of shipping containers used worldwide.

Table 23.1 ISO specifications for container

ISO Standard	Information Provided
668:2020	Freight Containers Classification, external dimensions, and ratings
830:1999	Freight Containers Terminology
6346: 1995	Freight Containers Coding, identification, and marking for freight containers
1161:1984	Freight Containers Specification of corner fittings for series 1
8323:1985	Freight Containers Air/Surface (intermodal) general purpose containers-specifications and tests
3874:1997	freight containers Handling and securing
1496-1:1990	Freight Containers Specifications and testing of series 1 freight containers

Source: Authors

3.3 Container Specifications

Standard dimensions of containers are listed below in Table 23.2, along with their weights.

Table 23.2 Details of standard dimensions of containers

Container Type	External Length (ft)	External Width (ft)	External Height (ft)	Internal Length (ft)	Internal Width (ft)	Internal Height (ft)	Door Opening Width (ft)	Door Opening Height (ft)	Approx. Tare Weight (kg)	Approx. Max Payload (kg)
10-ft Container	10' (3.05m)	8' (2.44m)	8'6" (2.59m)	9'2" (2.80m)	7'8" (2.34m)	7'10" (2.39m)	7'8" (2.34m)	7'6" (2.28m)	1, 300–1, 500	8, 000–10, 000
20-ft Standard	20' (6.06m)	8' (2.44m)	8'6" (2.59m)	19'4" (5.89m)	7'8" (2.34m)	7'10" (2.39m)	7'8" (2.34m)	7'6" (2.28m)	2, 200–2, 400	28, 000– 30, 000
20-ft High Cube	20' (6.06m)	8' (2.44m)	9'6" (2.90m)	19'4" (5.89m)	7'8" (2.34m)	8'10" (2.69m)	7'8" (2.34m)	8'6" (2.59m)	2, 300–2, 500	28, 000– 30, 000
40-ft Standard	40' (12.19m)	8' (2.44m)	8'6" (2.59m)	39'6" (12.03m)	7'8" (2.34m)	7'10" (2.39m)	7'8" (2.34m)	7'6" (2.28m)	3, 600–4, 000	26, 000–28, 000
40-ft High Cube	40' (12.19m)	8' (2.44m)	9'6" (2.90m)	39'6" (12.03m)	7'8" (2.34m)	8'10" (2.69m)	7'8" (2.34m)	8'6" (2.59m)	3, 800–4, 200	26, 000–28, 000
45-ft High Cube	45' (13.72m)	8' (2.44m)	9'6" (2.90m)	44'6" (13.56m)	7'8" (2.34m)	8'10" (2.69m)	7'8" (2.34m)	8'6" (2.59m)	4, 200–4, 800	25, 000– 27, 000

Source: ISO 668:2013

Shipping containers are typically used to transport goods internationally, and in building applications, shipping containers are referred to as Intermodal Steel Building Units (ISBU's), which are manufactured in various sizes. The typical sizes for shipping containers are 20' (6 m) and 40' (12 m), as shown in Figure 23.1 and 23.2 which are 8'-6" in height and 8' (2.4 m) wide. A taller version, High Cube, is 9' (2.77m) in height [6, 7, 20].

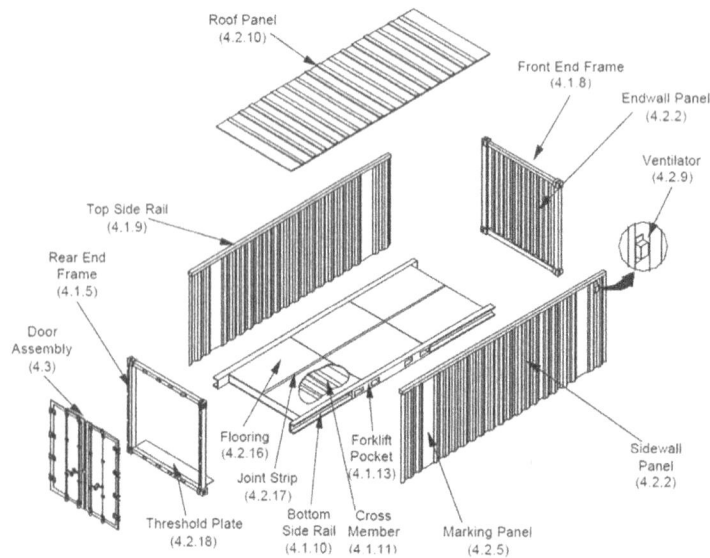

Figure 23.1 Busted view of container standard 20' shipping container

Source: Authors

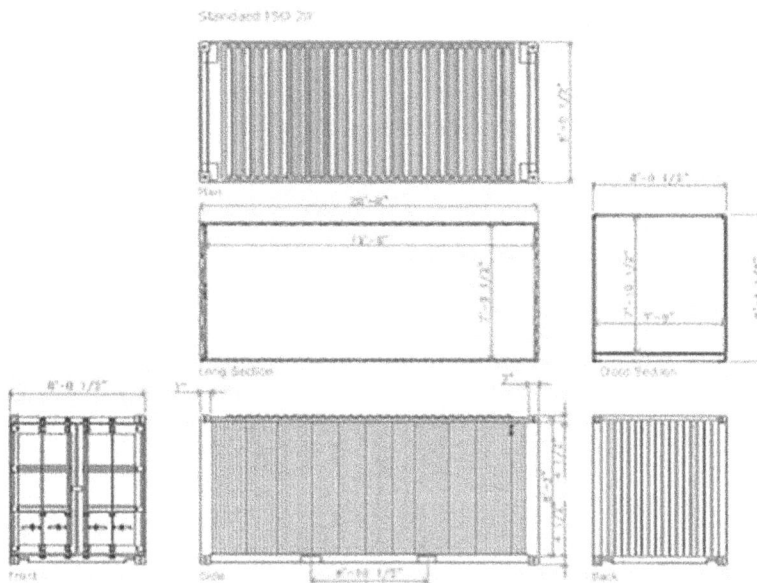

Figure 23.2 Plan of container

Source: Authors

3.4 Building Envelope – Insulation

Insulation is generally used to impede the flow of heat over a membrane. The choice of insulation is so crucial if it is to be habitable for human use. In addition to R-value, the materials' compatibility with the rest of the building envelope must be considered. (R-value is a measure of a material's ability to resist the transmission of energy; the greater the R-value, the better the insulation.) Doubling the thickness of an insulating material doubles its R-value and reduces energy transmission by half. However, the law of diminishing returns states that applying the same resources again yields half the net change. Comfort and durability are also important considerations. A complete insulation strategy considers cost, application procedures, product efficiency, and environmental impact. The only approach to strike the correct balance between construction cost, long-term energy savings, and total environmental impact is to examine a full wall assembly design and its energy analysis [18, 19].

4. PROJECT SELECTION: HOUSING PROJECT

- The project is constructed for 192 families who lost their houses to sea erosion and for 13 other families who had been living in rehabilitation camps for the past 6 years (case study).
- Ready to occupy 2 BHK flats in 24 blocks with 8 houses (540 square feet) in each.
- Site area: 3.47 acres.

4.1 Design Consideration

To design the selected housing project with the use of refurbished shipping containers, A detailed list of design considerations is derived from different sources, including:

- Site analysis

- Literature review
- Standards and norms

The design shown in Figure 23.3 is focused on one single block, and thus this block is the The focus of the study was to understand the feasibility of the structure, cost, and time taken to finish

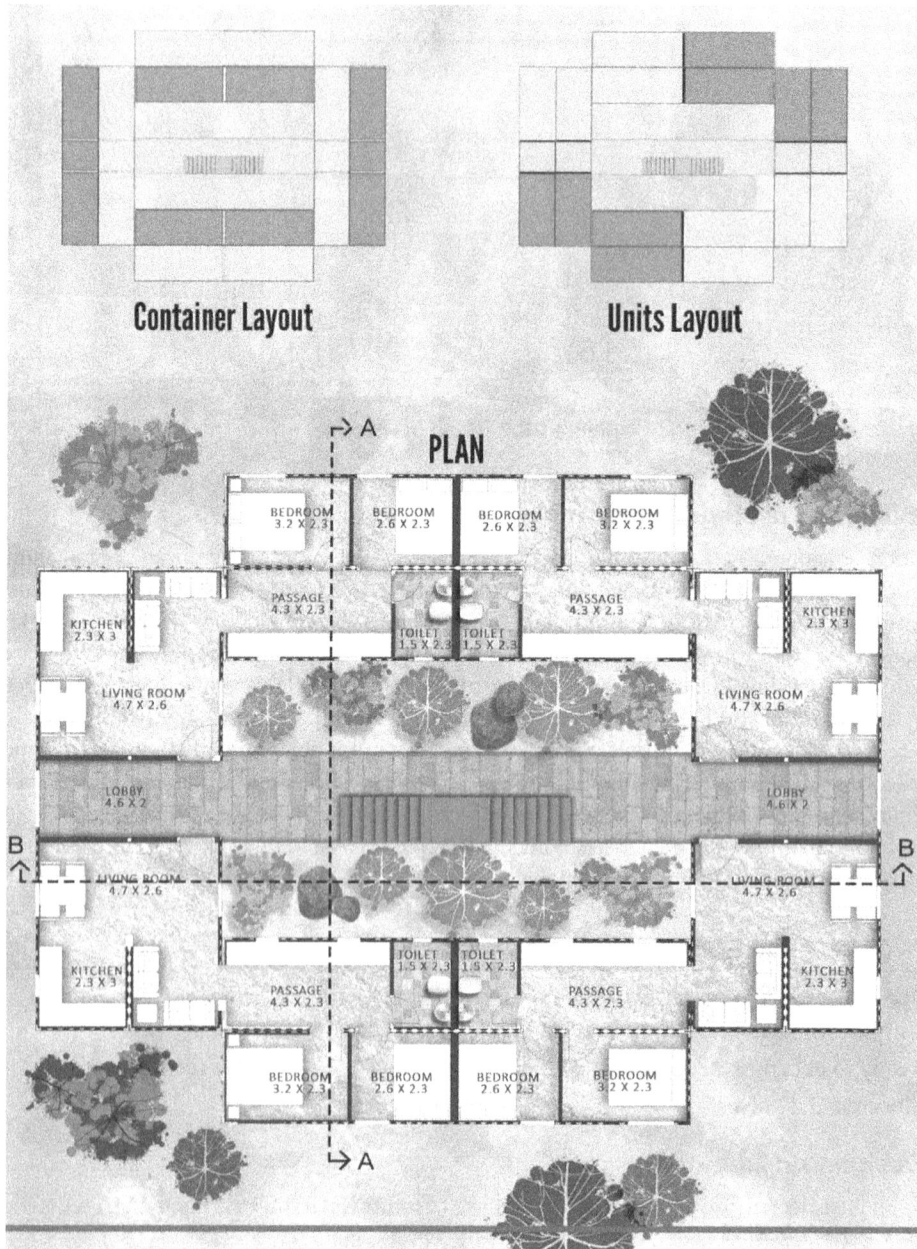

Figure 23.3 Floor plan of the proposed building

Source: Authors

this project. The design was finalized based on the parameters, making it the best fit for the area. requirements and providing more than the existing conventional plan. The main concept here is to provide a space that has a livelihood with containers, and making this possible is seen in the above design, which ensures total safety, a special arrangement, and a total sustainable architecture where materials such as cement, sand, blue metal, water, and intensive labour work have been mitigated. The exterior of the whole container architecture model is simple, and window shades have been added to the windows shown in the below Figs. 23.4.

Figure 23.4   3D View of the Building

Source: Authors

4.2 Area Statement

In the proposed design, we have allocated 192 units to the whole site, and one block contains 12 units, with 4 units on each floor, including the ground floor.

- Individual Unit Area: 54.2 sqm
- Single floor area (4 units): 269.13 sqm
- Built-up area of the block (12 units): 807.39 sqm

4.3 Analysis

To execute the analysis of the structure and design, the process is started with the help of Building Information Modelling (BIM). To initiate the process, the tool selected for execution is Autodesk Revit 2020. The whole structure is made in Revit, and thus this generates all the requirements such as visual view, structure, quantity, and estimation.

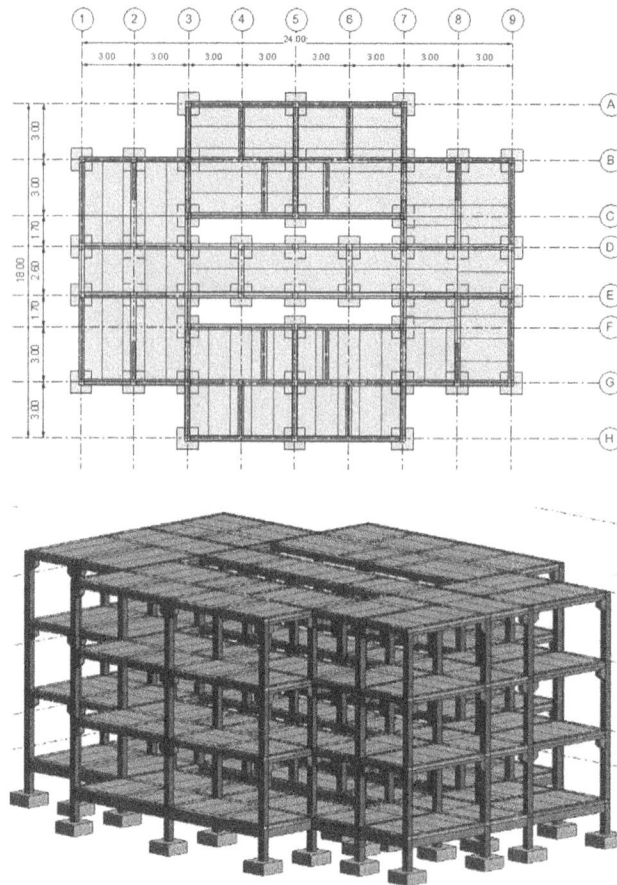

Figure 23.5 The snippets of the model

Source: Authors

4.4 Structure Analysis

Structural analysis of 40-foot containers was done. The software used for the analysis is Sim Scale, which is a cloud platform for structural stimulation. The maximum design load, as computed above, is applied to the structure with the base of the container fixed for analysis.

The following results are computed from the structural analysis:

- Von Mises Stress
- Cauchy stress
- Displacement
- Total strain

For structural analysis, two models were considered for the study, and the defined load combination was applied. The first model was a 40-foot container with no openings. The software used for the study is SimScale, which is a cloud platform for engineering stimulations (SimScale simulation software). The maximum load combination adopted was 1.25 DL + 1.5 LL, which gave a maximum design load of 4500 N/m^2.

The floor had a capacity of 180 psi, or 8600 N/m^2, which is more than the maximum design load of 4500 N/m^2. Hence, it can be concluded that the maximum design load adopted is safe and the structure is safe. On structural analysis of the structure using Simscale, the first model of the container (without opening) showed a maximum Cauchy stress of 8200 N/m^2 from Figure 23.6 and a Von Mises stress of 5514 N/m^2 from Figure 23.7, which is less than the floor capacity of 8600 N/m^2, hence the structure is safe against loading. Also, the maximum displacement on the structure is 6.9×10^{-8} m from Figure 23.8, which is a very low displacement, hence the structure is safe. Considering the second structure (container with openings), the maximum displacement of the structure was 2.19×10^{-6} m, which is very low.

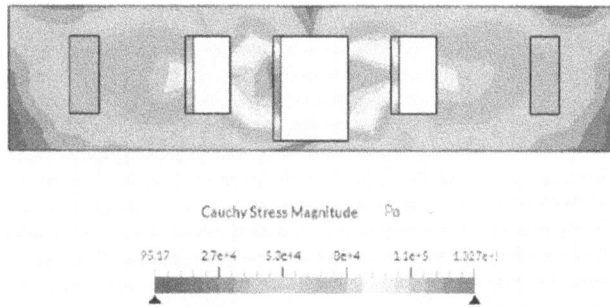

Figure 23.6 Cauchy stress on the wall

Source: Authors

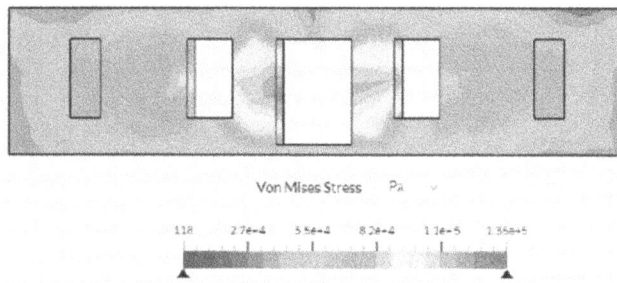

Figure 23.7 Von mises stress on the wall

Source: Authors

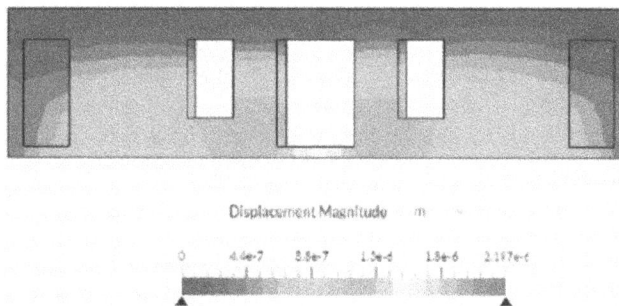

Figure 23.8 Displacement Magnitude on the wall

Source: Authors

4.5 Foundation

For the flood-prone area and the site on the radar of the airport and CRZ, it is addressed to have a stone or masonry foundation to hold the structure and make it flood-prone. This foundation will make the structure stable and flood-prone, as well as termite and corrosion-resistant, by increasing the life of the structure with isolated footing, as shown in Figure 23.9, and by reducing the life cycle cost of the project.

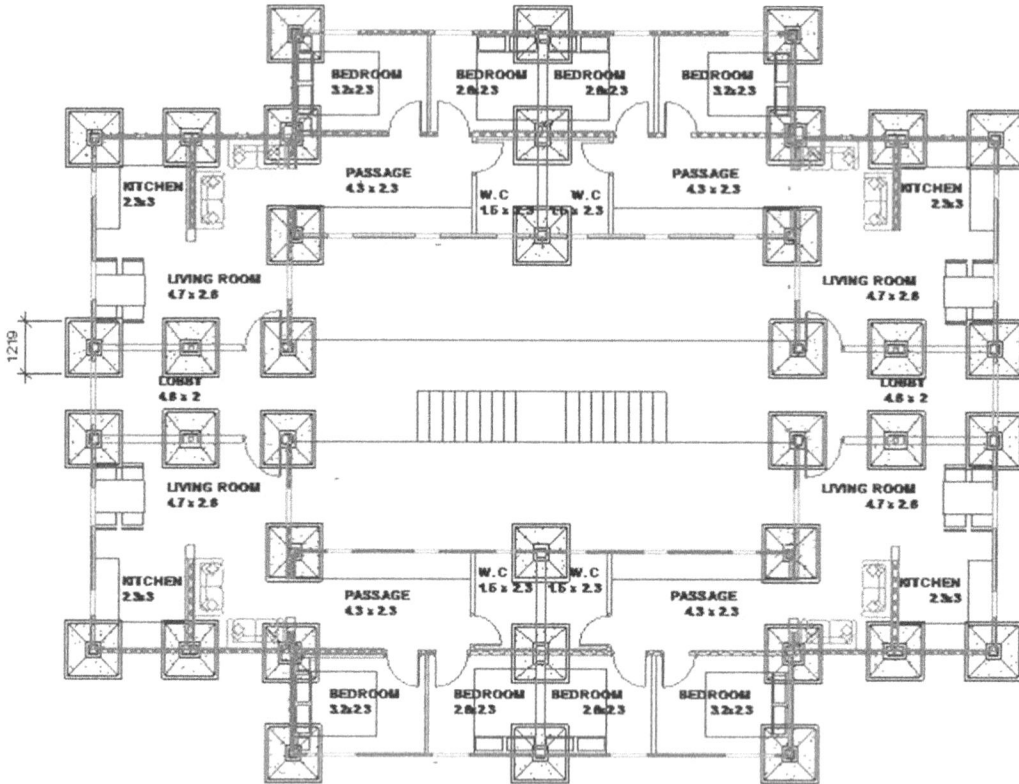

Figure 23.9 Foundation Layout

Source: Authors

4.6 Time Analysis

To calculate the duration of the project, the PERT method is preferred. So, in this method, the specific durations are calculated with different approaches, like pessimistic, optimistic, and most likely durations. This duration is based on the data that is provided by the architects and contractors. In the end, the total expected duration was found to be 101 days as shown in Table 23.3.

4.7 Cost Analysis

As per the planning stages data, the materials that are best suited for container architecture are identified, and calculated for the whole block of 12 units in Table 23.4, which will give us the cost of each unit at the end of the calculation.

Table 23.3 PERT analysis of the project

Task Name	Duration (Days)			Expected Duration
	Optimistic	Most Likely	Pessimistic	
Overall Project Duration	**73**	**93**	**123**	**101**
1. Planning of Project				
Capital allocation/financing	4	4	4	4
Develop initial home plan/design	7	9	20	11
Assemble site-related information and analysis	3	3	3	3
Material selection	1	2	3	2
2. Finalizing Design and Execution Approval				
Finalizing structure design and floor plans	5	8	10	9
Preparation of construction plans	5	5	5	5
Allocate final budget	2	2	2	2
Local structural engineering approval	5	7	15	8
Building permits	5	9	15	10
Identify general contractor	3	3	3	3
3. Site Preparation and Foundations				
Shipping container identification and purchase	5	5	5	5
Land clearance, drainage and access	5	5	5	5
Foundation layout and excavation	5	5	5	5
Run utilities to site	2	2	2	2
Lay foundations	7	10	16	11
4. Place and Modify Containers				
External air seal/spray foam insulation	1	1	1	1
Siting containers and attaching to foundation	2	3	3	3
Shipping container modifications, cutouts, and reinforcements	6	9	10	9
Structural framing and roofing	3	4	5	5
Install windows, doors, and siding	6	8	10	8
Utility services rough-in	5	5	5	5
Interior framing	4	5	5	5
5. Finish to Occupancy				
Insulation	4	8	8	8
Drywall	2	2	2	2
Fit flooring	2	2	2	2
Finish plumbing, electrical, and HVAC fixings	8	8	10	9
Finish fixtures, fittings, appliances and trim	10	12	15	13
Painting and decorating	3	3	3	3
External cladding	5	5	10	6
External landscaping	5	5	5	5
Final walk-through and cleanup	1	2	2	2
Receive final approval	2	2	4	3

Source: Authors

Table 23.4 Estimation of the container houses

Item	Unit Price	Quantity	Price
Used 40' high cube container	1, 50, 000/no	2	300000
Foundation	4500/m³		
Windows	2000/m²	7.98 m²	15960
Doors	4000/door	4 nos	16000
Insulations			
On wall	385/m²	64.688 m²	24904.8
On ceiling	75/sq ft = 800/m²	45.43 m²	36344
Wall finishing	80/m²	64.688 m²	5175
Partition walls	1600/m²	30.364 m²	48582.4
Toilet walls	323+1130.22=1453.22/m²	20.173 m²	29315.8
Flooring			
Inside home	105/sq ft = 1130.22/m²	45.43	51345.89
Inside toilet	323+1130.22=1453.22/m²	3.3647	4889.65
Lobby including staircase treads	105/sq ft = 1130.22/m²		
Steel columns	50000/ton		
Steel beams	50000/ton		
Staircase	100000/floor		
Flooring for passage	500/sq ft = 5400/m²	25.1 m²/floor	
Plumbing	100/sq ft = 1076.4/m²	54.2 m²	58340.88
Electrical	120/sq ft = 1291.68/m²	54.2 m²	70009.05
Contingency			5%

Source: Authors

On a rough comparison, container building cost less than the other two alternatives. The calculations might not give a clear picture suitable for all regions of India. The Muttathara project location was selected due to its recent development, close proximity to the port, and the abundance of available data. The feasibility seen in Table 23.5 may vary for different projects in different regions with different project scopes. There might be inaccuracies in the activity duration data since this is a new methodology of construction. The values collected from market surveys or online sources may not reflect real-world values because of new developments, like the pandemic situation. Data on the total availability of used containers in India might not be available because of a lack of statistical data.

Table 23.5 Cost comparative statement of different types of building

Building Type	Area of Each Unit (Sq. Ft.)	Cost of Each Unit (₹)	Cost Per Sq. Ft (₹)
Conventional Building	540	10, 50, 000	1944
Precast Building	710	10, 10, 000	1423
Container Housing	585	8, 10, 000	1385

Source: Authors

5. CONCLUSION

India's housing crisis has created a significant demand for affordable housing, particularly among urban migrants from the EWS and LIG categories. The economic idea of demand and supply is commonly used to explain the

Globally, there is a housing shortage. The housing crisis in India primarily stems from insufficient housebuilding, exacerbated by factors such as population growth, urban migration, increasing living costs, natural disasters, and various economic and sociological challenges.

These challenges significantly impede the government's capacity to offer suitable housing for individuals living in inadequate conditions. This research proposes employing container architecture as a cost-effective alternative for affordable housing by repurposing unused shipping containers. To assess the viability of deploying ISBUs, a comparison was made with 192 existing post-disaster shelter units constructed using traditional brick-and-mortar methods in a real-life scenario. To analyse the effectiveness of a planned ISBU shelter vs. the current building, numerous criteria were empirically assessed.

The study found that using containers for prefabricated buildings is a viable and cost-effective alternative to traditional methods. Other advantages include modularity, portability, and ease of transportation. Structural analysis evaluates the solution's constructability and soundness. The cost per square foot for conventional buildings was Rs. 1944, precast buildings were Rs. 1423, and containers were Rs. 1385.

REFERENCES

[1] Ataei, M. (2019). Design of a two-story iso shipping container building. Toronto, Ontario, Canada.

[2] Bernardo, L. F., Oliveira, L. A., Nepomuceno, M. C., & Andrade, J. M. (2013). Use of refurbished shipping containers for the construction of housing buildings: details for the structural project. *Journal of Civil Engineering and Management*, 19(5), 628–646.

[3] Cameron, D. C. (2019). A case study on the feasibility of shipping container homes as housing for disaster victims.

[4] Giriunas, K., Sezen, H., & Dupaix, R. B. (2012). Evaluation, modelling, and analysis of shipping container building structures. *Engineering Structures*, 43, 48–57.

[5] Hamilton, C. K. G. (2017). The benefits of incorporating shipping containers into the climate change adaption plans at the NASA Wallops Flight Facility.

[6] ISBU and Shipping Container Construction Approval Data. (n.d.). GreenCube ApprovalReport. International Organisation for Standardisation (ISO), International Maritima Organisation (IMO).

[7] ISO. (1990). ISO 1496-1: Series 1 Freight Containers—Specification and Testing—Part 1 General Cargo Containers.

[8] ISO/TC 104. (1995). ISO 668: 1995 Series 1 Freight Containers: Classification, Dimensions and ratings.

[9] ISO/TC 104. (1999). ISO 830: 1999 freight container vocabulary.

[10] ISO/TC 104. (1990). ISO 1496-1: 1990 Series 1 Freight Container Specification and Testing Part 1: General Cargo Containers for General Purposes.

[11] ISO/TC 104. (1997). ISO 3874: 1997 series 1 freight containers—handling and securing.

[12] Anagal, V., Karvenagar, P. I., & Dhongde, S. (2020). Container Housing–Challenges and Opportunities. In *First International Conference on Theory of Architectural Design: Global Practices Amid Local Milieu* (pp. 175–181).

[13] Islam, H., Zhang, G., Setunge, S., & Bhuiyan, M. A. (2016). Life cycle assessment of shipping container home: A sustainable construction. *Energy and Buildings*, 128, 673–685.

[14] Vijayalaxmi, J. (2010). Towards sustainable architecture–a case with Greentainer. *Local Environment*, 15(3), 245–259.

[15] Radwan, A. H. (2015). Reusing shipping containers in creating various architectural spaces. Helwan University Architecture Department, College of Fine Arts.

[16] Atmaca, A., & Atmaca, N. (2016). Comparative life cycle energy and cost analysis of post-disaster temporary housings. Applied Energy, 171, 429–443.

[17] Lingiardi, V., Gazzillo, F., & Waldron Jr, S. (2010). An empirically supported psychoanalysis: the case of Giovanna. Psychoanalytic Psychology, 27(2), 190.

[18] Li, Y., Li, G., Wang, T., Zhu, Y., & Li, X. (2019). Semicustomized design framework of container accommodation for migrant construction workers. *Journal of Construction Engineering and Management*, 145(4), 04019014.

[19] Moore, C. M., Yildirim, S. G., & Baur, S. W. (2015). Educational adaptation of cargo container design features. In the American Society for Engineering Education (ASEE) Zone III Conference.

[20] Oliveira, L. A., Nepomuceno, M. C., & Andrade, J. M. (2013). Use of refurbished Shipping containers for the construction of housing buildings: details for the structural project. *Journal of Civil Engineering and Management*, 19(5), 628–646.

Advances in Construction, Real Estate, Infrastructure and Project Management – Anil Kashyap et al. (eds)
© 2026 Taylor & Francis Group, London, ISBN 978-1-041-13433-6

24 Rethinking Waste in Construction: A Study on Reduction and Recycling Practices

Sudarshan D. Kore[1]

Assistant Professor,
School of Construction, NICMAR University
Pune

Anil Agarwal[2]

Professor,
School of Construction, NICMAR University
Pune

Babalu Rajput[3]

Assistant Professor,
School of Construction, NICMAR University
Pune

Amol D. Pawar[4]

Associate Professor,
School of Construction, NICMAR University
Pune

■ **ABSTRACT:** Construction waste is a significant environmental issue worldwide, contributing to adverse effects on the ecosystem and public health. The proper management of construction waste is crucial to promote sustainable development and minimize landfill waste. The project's objectives are to identify the sources of construction waste, catalog the types and quantities of waste generated throughout the project, compare the amounts estimated for the building's construction at the beginning and end to find the proportion of waste and provide recommendations for companies involved in the construction industry to lessen their garbage output. To fulfill the above goal the commercial project located Kalamassery, India, is considered for further analysis. The building project's material quantities lists were followed to gather information. The project's responsible construction business provided us with these lists. There was a comparison between the lists and the actual amounts of materials used in the construction. We calculated the proportion by which the initial and final quantities differed due to wastage. Appropriate strategies for waste reduction were suggested when the causes of these wastes were identified. According to the study of the building project, the main waste generators

[1]skore@nicmar.ac.in, [2]anilagarwal@nicmar.ac.in, [3]babalurajput@nicmar.ac.in, [4]apawar@nicmar.ac.in

DOI: 10.1201/9781003669814-24

were unused and surplus materials, trash from demolition, trash from packaging, and scrap from the building site. The amount by which material waste causes a disparity in quantity. The most common reasons for these wastes were improper substance storage and handling, inadequate communication between project stakeholders, inadequate planning, and an absence of suitable waste management methods.

■ **KEYWORDS:** Sustainability, Material, Waste, Reduction

1. INTRODUCTION

There will be a total of 9.8 billion individuals on the earth by the year 2050, an increase from a population of 7.6 billion in 2017 [12]. Several problems are exacerbated by the growing world population, including excessive water use, urbanization, increased chemical use, forest loss, and a rise in motor vehicles. The use of oil, gas, and coal is rapidly rising, which causes the quantities of methane and carbon dioxide in the environment to grow swiftly. By 2050, 2.5 billion individuals will live in cities, with Asia and Africa accounting for 90% of this increase [19]. The construction industry contributes to environmental degradation and is a crucial mechanism for developing infrastructure to promote city expansion. The building sector consumes a third of all energy produced and is responsible for 40% of all atmospheric CO_2 emissions [5]. Despite being the sector that consumes the rawest materials derived from natural resources. Furthermore, the ecology is harmed by the waste produced during construction. World Economic Forum anticipated that by 2025, there will be 2.2 billion tonnes of solid waste (SW), up from 1.3 billion tonnes in a study published in 2012 [4]. Almost half of the yearly SW created is made up of building and demolition waste [7].

Figure 24.1 Demolition of building

Source: Authors

Waste during construction adds no value and costs money directly and indirectly [13]. Building, remodeling, and demolition waste includes rubble and other unwanted goods [7, 8]. This waste may contain bricks, concrete, wood, glass, metal, and plastics. Construction waste includes any substance or product that must be removed from the building site and shipped elsewhere, must be used on the site but not for the project's intended purpose due to harm, excess, or non-use, can't be used because it doesn't meet requirements, or is a byproduct of construction [1]. Pollution, deforestation, and greenhouse gas emissions from building debris harm the ecosystem [2]. Construction waste may also fill landfills, damaging the environment. Industry and human activity create waste that cannot be sold. Hazardous CDW disposal and dumping harms society and the environment [3].

1.1 Causes of Construction Waste

Waste generated during construction can be attributed to several different processes. Without a doubt! The term "construction waste" refers to any material that is produced during construction activities but is not utilized for the purpose for which it was intended and is then discarded accordingly. Several variables contribute to the production of garbage from construction [4]. The below are major factors that cause construction waste are discussed.

- Design and Design Changes
- Material and Procurement
- Construction Methods and Planning
- Human Resources
- Effect of Weather
- Wrong Material Storage

1.2 Environmental Impact of Construction Waste

Construction waste has significant environmental impacts at various stages of its lifecycle, from extraction of raw materials to disposal. Here are some of the key environmental impacts:

- Reducing Available Resources: Timber, minerals, and aggregates are all essential building materials, but their extraction can deplete natural resources, destroy habitats, and cause biodiversity to vanish.
- Energy Utilization: Greenhouse gas emissions and the acceleration of climate change are exacerbated by the massive quantities of energy needed for the production, transportation, and processing of building materials.
- Air Pollution: Negative impacts on air quality and human health can be caused by construction-related air pollutants, which include particulate matter, volatile organic compounds (VOCs), and nitrogen oxides (NOx).
- Water Pollution: When building materials like paints, chemicals, and solvents are not disposed of properly, they can contaminate water bodies through runoff. This can cause water pollution and endanger aquatic ecosystems.
- Landfill Space: A large amount of trash is sent to landfills, and a lot of it is construction debris. This takes up precious area and causes soil to be contaminated and deteriorate.
- Destroying Habitats: Illegal dumping in natural areas, which results from improper construction waste disposal and management, destroys habitats and disrupts ecosystems.
- Greenhouse Gas Emissions: As the organic elements in building waste decompose anaerobically, landfills release methane, a powerful greenhouse gas. Warming the planet and altering its climate are both aided by methane.
- Construction Debris Pose Serious Health Dangers to Workers, Neighbors, and Wildlife from Asbestos, Lead-Based Paints, and Chemical Pollutants.

- The Visual Impact: Construction trash, including abandoned materials and litter, can diminish the visual quality of both urban and natural areas and take away from the beauty of landscapes if not effectively handled.
- Impact on Ecosystems: Excavation and other forms of land clearing in the construction industry have the potential to alter patterns of local biodiversity, fragment habitats, and disrupt natural ecosystems.

Construction debris has far-reaching environmental consequences, and effective waste management solutions—including reduction, reuse, recycling, and correct disposal—are necessary to mitigate these effects. Minimizing the environmental impact of construction operations can be achieved by the use of sustainable practices. These practices include resource-efficient construction technologies, sustainable design, and the promotion of the circular economy. To further lessen the environmental toll of construction debris, we need legislative frameworks, education initiatives, and financial incentives for green building.

2. LITERATURE REVIEW

The environmental impact of waste advancement is substantial. The development of waste has a major effect on the natural world. Construction mechanical and industrial zones provide a great deal of output to meet the demands of India's many businesses and hotels as well as the country's mandatory foundation extension projects. As a result, proper disposal of building site garbage is essential [16]. Here, waste minimization plans are the way to go for cutting down on-site waste. Owners and contractors suffer enormous financial and administrative losses due to waste in construction projects [11]. Therefore it is essential to plan and control the waste, how it can be reduced and reused for other uses was discussed by Janani & Lalithambigai, [9] in their study. Loss of leadership, inadequate storage facilities, mistakes in layout, changes in layout, employee error, poor interpersonal skills, an absence of ideas for recycling, and an ineffective work style are eight more aspects that they see as contributing to increased waste. The use of bar codes for effective waste reduction in the construction industry, appropriate storage facilities, the recruitment of staff to oversee waste generation on the project site, and numerous other recommendations for waste reduction were also part of the study.

According to Mhaske et al., [14] one of the most common issues discovered in building projects involves not enough preparation of the construction waste. The study's findings suggest that identifying the main contributors of construction waste can be accomplished by methodically analyzing the activities that generate trash and the kinds of waste that happen frequently on building sites. Once it is recognized, it can be reduced or avoided, which will have a significant positive financial impact on the company. It is possible to save money, conserve resources, and protect the environment if building waste is collected from construction sites and recycled. The construction and demolition waste in the UK is expected to be 120 million tons annually, comprising approximately 13 million tonnes of unusable material, according to recent estimates released by the UK Government [15]. In addition, new laws being passed, innovative methods and technology emerging for recovering and disposing of garbage, and an increase in public awareness are all working together to transform the waste management landscape. It is the duty and opportunity of clients, suppliers, contractors, and designers to minimize construction waste. But to effectively and sustainably reduce waste, all parties involved in the building supply chain must take a more proactive stance toward waste management, or "design out" waste. Reengineering present practices to reduce waste while contributing to a healthier environment through effective and affordable sustainable waste minimization solutions is necessary. According to Thongkamsuk et al., [18]. 10–30% of the type of waste which is dumped at different disposal sites worldwide is often made up of construction and demolition debris. In the paper, it is stated that 23% of all C&D trash is made up of concrete, 18% is flooring, 13% is steel, and 6% is wood. Furthermore, over 60% of all rubbish produced can be attributed to just these four items. A lack of legal regulations and guidelines for C&D waste dealing with, such as minimizing waste at the source, treating waste, recycling, reuse, shipping, and disposal, as well as a shortage of operational staff with these areas of expertise, are the main causes of this generated waste. These issues also include poor planning during the building design phase. A proper and effective waste reduction strategy starts at the source with trash organization, reuse, recycling, transportation, and burial. Apart from the planning stage, factors that impact the generation of large-scale waste include the storage of building supplies and damage caused by transportation. A large quantity of waste is produced by the building industry as a result of its fast expansion and improvement. The researchers Kalilurrahman & Janagan [10] set out to determine what exactly causes trash to accumulate. A comprehensive survey and many interviews with individuals working in the construction business have already been carried out. They have proposed a land disposal strategy with zero waste based on the 3Rs: reduce, reuse, and recycle, as well as a strategy for dealing with building and demolition debris. The timely and cost-effective procurement, storage, and management of the right kind and amount of construction materials is the hallmark of any project that employs a material management system [17]. Construction sites are becoming more efficient and less prone to human mistake as a result of the use of new and emerging technologies including radio frequency identification (RFID), bar-code readers, and wireless communication systems. Implementing a waste management system into the project could help cut down on construction waste. To lessen material waste, one can apply the 3R and 4R principles of reduction, reuse, recycling, and recovery. These have several uses over a product or service's lifetime. The study conducted by Ismail et al [6] provides a comprehensive account of the CDW that is generated throughout a construction project. We must develop a safe disposal alternative for CDWs because Egypt's problem with this has become a big concern. From in-depth interviews with industry personnel, three critical aspects influencing safe disposal of CDW (SDCDW) were identified. The first three include tracking and scheduling of construction debris, locating illicit disposal sites, and managing routes, collections, and transportation. "Detection of CDW illegal disposal sites" was shown to be the most effective parameter, while "Selection of shortest path transport route" was known to have the highest level of application.

Figure 24.2 DCP quarters Kalamassery, Kerala, India

Source: From the case study evaluated by the author

To understand the cause of the waste generation in constructin project a real time case has been considered for the study. The USQ 2 Kalamassery project which includes the construction of the D.C.P. quarters for the Kerala police. The Kerala Police Housing Construction Corporation Limited acted as the project's consultant, and E.M. Francis Antony finished the work.

Table 24.1 Percentage difference in quantity

S no	Material	Difference in %
1	Centering, etc., and removing of form for mass concrete footings, column bases, etc.	14
2	Straightening, cutting, bending, positioning, and binding of steel reinforcement for R.C.C. work is all finished up to plinth level. Bars with a thermo-mechanical treatment grade of at least Fe-500D	11
3	For superstructures up to floor two level with a thickness of 15cm, solid masonry utilising precast solid blocks (factory produced) of dimension 30x20x15cm or the nearest size available in accordance with IS 2185-part I of 1979 Cement to coarse sand ratio of 1:6, etc. complete	11
4	1:3 (1 cement:3 fine sand) 1:6 mm cement plaster)	12
5	Supplying and spreading white cement-based putty with an average thickness of 1 mm from a manufacturer and brand you can trust to prepare the surface evenly and completely.	14
6	Using Premium Acrylic to finish walls Smooth exterior coating with the necessary shade of silicone additives: New work (two or more coats applied at 1.43 litres per 10 square metres over and including the external primer priming coat applied at 2.20 kilograms per 10 square metres)	17
7	Supplying and installing vitrified floor tiles in various sizes (the thickness will be determined by the manufacturer), adhering to IS: 15622, of approved make, in all colours and shades, laid on a 20 mm thick cement mortar that is 1 cement to 4 coarse sands, complete with grouting the joints with white cement and matching pigments, etc. 600 × 600 mm tile size.	14
8	Supplying and placing M-25 grade machine-batched and machine-mixed design mix cement concrete for reinforced cement concrete work, utilising cement content in accordance with authorised design mix,	10

Source: Compiled by authors

Table 24.2 Allowable wastage as per BIS (Bureau of Indian Standards)

Item No	Material	Allowable wastage
1	Cement	0.01
2	Sand	0.06
3	Coarse Aggregates	0.025
4	Tile	0.07
5	Granite	0.08
6	Reinforcement Steel	0.03
7	Structural Steel	5–10%
8	Bricks	0.05
9	Filling Soil	4% to 5%
10	Paint	3% to 4%
11	Wood	0.05
12	Concrete	0.05
13	Mortar	0.03
14	Shuttering Board	6% to 7%

Source: Taken from BIS and made by author

3. RESULTS AND DISCUSSION

According to the information we received from the project's site engineer, there are a number of reasons why the final bill of quantity differs from the quantity expected, including changes to the plan, wastage and measurement problems. Since the goal of our study is to determine how different waste quantities affect the project, we questioned the site engineer about the variables that led to the production of so much waste. Using the information he provided, we were able to identify the elements that led to wastage in various goods, and they are as follows:

3.1 Steel Reinforcement

As per BIS standard the allowable wastage was 3% in our project it was 11% from the site information collected it shows that remaining 8% of difference are due to following reasons, .

- Cutting and shaping: Steel was shaped and cut to meet particular project locations. The residual steel from this operation is too tiny to be useful, therefore it is discarded.

- Overordering: To guarantee that there is sufficient material to finish a job, extra steel was ordered. Due to incorrect storage of the surplus steel, which led to corrosion, they had to be scrapped.

- Labor errors: Because workers lacked the necessary steel working skills, they wasted steel by cutting it to lengths that were longer than those specified in the drawing.

Figure 24.3 Steel waste in in site

Source: Authors

3.2 Bricks and Blocks

Based on the data received from the site, it appears that the remaining 6% of the difference is attributable to the following reasons: our project had 11% wastage, which is higher than the 5% allowed by BIS standards.

Figure 24.4 Brick waste in site

Source: Authors

- Cutting and trimming: To fit into particular spaces or attain particular forms, bricks and blocks were cut and trimmed. Due to the frequent discarding of smaller parts, this procedure produces a lot of trash.

- Damage: Bricks and blocks are damaged during handling and shipping, and these damaged parts are not utilized in the building project, leading to waste.

- Whether the following occurred as a result of inappropriate storage: Bricks were exposed to rain, which reduced their strength, and they were not utilized.

- Quality control: A batch of 1, 000 bricks had poor quality, and a wall made of them cracked when electrical work was being done. The quality control methods for these bricks also didn't satisfy the necessary criteria. This had the effect of rejection of brick pieces.

3.3 Flooring Materials

As per BIS standard the allowable wastage was 7% in our project it was 14% from the site information collected the following reasons;

- Cutting and shaping: To suit particular parts of a project, such as around corners or along walls, tiles are cut and shaped. This could lead to unused tile pieces being thrown because they are too little to be used.
- Installation mistakes: Tiles that were uneven due to improper installation had to be taken out and replaced. This led to more garbage being produced.
- Electrical change: After the floor work was finished, the electrical design underwent some adjustments, and as a result, a new drawing showed some electrical cables running through the floor, necessitating the replacement of the flooring from an entire room.
- Improper room design: Most of the rooms in these buildings were circular rather than the correct square or rectangular shape, which led to additional cutting waste.

Figure 24.5 Tiles waste on site
Source: Authors

Figure 24.6 Plastering waste
Source: Authors

3.4 Plastering

As per BIS standard the allowable wastage was 3% in our project it was 12% from the site information collected it shows that remaining 9% of difference are due to the following reasons;

- Uneven wall form: The walls were not plumb, which needed more application than necessary in some areas to get a precise shape.
- faulty curing; Some plaster cracked as a result of faulty curing, which was corrected.
- Lack of knowledge or expertise: A plasterer with less experience made mistakes that resulted in waste, such as applying too much or too little material or improperly smoothing the surface.
- Disposal of surplus material: Plastering is a dirty task, and the plasterer did not adequately clean up after finishing the job, leaving extra material that needed to be thrown away.

3.5 Paint

As per BIS standard the allowable wastage was 4% in our project it was 14% from the site information collected the following reasons;

- Application mistakes: Improper application of paint left it uneven or sloppy, necessitating removal and reapplication.

Figure 24.7 Paint waste
Source: Authors

Figure 24.8 Timber waste

Source: Authors

Figure 24.9 Concert waste

Source: Authors

- Color adjustment. After the first coat was applied, the colour was adjusted and the whole area where the initial colour was applied was redone.
- Spillage: Since paint is a liquid, it was very likely to spill while being transported, mixed, or applied. This led to waste creation and paint loss.
- Rain: The external paint was harmed by rain, which caused the colour to fade, and new work had to be done.

3.6 Timber

As per BIS standard the allowable wastage was 7% in our project it was 14% from the site information collected it shows that remaining 7% of difference are due the following reasons;

- Cutting and shaping: To fit into particular sections of a project, such as around corners or along walls, timber has to be cut and shaped. As a result, there are residual bits of lumber that are too tiny to be useful and are thrown away.
- Incorrect installation of the wood results in it being uneven and not fitting the criteria, necessitating removal and replacement.
- Insufficient upkeep. Some timbers could not be used because of poor care.

3.7 Concrete

The following factors contributed to the remaining 5% of the gap between our project's 10% and the 5% allowed by BIS standards:

- Inadequate batching: Because the hand-mixed concrete mix was not prepared properly, there was a lot of waste.
- Inaccurate measures: inaccurate material measurements result in an inaccurate mix, which wastes concrete.
- Concrete spills: Mistakes or accidents led to concrete spills, wasting it.
- RMC pump waste: When RMC was utilized, some concrete was lost during the inspection of the RMC pump and pipe line. At the conclusion of the job, some concrete was also left in the pipe.

4. REMEDIAL MEASURES

In order to protect the environment and the economy, waste from construction supplies such as cement, steel, concrete, tiles, paint, and wood must be reduced. Here are a few strategies for minimizing trash during construction:

- Good Planning
- Utilizing Recycled Products
- Utilizing Prefabricated Materials
- Recycling and Reusing Materials like Concrete and steel
- Adopt lean construction principles
- Proper Storage
- Education and Training

The environment and industries profits will both gain from the implementation of these tactics in building projects that decrease waste on cement, steel, concrete, tiles, paint, and wood. The detailed strategy points were highlighted in the below table to reduce the wastage of materials.

Table 24.3 Strategy for waste reduction in the construction industry

S. No	Mtatrial	Strategy
1	Cement	Careful preparation
		Accurate measurement:
		Making use of batching plants
		Appropriate Storage of cement
		Use of substitute materials
		constant monitoring
2	Steel	Careful preparation
		Accurate assessment of the amount of material needed
		Steel reinforcement optimisation
		Use of prefabrication
		Reuse and recycle steel:
		Regularly monitor
3	Concrete	Accurate measuremnt of quantity
		Use of batching plants
		Maximize material usage
		order or preparation of required quantity of concrete
4	Tile	Use of modular tile sizes
		Accurate measureemnt of quantity
		Creation of minimum cutting
		Proper storage
5	Paint	Only buy the paint you actually need.
		Select the proper weather
		Should you prime or not?
		paint waste management

S. No	Mtatrial	Strategy
6	Timber	Plan ahead
		Reuse and recycle steel:
		Selection of proper sturdy wood
		If possible avoid the use of wood
		use of new constrction methodology

Source: Authors

5. CONCLUSIONS

In a nutshell achieving sustainable growth requires a reduction in waste throughout construction projects. It is well-known that the building industry's massive waste has negative impacts on ecosystems and human health. Recycling, reusing, and cutting down on materials used in construction can help construction companies save money, reduce their environmental impact, and build a more sustainable future. The significance of well-managed construction waste has been brought to light by this study. We have determined the origins and amounts of construction waste produced during the construction of a commercial building, and we have also offered solutions to lessen this waste.

The following are typical of such factors: insufficient batching and measuring; excessive material orders; mistakes in labor; injury during management and shipment; exposure to environmental variables such as rainfall; inadequate quality control; incorrect application errors; and spills. According to the findings, insufficient planning, miscommunication, improper material storage and management, and improper trash disposal are the main reasons for construction waste. Implementing lean building principles, having sufficient storage, getting education and training, using recycled and prefabricated components, recycling and reusing resources, and meticulous planning are all effective ways to decrease waste during construction. Both the environment and businesses benefit from construction projects that minimize trash. A more sustainable future is within reach when businesses cut down on waste, which has financial benefits, positive PR, and other benefits.

REFERENCES

[1] Ajayi, S. O., Oyedele, L. O., Akinade, O. O., Bilal, M., Owolabi, H. A., Alaka, H. A. et al. (2016). Reducing waste to landfill: A need for cultural change in the UK construction industry. *Journal of Building Engineering*, 5, 185–193. https://doi.org/10.1016/j.jobe.2015.12.007

[2] Ajayi, S. O., Oyedele, L. O., Akinade, O. O., Alaka, H. A., & Owolabi, H. A. (2017). Optimising material procurement for construction waste minimization: an exploration of success factors. *Sustainable Materials and Technologies*, 11, 38–46. https://doi.org/10.1016/j.susmat.2017.01.001

[3] Utsev, J. T., Imoni, S., Onuzulike, C., Akande, E., Orseer, A. M., Tiza, M. T. (2024). Strategies for sustainable construction waste minimization in the modern era. *Bincang Sains dan Teknologi*, 3(1), 1–10. https://doi.org/10.56741/bst.v3i01.506

[4] Botchway, E. A., Asare, S. S., Agyekum, K., Salgin, B., Pittri, H., Kumah, V. M. A. et al (2023). Competencies driving waste minimization during the construction phase of buildings. *Buildings*, 13. https://doi.org/10.3390/buildings13040971

[5] Crawford, R. H., Mathur, D., & Gerritsen, R. (2017). Barriers to improving the environmental performance of construction waste management in remote communities. In *Procedia Engineering. Elsevier Ltd*, pp., 830–837

[6] Ismail, E. H. R., El-Mahdy, G. M., Ibrahim, A. H., & Daoud, A. O. (2023). Analysis of factors affecting construction and demolition waste safe disposal in Egypt. *Alexandria Engineering Journal*, 70, 515–523. https://doi.org/10.1016/j.aej.2023.03.012

[7] Jain, M. S. (2020). Possible ways of re-utilization of the construction and demolition wastes in various construction sectors. Sustainable Development and Environment (pp. 99–109).

[8] Jain, S. (2016). A case-study on use of precast technology for construction of high-rise buildings.

[9] Janani, R., & Lalithambigai, N. (2020). A critical literature review on minimization of material wastes in construction projects. In *Materials Today: Proceedings. Elsevier Ltd* (pp. 3061–3065).

[10] Kalilurrahman, M. M., & Janagan, M. S. S. (2015). Construction waste minimization and reuse management. *International Research Journal of Engineering and Technology*, 2(8).

[11] Patel, K. V., & Vyas, C. (2011). Construction materials management on project sites. *National Conference on Recent Trends in Engineering & Technology*.

[12] Laovisutthichai, V., Lu, W., & Bao, Z. (2022). Design for construction waste minimization: guidelines and practice. *Architectural Engineering and Design Management*, 18, 279–298. https://doi.org/10.1080/17452007.2020.1862043

[13] Mahinkanda, M. M. M. P., Ochoa Paniagua, J. J., Rameezdeen, R., Chileshe, N., & Gu, N. (2023). Design decision-making for construction waste minimisation: a systematic literature review. *Buildings*, 13(11), 2763.

[14] Mhaske, M., Darade, M., & Khare, P. (2017). Construction waste minimization. *International Research Journal of Engineering and Technology*, 2(8).

[15] Osmani, M. (2012). Construction waste minimization in the UK: current pressures for change and approaches. *Procedia Social and Behavioral Sciences*, 40, 37–40. https://doi.org/10.1016/j.sbspro.2012.03.158

[16] Phu N. L., & Cho, A. M. (2014). Factors affecting material management in building construction projects. *International Journal of Scientific Engineering and Technology Research*, 3(10), 2133–2137.

[17] Singh, T. S. (2015). Management of construction waste materials: a review. *International Journal of Geology, Agriculture and Environmental Sciences,* 3(4).

[18] Thongkamsuk, P., Sudasna, K., & Tondee, T. (2017). Waste generated in high-rise buildings construction: a current situation in Thailand. In *Energy Procedia. Elsevier Ltd*, pp. 411–416

[19] Thomas, A. J., Akande, I., Oso, O. M., Oke, A. B., & Akindamini, T. (2024). Global Waste Management Practices in Abattoir: Challenges for Implementation in Nigeria. *Journal of Environmental Protection*, 15(12), 1022–1034.

Advances in Construction, Real Estate, Infrastructure and Project Management – Anil Kashyap et al. (eds)
© 2026 Taylor & Francis Group, London, ISBN 978-1-041-13433-6

25 Analyzing the Causes of Claims in Indian Construction Industry— Stakeholder's Perspective

Abhishek Shrivas[1]
Assistant Professor,
NICMAR University, Pune

Navya Sankar[2]
Student,
NICMAR University

■ **ABSTRACT:** On a global level, the present situation has shown a significant rise in claims' numbers and frequency as regards complicated and costly projects. These occurrences have an adverse effect on the financial requirements of project owners leading to liquidity losses and cash flow problems for contractors. This study endeavours to identify the primary causes and origins of claims within the Indian construction industry from 3 major stakeholder's perspective (client, consultant and contractor). Furthermore, it aims to ascertain the prevalence of various types of claims in Indian construction projects. The study's objectives were realized through the administration of a meticulously designed questionnaire, disseminated among various stakeholders operating within the Indian construction sector and provide appropriate suggestions to avoid/ mitigate claims of such cause. Relative importance indexing and spearman's correlation methods were used to rank the causes of claims and found out the correlation among the stakeholders. The findings underscore several pivotal factors contributing to claim generation, on an overall aspect Change in scope of project by client, Specification changes, Excessive change orders, Design revisions during construction phase and Stoppage of work by employer are the top 5 ranked causes of claims. It can also be concluded from this study that change in scope of the project is the most common claim in construction industry in context of India from an overall stakeholders' perspective. This study will help, construction managers to meticulously plan to avoid claims to a certain extent.

■ **KEYWORDS:** Claims, Stakeholders, Relative importance indexing, Spearman's correlation methods, Construction sector

---

[1]sabhishek@nicmar.ac.in, [2]navyasankar5350@gmail.com

DOI: 10.1201/9781003669814-25

1. INTRODUCTION

One of the main drivers of India's economic expansion is the construction sector. About 9% of India's GDP came from the construction sector in 2020. The construction industry contributes INR 2.49 lakh crore to the country's GDP, and the government of India has pledged to making significant investments in the infrastructure sector totalling about INR 5.97 lakh crores in the current fiscal year. Furthermore, it is projected that in order to maintain the nation's development, an investment of almost INR 304 lakh crore will be needed in the Indian infrastructure sector through 2040. However, the Indian construction industry is prone to inherent challenges related to time, cost, and quality management. Hence leading to claims in every construction project. The construction sector is facing a growing number of challenges due to the introduction of new standards, sophisticated technologies, and owner-preferred modifications. Although it is generally believed that the cooperation of the contractor, consultant, and client is indispensable for the successful completion of projects, disagreements over planning and construction details are a constant source of issues and conflicts. There are never any claims-free building projects; all have quantities that change over time. Decreased claims mean fewer disagreements, according to claims management. Concerns over construction claims have grown among the various stakeholders involved in the building sector in recent years. This is explained by the fact that nearly all projects have construction claims. According to a survey commissioned by the CII, (i) pending claims from government agencies are a major contributor to construction companies' growing debt, making about 150% of the total. (ii) Of the more than 85% of claims that have been filed, 11% are at the employer level, 64% are at the arbitrators' level, and 8.5% are in the courts: (iii) It takes 7.5 years on average for settlement In FY15, NHAI agreed to and paid out only about 8% of the total amount claimed (among settled claims); (iv) "Awarded claims" do not settle following the arbitrator's decision; (v) almost all court orders uphold the arbitrators' decisions, and referring claims to courts causes a pay-out delay of about 2.5 years. The problems facing the industry have been discussed with representatives of banks, NHAI, construction companies, PSUs, and relevant departments and ministries. The following stance is reached after considering the feedback from all parties involved: (a) The majority of the outstanding claims are legacy issues pertaining to previous item-rate contracts; (b) significant claims made by contractors/concessionaires against PSUs are still pending in arbitration proceedings or in courts. The numbers of claims are likely to come down by adoption of Trunkey projects in NHAI departing from old item rate contracts Unavoidable in the construction sector, they are becoming more and more so now that complicated projects, the industry's cost structure, and the owners' and operators' legal frameworks all play a major role in the notable rises in capital requirements. Employers frequently petition the courts for postponed payment, and the courts affirm their rulings. Both wealthy and developing nations do a lot of research. Unfortunately, not enough research articles on the aforementioned topic have been published in India. The study's goals were achieved by administering a carefully crafted questionnaire to a range of stakeholders involved in the Indian construction industry, and by offering suitable recommendations to prevent or lessen claims of this kind.

2. LITERATURE REVIEW

Arditi et al. [4] established practices to measure the impact of delays and resolution of disputes resulting in litigation and unions. Design and Engineering explores the advantages and disadvantages of various deferred analysis methods. Fifty-eight state job description cases between 1992 and 2005 are analysed to understand common issues and corrective actions. The choice of lagged analysis includes, for example, political characteristics and approaches that affect knowledge. Elhag et al. [8] work explores the possibilities of collaborative procurement to address the growing demands and disputes of the UAE construction industry. Conflict, Ansari states, is due to the many human relationships involved in construction projects, which can lead to compensation claims and subsequent delays and cost overruns. Effective project management is critical to project success as it affects budget, schedule and overall performance. This study uses a combined AHP-TOPSIS method and system dynamics to classify the causes of claims, assess their impact on Key Performance Indicators (KPIs), and simulate the interactions between KPIs over time, highlighting delays as the main cause of claims significant impact on various project metrics. The suggested model simulates various management strategies, which aids in understanding and reducing the effects of subpar performance. According to Al-Qershi et al. [1], the Indian government annually allocates a considerable sum of money to the construction of infrastructure. Projects with inadequate planning, scheduling, administration, control, and client coordination give rise to claims, and the construction sector is especially vulnerable to them. According to Mukilan et al. [12], disputes between contractors and clients are caused by difficulties with the appeals process, which emphasizes the need for better construction claims management techniques to handle potential claims and reduce disputes. India's intricate, high-budget projects have numerous claims, which affect clients' budgets and entrepreneurs' cash flows, leading to liquidity constraints. In addition to highlighting elements like scope revisions and delays in site handover, this report analyses the key reasons and sources of claims in the Indian construction sector.

The recommendations to reduce benefits are based on survey responses from various contractor organizations and highlight the prevalence of additional labour requirements in construction projects in India. Almutairi et al. [3] explores that the COVID-19 crisis has brought unprecedented challenges to the construction industry, including reduced productivity, increased costs, increased conflicts and delays, leading to reduced profit margins. Open surveys show an 80% increase in disputes and claims, and project costs have increased by more than 40% due to the pandemic. Material shortages, compliance issues, social distancing and labour shortages are key drivers of conflict that require changes in construction management practices and strategies. Asset orders, discussions, and contract clauses are useful in settling COVID-19-related disputes. While the construction industry has a long history of adjudication, a study done in India suggests that legal competence does not appear to have a substantial impact on daily operations of contracts [10]. Even while the majority of professionals understand its significance, many do not have access to case law libraries and few regularly receive case law training. This lack of understanding drives up the expense of resolving disputes and emphasizes the necessity of an easily readily available legal education program to enhance contract management procedures in the building sector. According to Ajibade, calls for the suspension or delay of building projects frequently result in arguments and confrontations that have a negative impact on project funding. This study looks at how contract negotiations can help to mitigate these kinds of conflicts and discovers that agreements around the measurement and assessment of delays can help to lessen the intensity of the disagreements. Detailed contract negotiations covering a range of topics, including evidence rules, document retention policies, and compensation conditions, are among the recommendations. These negotiations can enhance decision-making and lessen the likelihood of conflicts during the project's implementation, which will ultimately result in a smoother project implementation.

Chaphalkar et al. [7], examining a total of 52 arbitral awards in India, found that 38 out of 52 arbitral awards related to delay-related claims. Chaphalkar et al. [6], examining a total of 23 arbitration awards on construction projects in India, analysed a total of 419 claims that resulted in construction disputes. According to their study, 19% of claims were variable claims, 17% were additional work requirements, and 11% were escalation claims of the total number of claims studied. Mohamed et al. [11] studied 31 loss factors in the Egyptian industry of construction and found that contentious compensation factors are time extension, late client interim payment, teamwork competence, owner changes, poorly written contracts, delayed procurement of equipment and material, incomplete drawings and specifications, and the nature of collaboration and communication. Al-Mohsin (2012) examined 26 building industry regulations and gathered information on 45 Omani projects that met particular standards. Based on where the allegations originated, he separated these elements into four groups: factors relating to the owner, consultant, contractor, and contractual agreement. Adrian distinguished four primary categories of compensation claims.: delay claims, scope of work claims, acceleration claims and job change claims. Asem et al. [5] states that construction requirements arise from various events that require contractors to strictly adhere to contract terms and provide detailed documentation of additional costs and time additions. Project owners must implement a comprehensive requirements management process, including effective tracking and management of contractor requirements. The document outlines key milestones and compliance checkpoints and emphasizes the importance of using tools such as simulation and financial analysis to assess the validity and reasonableness of requirements, facilitating solutions for all parties involved in construction projects. Enshassi et al. [9] examined 41 factors that led to claims in the Gaza Strip in three categories: owner-related factors, planning, and cost estimates.

3. RESEARCH GAP

After the thorough literature review researchers have identified few research gaps mentioned below. The most prominent research gap is that there is limited research on the causes of claims in regards to stakeholders' perspective. Not many papers are available on the interrelationship among the major stakeholders' (client, consultant and contractor) and their perspective with respect to causes of claims in the Indian construction industry. In this study researchers focused on, causes of claims in regard to stakeholders' perspective

4. RESEARCH METHODOLOGY

4.1 Research Design

See Figure 25.1 and Figure 25.2.

4.2 Data Collection

The data was collected through questionnaire. There are many causes of claims in Indian construction industry. Among which some have more importance, some have less importance, and some may be the root cause of the overrun and delay in completion of the project. A total of 50 common causes of claims were first identified through a thorough study of literature. Then these factors were shared with a panel of experts, (legal and industry) who had rich working experience in the

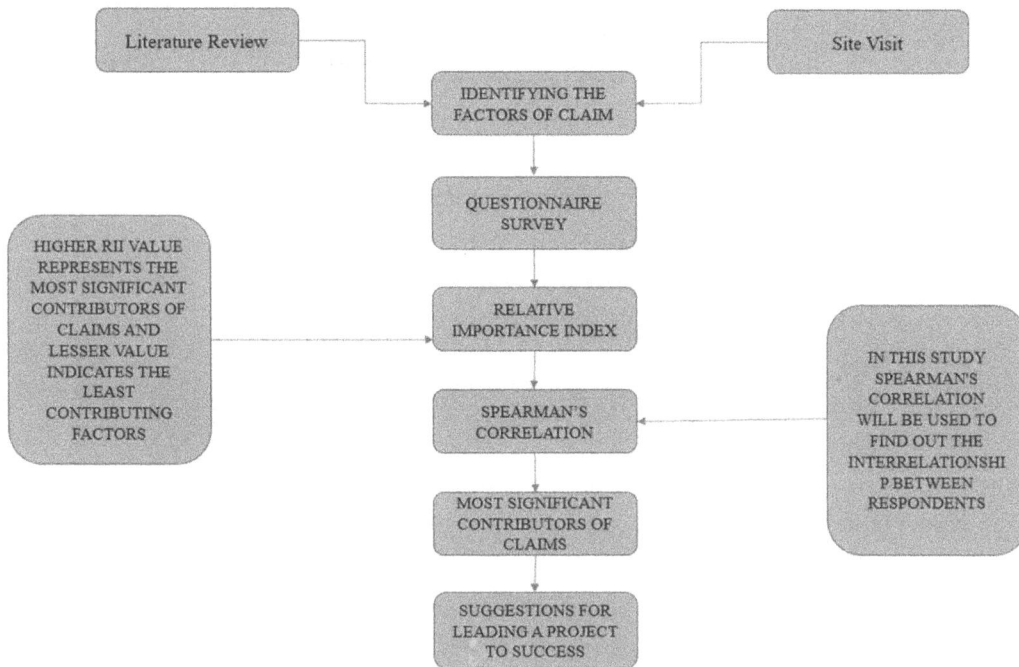

Figure 25.1 Research design

Source: Authors

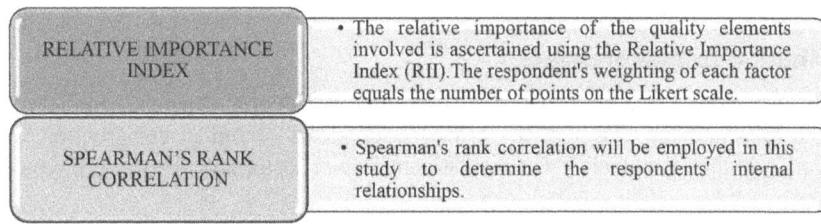

Figure 25.2 RII and spearman's rank correlation

Source: Authors

construction sector. These experts identified 15 most critical repeated causes of claims to be the part of questionnaire survey. Respondents were asked to rate each of 15 factors on a five-point scale, where one mean strongly disagree and five means strongly agree. The questionnaire was sent through mail to the respondents. The respondents were from all three stakeholders of the project, i.e. client, contractor, and consultant. All of them had a minimum work experience of 5 years in the Indian construction. Table 25.1 gives a summary of the respondent's brief profile. Out of the 120 respondents who completed the questionnaire, 84 were selected for further examination based on their relevance to the Indian construction sector. 70 percent of respondents have responded. Clients, consultants, and contractors comprise the respondents who contributed to the study. A total of fifteen common causes of claims in the Indian construction industry were included in the questionnaire. A survey was used to collect the data, which was subsequently analysed using the Spearman rank correlation and the Relative Importance Index while accounting for varying degrees of experience.

Table 25.1 Response table

	Client	Consultant	Contractor
No of relevant Responses	31	20	33
No of responses received	40	40	40
Responsive rate in percentage	77.5%	50%	82.5%

Source: Author's compilation

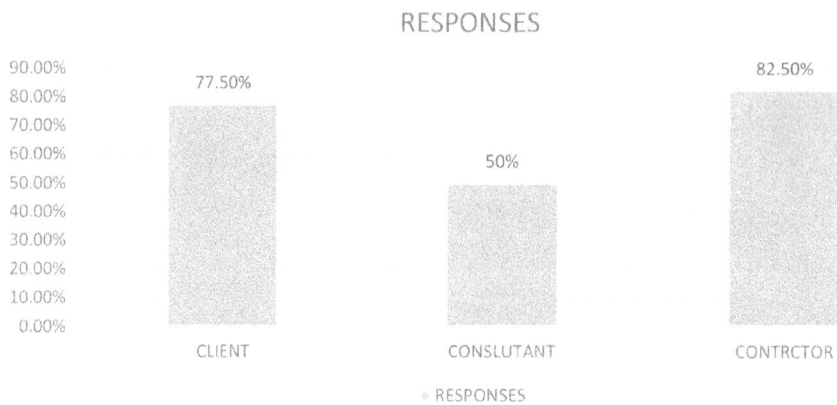

Figure 25.3 Bar chart regarding questionnaire response

Source: Authors

4.3 Distribution to Stakeholders

Questionnaire consists of total of 15 factors that are responsible for the claims in Indian construction industry. The questionnaire was sent to various respondents of Indian construction sectors like transportation, building, heavy civil, water treatment, power sector and other minor sectors.

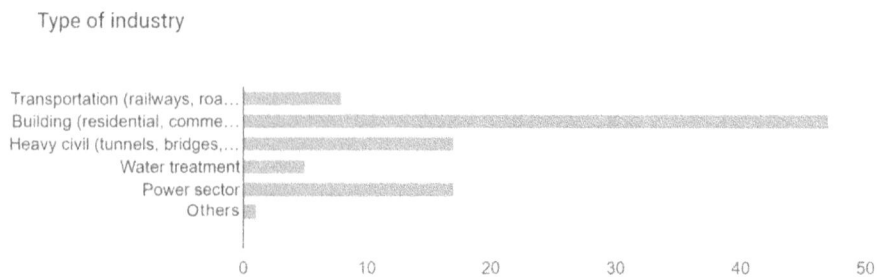

Figure 25.4 Distribution of respondents

Source: Authors

4.4 Objective

The objective of the study are as follows, identify and rank the most common causes for claims in Indian construction industry. To assess which causes for claims are more critical from the stakeholder's point of view, i.e. client, contractor, and consultant.

5. DATA ANALYSIS

The data analysis is divided into two parts. In section 4.1, Data analysis is done using RII and in section 4.2, Spearman's correlation is done to find out the degree of association between client, contractor and consultant.

5.1 Relative Importance Index

The Relative Importance Index (RII) is calculated using the following formula:

$$RII = W/(A \times N)$$

where,

W – The weight given to each factor by the respondents and ranges from 1 to 5.

A – The highest weightage (i.e., 5 in this case) and;

N – The total number of respondents.

5.2 Spearman Rank Correlation

The Spearman Rank Correlation is calculated using the following formula:

$$\rho = 1 - \frac{6\sum d_i^2}{n(n^2 - 1)}$$

Where   d = difference in the ranking of a two set of causes; and

n = number of data in one set of causes

Relationship between two categories is classified according to the value of "ρ".

±0.1 to ±0.3 → Low degree of correlation

±0.3 to ±0.5 → Moderate degree of correlation

±0.5 to ±1.0 → High degree of correlation

± 1.0 → Ideal correlation (perfect)

± 0.0 → No correlation

6. FACTORS OF CLAIMS FROM OVERALL STAKEHOLDER'S PERSPECTIVE

The responses given by the stakeholders are as follows. These top five factors are most recurring causes of claims from overall stakeholder's perspective,

1. Change in scope of project
2. Specification changes
3. Excessive change orders
4. Design revisions during construction phase
5. Stoppage of work by employer

Table 25.2 RII and ranking for overall stakeholder's perspective

Causes for Claims	RII	Rank
Do you think contractual ambiguities leads to claim?	0.7902	6
Are excessive change orders a reason for claim?	0.7976	3
Is delay in obtaining site clearance a cause?	0.7780	8
Is delay in obtaining permit to start work a cause?	0.7756	9
Can specification changes lead to claim?	0.8341	2
Are variation in quantities a reason for claim?	0.7829	7
Can delay in certification of bills be a cause for claim?	0.6976	14
Is change in scope of project a reason for claim?	0.9415	1
Can design revisions during construction phase be a cause for claim?	0.7976	4
Will delay in final payment be a reason for claim?	0.6683	15
Will stoppage of work by the employer lead to claim?	0.7976	5
Can delay in land acquisition lead to claim?	0.7659	10
Do you think delay in obtaining environmental clearances lead to claims?	0.7049	13
Can delay in removal of land encroachments can cause a claim?	0.7512	11
Can price escalation be a cause for claim?	0.7293	12

Source: Author's compilation

7. Factors of Claims from the Client Perspective

The responses given by the clients are as follows. These top five factors are most recurring causes of claims from client's perspective,

1. Change in scope of the project
2. Excessive change orders
3. Variation in quantities
4. Delay in removal of land encroachments
5. Delay in land acquisition

Table 25.3 RII and ranking from client's perspective

Causes for Claims	RII for Client	Rank
Do you think Contractual ambiguities leads to claim?	0.7400	10
Are excessive change orders a reason for claim?	0.8200	2
Is delay in obtaining site clearance a cause?	0.7733	6
Is delay in obtaining permit to start work a cause?	0.7733	6
Can specification changes lead to claim?	0.7400	10
Are variation in quantities a reason for claim?	0.7933	3
Can delay in certification of bills be a cause for claim?	0.7133	13
Is change in scope of project a reason for claim?	0.9133	1

Causes for Claims	RII for Client	Rank
Can design revisions during construction phase be a cause for claim?	0.7733	6
Will delay in final payment be a reason for claim?	0.6600	15
Will stoppage of work by the employer lead to claim?	0.7733	6
Can delay in land acquisition lead to claim?	0.7800	5
Do you think delay in obtaining environmental clearances lead to claims?	0.7400	10
Can delay in removal of land encroachments can cause a claim?	0.7933	3
Can price escalation be a cause for claim?	0.7133	14

Source: Author's compilation

8. FACTORS OF CLAIMS FROM THE CONSULTANT PERSPECTIVE

The responses given by the consultants are as follows. These top five factors are most recurring causes of claims from consultant's perspective,

1. Change in scope of the project
2. Specification change
3. Contractual ambiguities
4. Excessive change orders
5. Design revisions during construction phase

Table 25.4 RII and ranking in consultant's perspective

Causes for Claims	RII for Consultant	Rank
Do you think Contractual ambiguities leads to claim?	0.81	3
Are excessive change orders a reason for claim?	0.79	4
Is delay in obtaining site clearance a cause?	0.73	7
Is delay in obtaining permit to start work a cause?	0.72	8
Can specification changes lead to claim?	0.83	2
Are variation in quantities a reason for claim?	0.74	6
Can delay in certification of bills be a cause for claim?	0.65	13
Is change in scope of project a reason for claim?	0.95	1
Can design revisions during construction phase be a cause for claim?	0.76	5
Will delay in final payment be a reason for claim?	0.67	12
Will stoppage of work by the employer lead to claim?	0.72	8
Can delay in land acquisition lead to claim?	0.7	11
Do you think delay in obtaining environmental clearances lead to claims?	0.63	14
Can delay in removal of land encroachments can cause a claim?	0.63	14
Can price escalation be a cause for claim?	0.71	10

Source: Author's compilation

9. FACTORS OF CLAIMS FROM THE CONTRACTOR PERSPECTIVE

The responses given by the contractors are as follows. These top five factors are most recurring causes of claims from contractor's perspective,

1. Change in scope of the project
2. Specification change
3. Stoppage of work by employer
4. Design revisions during construction phase
5. Contractual ambiguities

Table 25.5 RII and ranking in contractor's perspective

Causes for Claims	RII for Contractor	Rank
Do you think Contractual ambiguities leads to claim?	0.825	5
Are excessive change orders a reason for claim?	0.78125	11
Is delay in obtaining site clearance a cause?	0.8125	6
Is delay in obtaining permit to start work a cause?	0.8125	6
Can specification changes lead to claim?	0.925	2
Are variation in quantities a reason for claim?	0.8	8
Can delay in certification of bills be a cause for claim?	0.7125	14
Is change in scope of project a reason for claim?	0.9625	1
Can design revisions during construction phase be a cause for claim?	0.84375	4
Will delay in final payment be a reason for claim?	0.675	15
Will stoppage of work by the employer lead to claim?	0.86875	3
Can delay in land acquisition lead to claim?	0.79375	9
Do you think delay in obtaining environmental clearances lead to claims?	0.71875	13
Can delay in removal of land encroachments can cause a claim?	0.7875	10
Can price escalation be a cause for claim?	0.75625	12

Source: Author's compilation

10. DEGREE OF ASSOCIATION BETWEEN CLIENT, CONTRACTOR AND CONSULTANT

After the individual analysis of each category, the degrees of association between each two categories were carried out by using spearman's rank correlation. There is a moderate degree correlation between (client and consultant) and (client and contractor). There is a high degree of correlation between (consultant and contractor). All the details regarding the spearman's rank correlation given in the following table.

Table 25.6 Details of correlation between two categories

	Client	Consultant	Contractor
CLIENT	----	0.42	0.48
CONSULTANT	0.42	----	0.75
CONTRACTOR	0.48	0.75	----

Source: Author's compilation

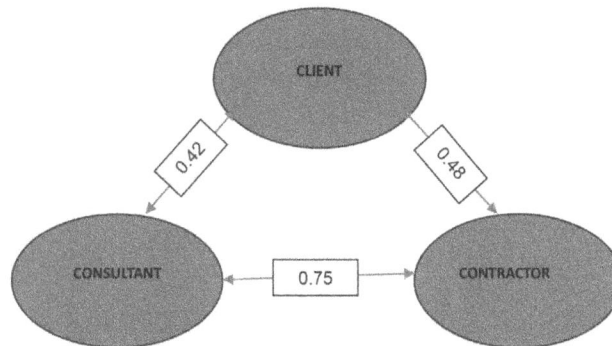

Figure 25.5 Representation of correlation between categories

Source: Authors

11. CONCLUSION AND RECOMMENDATION

The study looked on the origins and prevalence of various claim kinds in the Indian construction sector. A questionnaire was designed and distributed to the stakeholders. The respondents for the questionnaire were from different contractors, consultants and clients. From the results, it is found that in an overall aspect Change in scope of project by client, Specification changes, Excessive change orders, Design revisions during construction phase and Stoppage of work by employer are the top 5 ranked causes of claims. It can also be concluded from this study that change in scope of the project is the most frequent claim in Indian construction industry from an overall stakeholders' perspective.

The research's suggestions have been developed in a way that makes them realistic, observable, and obtainable in the form of steps and procedures, in line with the findings. Building industry experts may adhere to in order to prevent, lessen, and handle their claims. The crucial actions project stakeholders can take to address the previously suggested reasons and reduce or mitigate claims are as follows:

- Prior to submitting the paperwork for tender, the client and the representative they appointed need to specify and approve the scope of the task. A comprehensive set of plans and specifications that provide sufficient information outlining the project's scope minimizes the likelihood of recurrent revisions, supplementary work claims, and exorbitant change order claims.

- It is recommended that clients not rush the design. The engineering firm and design professionals should be given sufficient time to coordinate, verify, and finish the design, drawings, and specification.

- All parties should work together to ensure outstanding coordination and to establish effective methods of communication with all stakeholders during the project.

- At the initial stage of the project, a Dispute Adjudication Board (DAB) should be established.

- It is suggested that clients recognize and agree to the contractor's valid claims as soon as possible to avoid any further delays, as this will impact the contractors' cash flow.

This study attempts to shed light on the relationships between the main players in the Indian construction industry—the client, consultant, and contractor—as well as their perspectives on the reasons behind claims. With the goal to gain a greater understanding the relationships between the various stakeholders about the reasons behind claims in building projects, a conceptual framework has been established. The project managers who oversee building projects can utilize this study as a foundation for analysing the connections and interdependencies between the various claim reasons. Researchers may use this study as a foundation for more studies in this area.

12. LIMITATIONS

The analysis of construction project claims encounters various challenges stemming from data availability, in this case only 3 main stakeholders' (client, consultant &contractor) perspectives was taken into consideration as there was time constraint. The contractual ambiguities, cultural differences, human factors, accessibility and ethical considerations limited access to comprehensive and accurate data; which inhibits a thorough examination of construction projects hindering the identification of underlying causes of claims. Conflicting viewpoints among stakeholders further complicate matters making it difficult to establish consensus on the issues at hand. Complex contractual language adds another layer of complexity leading to differing interpretations among stakeholders and subjectively examined disputes. Adopting an official language can help to reduce these issues by streamlining the study process and promoting clearer communication. Furthermore, geographical and cultural variations in national customs can have a substantial impact on how claims are interpreted and resolved, making it difficult to draw broad generalizations. Furthermore, human factors, financial circumstances, and outside influences add to the complex character of claims, which could fail to be adequately reflected in assessments. Another challenge is accessibility, since not all parties to a project issue may be easily reached for cooperation or consultation. The study becomes more complex by ethical considerations, which may lead to limits resulting from privacy concerns, informed consent challenges, or participant harm risk. Last but not least, time restraints placed restrictions on the gathering and analysis of data, sometimes causing important factors or linkages to be overlooked and limiting the study's breadth and depth. In order to overcome these obstacles, a multimodal strategy that includes tactics to improve data accessibility, encourage stakeholder collaboration, and make contractual language more culturally sensitive must be used. Recognize human aspects to maintain accessibility, respect moral principles, and efficiently manage time.

REFERENCES

[1] Al-Qershi, M. T., & Kishore, R. (2017). Claim causes and types in Indian construction industry – contractor's perspective. American Journal of Civil Engineering and Architecture, 5(5), 196–203.Arditi, D., & Pattanakitchamroon, T. (2008). Analysis methods in time-based claims. *Journal of Construction Engineering and Management*, 134(4), 242–252.

[2] Al-Mohsin, M. A. (2012). Claim analysis of construction projects in Oman. International Journal on Advanced Science, Engineering and Information Technology, 2(2), 186–191.

[3] Almutairi, S., Bakri, M., AlMunifi, A. A., Algahtany, M., & Aldalbahy, S. (2023). The Status of the Saudi Construction Industry during the COVID-19 Pandemic. Sustainability, 15(21), 15432. https://doi.org/10.3390/su152115432

[4] Arditi, D., & Pattanakitchamroon, T. (2008). Analysis methods in time-based claims. Journal of Construction Engineering and Management, 134(4), 242–252.

[5] Asem, M., Abdul-Malak, U., Mustafa, M. M. H., El-Saadi, H., & Abou-Zeid, M. (2002). Process model for administrating construction claims. Journal of Management in Engineering, 18(2), 84–94.

[6] Chaphalkar, N. B., & Sandbhor, S. S. (2015). Application of neural networks in resolution of disputes for escalation clause using neuro-solutions. *KSCE Journal of Civil Engineering*, 19(1), 10–16. https://doi.org/10.1007/s12205-014-1161-3.

[7] Chaphalkar, N. B., & Iyer, K. C. (2014). Factors influencing decisions on delay claims in construction contracts for Indian scenario. Australasian Journal of Construction Economics and Building, 14(1), 32–44.

[8] Elhag, T., Eapen, S., & Ballal, T. (2020). Moderating claims and disputes through collaborative procurement. *Construction Innovation*, 20(1), 79–95.

[9] Enshassi, A., Mohamed, S., & El-Ghandour, S. (2009). Problems associated with the process of claim management in Palestine: Contractors' perspective. Engineering, Construction and Architectural Management, 16(1), 61–72.

[10] Jagannathan, M., Nawle, V., Malla, V., & Delhi, V. S. K. (2022). Role of case laws in claim management and contracts. Journal of Legal Affairs and Dispute Resolution in Engineering and Construction, 14(4), 04522025.

[11] Mohamed, H. H., Ibrahim, A. H., & Soliman, A. A. (2014). Reducing construction disputes through effective claims management. American Journal of Civil Engineering and Architecture, 2(6), 186–196.

[12] Mukilan, K., BalaNivetha, M., Velumani, P., & Christopher Gnanaraj, S. (2019). A qualitative study and analysis of causes and disputes in claims in construction industry. International Journal of Civil Engineering and Technology, 10(1), 951–957.

Advances in Construction, Real Estate, Infrastructure and Project Management – Anil Kashyap et al. (eds)
© 2026 Taylor & Francis Group, London, ISBN 978-1-041-13433-6

26

A Systematic Literature Review of Critical Enablers and Barriers of Transport Corridor Megaproject Delivery

Shreya Bandekar[1],
Sankhipta Das[2], Shrunkhala Shinde[3],
Bhagyashree Bangde[4]
MBA APM Student, NICMAR University,
Pune, India

Aritra Halder[5]
Assistant Professor at NICMAR University,
Pune, India

■ **ABSTRACT:** Transport Corridor Megaprojects (TCM) play a crucial role in economic development and regional connectivity. However, such large-scale projects also suffer from cost and schedule overruns, land acquisition issues, regulatory and administrative hurdles, the presence of pressure groups, technological instability, etc, throughout their life cycles. Therefore, it is important to identify the enablers of and barriers to the efficient delivery of TCMs. Enablers are those elements that promote a seamless and effective delivery and barriers are impediments that prevent the successful completion of such megaprojects.

Existing research on TCM is contextually diverse and fragmented. Despite the growing interest in the development of TCM, current research does not provide a consolidated view of the major facilitating factors and inhibiting conditions that affect the successful delivery of them. This study tries to address this gap by conducting a systematic literature review and by proposing a conceptual framework linking the enablers and barriers of TCMs with the key performance indicators (KPIs) of their successful delivery. First, 34 articles were shortlisted from the Scopus database using the systematic method of PRISMA (Preferred Reporting Items for Systematic Reviews and Meta-Analyses) approach. Then the TCCM (Theory, Context, Characteristics and Methodology) approach was adopted to Identify the enablers, barriers, and KPIs of TCM through a comprehensive literature review.

The PESTEL (Political, Economic, Social, Technological, Environmental, and Legal) framework was adopted to characterize the external environmental factors contributing as enablers and barriers. The internal environmental factors were broadly characterized as stakeholder and operational management issues. Finally, a conceptual framework was established linking the enablers,

[1]shreyavb.arch@gmail.com, [2]sankhipta20.das@gmail.com, [3]shrunkhalashinde@gmail.com, [4]bhagyashreebangde15@gmail.com, [5]aritra.nicmar@gmail.com

DOI: 10.1201/9781003669814-26

barriers, and key performance indicators of TCMs. The findings can help researchers develop future models of TCM success frameworks and can help practitioners improve the efficiency of TCM delivery by fostering the enabling factors and eliminating the inhibiting elements.

■ **KEYWORDS:** Transport corridor, Megaprojects, TCCM, PRISMA, PESTEL

1. INTRODUCTION

Megaprojects can be defined as complex operations that need many years to plan and execute, include several public and private stakeholders, and have the potential to significantly change society [39]. Success in megaproject performance comes from effectively managing enablers and barriers. These enablers and barriers in turn form the key performance indicators (KPI) for TCM. Enablers serve as catalysts throughout the project's lifespan facilitating progress while barriers may hinder goal achievement resulting in obstructed progress. Therefore, enablers must be used to boost the project while barriers must be controlled and managed.

TCMs play a crucial role in economic development and regional connectivity [47]. A Transport Corridor (TC) integrates multiple modes of transportation, such as roads, railways, and waterways, and plays an important role in facilitating trade [44]. Currently, India (National Industrial Corridor Development Programme) is working on 11 Transport corridor megaprojects (TCM). However, TCMs have many intricate interdependencies throughout their lifecycle that can make project delivery extremely difficult. One of the issues with the current TC is a lack of infrastructure, environmental impacts, increased greenhouse gas emissions, habitat destruction, and air and noise pollution pose a challenge in the delivery of TCM. Cost and schedule overruns and substandard works are major factors failing the objectives of the project performance.

Researchers have been looking into enablers; however, few studies have specifically addressed enablers and barriers with respect to the project management of TCM. This creates a research gap. Specifically, three questions remain unanswered: 1) What are the unidentified key enablers and barriers to TCM? 2) What are the key performance indicators for TCM across the globe? 3)What best practices are needed throughout the lifecycle of the project to address challenges and maximize potential in TCM?

This gap is addressed with the help of a comprehensive framework of the internal environment (stakeholder and operational management) and external environment (PESTEL) to identify enablers and barriers. These enablers and barriers are assembled in an implementation framework that suggests the KPI of TCM's success. To make the project scope more focused and manageable, TCCM and PRISMA methodologies are used. The study area is also limited to Road TCM. Projects involving sea and air transportation are excluded.

2. LITERATURE REVIEW

2.1 PRISMA Method

As mentioned in [46] a literature review is conducted through the usage of a comprehensive search that scans the relevant body of literature with clearly stated and comprehensible search choices and selection criteria.

Publications that are relevant to TC are retrieved from Scopus databases. The search criteria are defined, on Scopus, as keywords (Topic: Transport Corridors, Road Projects, Megaprojects, Success in Megaprojects, Economic Corridors, Highway Projects in India) plus publication year ranging from year 2005 to year 2024 (up to the date of data collection – 2024 March). The initial search returned 1755 journal articles. This number was reduced by filtering to specific fields like Document type as "Article" with Source type as "Journals", and "English" Language. The selected subject areas were "Construction Industry" and "Business Management and Accounting" (as defined in Scopus). There were 298 items left after filtering. The search was then done by examining each of the 298 papers' abstract, introduction, and conclusion parts. A stage of the manual selection process looked at the articles' relevancy. The journal article was disregarded which was irrelevant or incompatible with Road TCM. Thereafter, 34 journal papers were ultimately chosen for a thorough review. The screening was then followed by the PRISMA Statement, as indicated in Figure 26.2. Similarly, Figure 26.1 illustrates the distribution of the 34 publications by year where most of the study regarding TCM was done after 2011. As a result, each publication was specifically assigned to one of the enabling categories and some with barring categories. In this review, each article was associated with a single enabler or barrier.

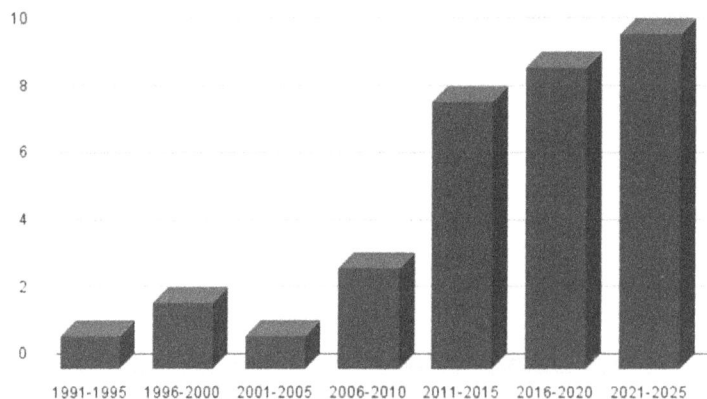

Figure 26.1 Publication by year

Source: Compiled by author

2.2 Research Agenda: TCCM Method

Based on a simple structure called TCCM—where T stands for theory development, C for context, C for characteristics, and M for methodology—we have identified the knowledge gaps and analysed the final 34 research papers. The TCCM method is a useful method that helps with giving detailed knowledge about research analyses and categorizes the papers based on theory, context, characteristics, and methods. The TCCM method is adopted from [26].

Theory Development

In this review, we note the use of theories on TCM that can open up the uncovered areas through theoretical lenses. These areas might result in methods to achieve superior project performance. There is a need for theoretical and analytical frameworks that clarify the relationship among success

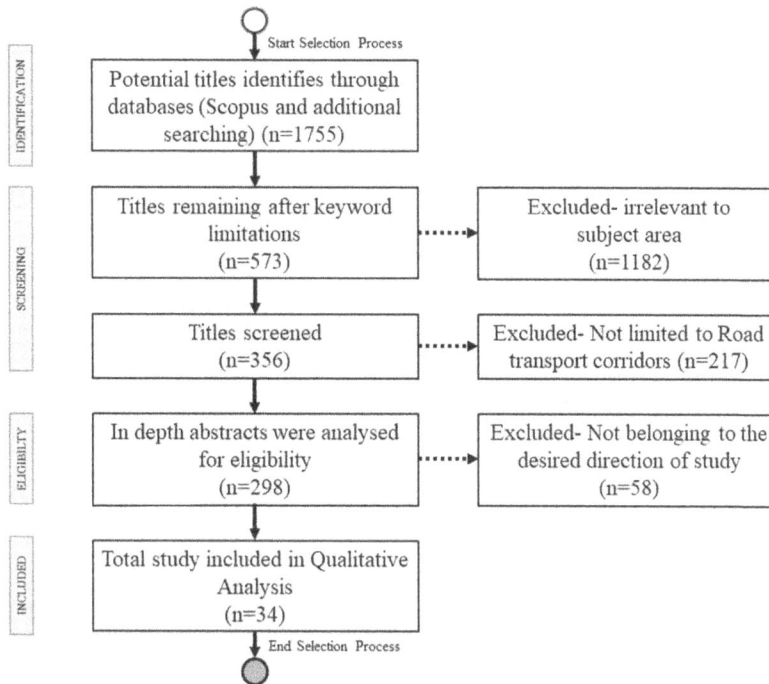

Figure 26.2 PRISMA method for paper screening (PRISMA method)

Source: Adapted and modified from [37]

factors, enablers, and barriers which in turn enhance the project performance. The list of theories used in 34 papers is given below and is broadly classified in Figure 26.3.

1. Three-dimensional theory of Project Management [20]
2. Complexity Theory [32]
3. Transition Management Theory [32]
4. Legitimacy Theory [5]
5. Transaction Cost Theory [5]
6. Governance Theory [32]
7. Production Theory [8]
8. Probability Theory (Fuzzy set Theory) [13]
9. Risk-based normative decision Theory [36]
10. Theory of Integrated Transport Networks Servicing [3]
11. Supply Chain Management Theory [10]
12. Social Theory [32]

Context

The Context for 34 papers included the origin of the paper, location of TCM and broad context of practice that were adopted by various regions. This was then grouped as per continents and was

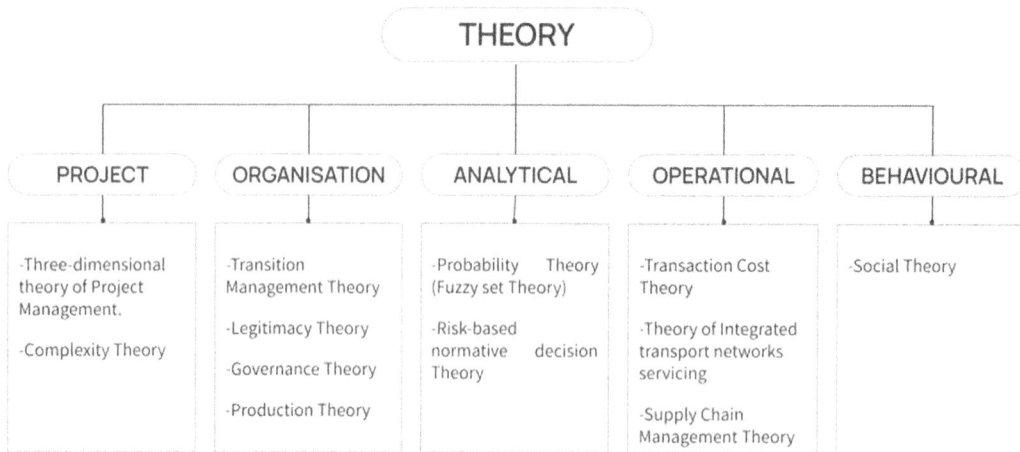

Figure 26.3 Theories from 34 papers (Theories from TCCM)

Source: Compiled by author

analysed (Figure 26.4). It was noted that less amount of work is done in developing countries. Hence, there are opportunities to carry out research in the context of firms from developing countries.

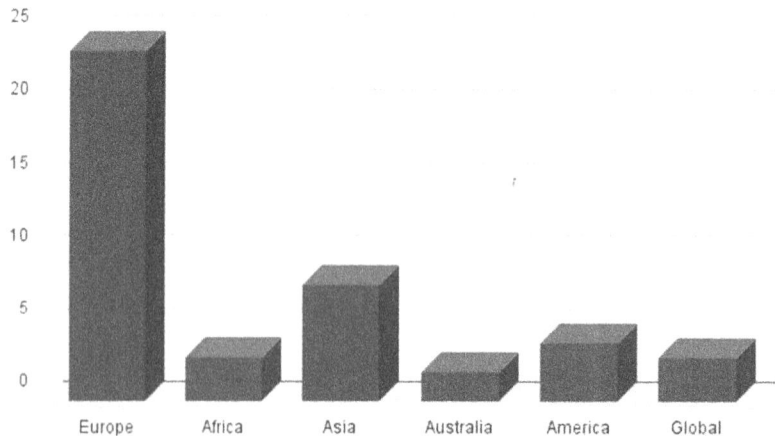

Figure 26.4 Contextual study of 34 papers (Context for TCCM)

Source: Compiled by author

Characteristics

We have categorized the characteristics of TCM into 11 main areas. These include various factors that are listed in Table 26.1. By examining and analysing these research papers we can identify the causes and consequences of TCM, contributing to a more complete understanding of the subject.

Table 26.1 Characteristics from 34 Papers (Characteristics of TCCM)

Characteristics	No. of Articles	Citations
Risk management	2	[33, 36]
Connectivity	3	[24, 27], P32
Spatial Planning	2	[13, 35]
Sustainability indicators	3	[16, 31], P20
Political	3	[20], P8, [23]
Finance	7	[4, 9, 10, 13, 23, 34, 42]
Community engagement	6	[4, 11, 13, 20, 27, 31]
Technological integration	8	[12, 15, 18, 28, 45], P20, P9
Environmental considerations	7	P32, [1, 2, 19, 24, 31]
Policy and regulation compliances	6	[10, 18, 23, 24, 27, 42] P7
Internal Environment	9	[2, 5, 7, 14, 15, 35] P9, [10, 12]

Source: Compiled by author

Characteristics

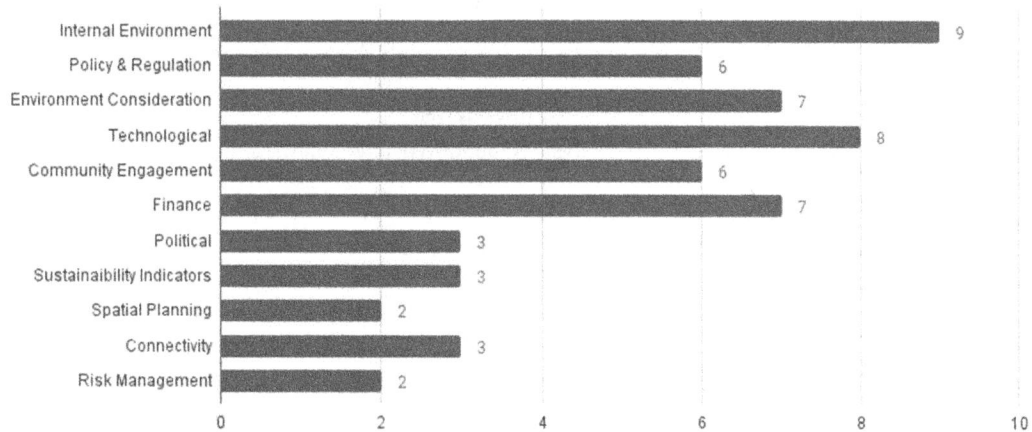

Figure 26.5 Characteristics from 34 Papers (Characteristics of TCCM)

Source: Compiled by author

Methods

After examining previous research, we observed that several methodologies are utilized to evaluate the characteristics that contribute to the success of TCM. These techniques include a variety of other approaches are defined in Table 26.2.

Table 26.2 Methods from 34 papers (Methods of TCCM)

Qualitative Analysis			Quantitative Analysis	Mixed methods (Qualitative + Quantitative)	
Case study	Interview	Both (Case study & Interview)			
[27]	[30]	[20]	[31]	[12]	[17]
[34]			[8]	[21, 41]	[42]
[6]	[23]		[5]	[9]	[32]
[7]			[5]	[9]	[32]
[35]			[19]	[10]	[13]
[2]				[28]	[15]
				[33]	[18]
				[36]	[1]
				[14]	[24]
				[25]	[15]
					[11]

Source: Compiled by authors

The aim is to find which methods are used most effectively for investigating the outcomes for TCM. The most commonly used methods of analyzing the success factors of TCM include:

1. Quantitative analysis (11.8%): Regression Analysis, Pearson correlation Analysis, PERT Analysis, Probability Analysis,
2. Qualitative analysis (29.4%): Case Study, Interviews,
3. Mixed Methods (58.8%): Hybrid Analysis (Qualitative and Quantitative Analysis).

3. REVIEW OF THE RESULTS

The thorough analysis of 34 research papers summarised the major challenges that were faced in TCM. Using the PRISMA and TCCM methods a range of enablers and barriers were identified, which were then organized into a framework depicted in Figure 26.7. By combining the insights from these studies, it was easy to identify the significance of the PESTEL framework, which was consistently emphasized across the analysis. Understanding these enablers and barriers is important to pinpoint the key performance indicators (KPIs) that signify the successful implementation of TCM.

3.1 Identification of Enablers

Table 26.3 shows the 39 particular enablers that were identified from the papers. The PESTEL framework is critical as it evaluates external factors such as political, economic, social, technological, environmental, and legal issues, operational

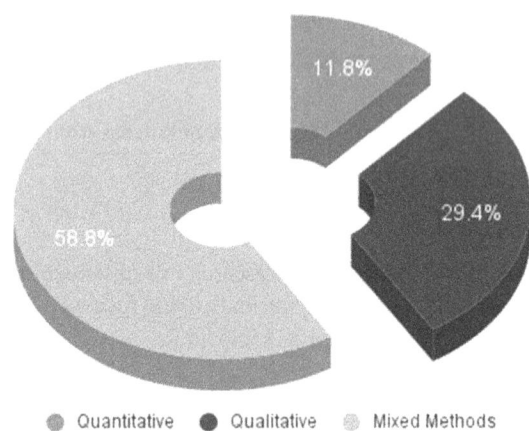

Quantitative ● Qualitative ● Mixed Methods

Figure 26.6 Methods from 34 papers (Methods of TCCM)

Source: Compiled by author

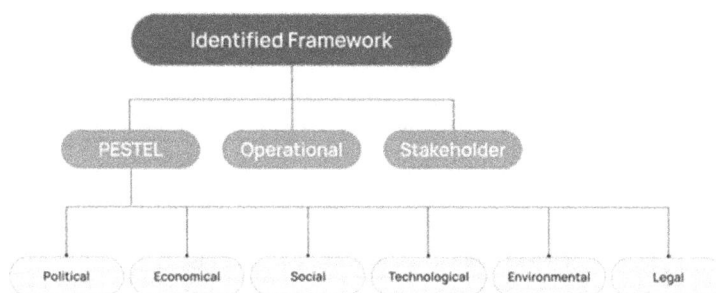

Figure 26.7 Identified framework for classification of enablers and barriers

Source: Compiled by author

and stakeholder management. It helps in identifying the enablers and barriers in TCM [43]. The developing keywords that are closely linked to a certain enabler were chosen from the articles to create the key concepts listed in Table 26.3.

Table 26.3 List of enablers of TCM

Sr. No.	Categories	Key enablers	Reference Articles
External Environment			
E1	Political	Strong commitment and will	[9, 12]
		Government involvement	[9, 32]
		Enhanced political stability	[32]
E2	Economical	Stability of price	[21]
		Market competitiveness	[9]
		GDP per capita-economic growth	[20, 35]
		Diverse funding sources	[32]
E3	Social	Encouraging community involvement	[9, 35]
		Committed public agency	[9]
E4	Technological	Advancement of technology	[40, 8, 15]
		Use of Intelligent Video Gates (IVGs) and Digital Automatic Couplers (DACs)	[8]
		BIM in construction	[8, 32]
E5	Environmental	Environmental sustainability performance	[16, 32]
		Use of Natural Materials	[16, 35]
		Geographical Information System (GIS)	[1, 22]
E6	Legal	Preparation of an intact contractual agreement	[9]
		Dispute resolution mechanism	[9]
		In-time land acquisition	[9, 32]
Internal Environment			
E7	Operational	Efficiency of resource allocation	[10, 14, 35]
		Quality management	[32]
E8	Stakeholders	Horizontal collaboration	[12]
		Good governance by top management	[9, 32]

Source: Compiled by author

Table 26.4 List of barriers of TCM

Sr. No.	Categories	Key barriers	Reference Articles
External Environment			
B1	Political	Corruption	[5, 10, 36]
		Political interference	[9, 25]
		Low government support	[18]
B2	Economical	Scarcity of public funds	[9]
		Zoning challenges	[10]
		Financial constraints	[2, 34, 36]
		Competitive demands from other sectors	[2]
		Market demand risk	[36]
		Debt servicing risk	[36]
		Intense competition	[12, 13]
		Lower employment rate	[13]
B3	Social	Pressure groups	[11, 27]
B4	Technological	Technical complexity	[6, 20, 35]
B5	Environmental	Complex geography of the site	[25]
		CO2 emissions	[31]
		Water scarcity	[31]
B6	Legal	Mistakes in the tendering process	[5]
		Delays in awarding contracts	[10]
		Delay due to new regulations	[20]
		Land acquisition	[9, 10, 23, 27, 36, 42]
		Regulatory challenges	[9, 35]
Internal Environment			
B7	Operational	Absence of a standardized procedure SOP	[19]
		Limited resources. (labour, equipment, raw material)	[14, 35]
		Schedule overrun	[20, 23]
B8	Stakeholder	Mental barrier	[12]
		Poor decision making	[5]
		Complexity in coordination	[7, 18]
		Lack of trust	[13]

Source: Compiled by author

3.2 Identification of Barriers

In addition to enablers and their potential, research has shown that difficult problems nevertheless arise while initiatives are developing. Controlling barriers created during the project's lifecycle is necessary, even if the enablers are implemented. The report goes on to identify further barriers in Table 26.4.

3.3 Key Performance Indicators

The above-listed Enablers (Positive Impact) and Barriers (Negative Impact) together are seen in the life cycle of transport corridor projects. However, the question arises how to measure the successful completion of projects. A key aspect that makes significant contributions to the project's completion is used to characterize a success factor. The performance of construction has been predicted and assessed using a variety of methods and strategies in recent years, some of which are based on key performance indicators (KPIs). The analysis of prior studies indicates that there hasn't been much focus on the KPIs of TCM. Hence, integrating KPIs and determining their importance in projects is done [38]. Here are some major KPIs listed below in Table 26.5.

Table 26.5   List of key performance indicators of TCM

Sr. No	Categories	Key Performance Indicators	Reference Authors
		External Environment	
1	Political	Satisfactory government assistance	[32]
2	Economical	Control cost overrun	[32]
3	Social	Satisfaction of community	[35]
4	Technological	Use of latest technology	[28]
5	Environmental	Efforts on emission reductions	[16]
6	Legal	Low contractual ambiguity	[27]
		Internal Environment	
7	Operational	Quality standards maintained	[8]
8	Stakeholder	Conflicts management	[8]

Source: Compiled by author

4. SYNTHESIS OF REVIEW OF RESULTS

By synthesizing the review of results, an interrelationship framework is derived for future project practice. The framework is characterized by both the internal and external level environment of the project. The categorization is done using PESTEL which helps in understanding the impact of the external environment, while the stakeholder and operational management categories help in understanding the impact of internal environments on TCM. By thoughtfully addressing these barriers, stakeholders may improve project performance, reduce risks, and assure the transportation corridor projects' long-term viability and positive impact on economies and communities.

5. CONCLUSION

This paper presents the extent of the literature on TCM and its success factors. Through a systematic literature review of carefully selected publications, a comprehensive framework was developed. This framework not only takes into consideration the internal and external environmental effects throughout the lifecycle of the project but also it fills in the gaps that exist between the current best practices that are required for TCM throughout its lifecycle. This research provides valuable insights for construction researchers and practitioners, enhancing their understanding of managing future TCM. Future studies can be taken forward based on this framework and researchers can create

Enablers	
Political :	• Government involvement • Enhanced political stability
Economical:	• Diverse funding sources • Stability of price
Social:	• Encouraging community involvement • Public acceptance and participation
Technological:	• Project technical feasibility • Noise barrier systems
Environment:	• Use of Natural Materials • Geographical info. System(GIS)
Legal :	• Dispute resolution mechanism • Transparent procurement
Operational:	• Transparent procurement • Quality management
Stakeholder:	• Strong commitment and will • Horizontal collaboration

Barriers	
Political :	• Corruption • Political interference
Economical:	• Scarcity of public funds • Financial Constraints
Social:	• Pressure groups • Insufficient public funding
Technological:	• Technical Complexity
Environment:	• Loss of Biodiversity • Deforestation due to road fragmentation
Legal :	• Delays in awarding contract • Mistakes in Tendering Process
Operational:	• Low supply of raw materials • High Transhipment cost
Stakeholder:	• Mental barriers • Complexity in coordination

Enhance Positive Impact → **Key Performance Indicators** ← Control Negative Impact

Political :	Satisfactory Government Assistance	Environment:	Efforts on emission reduction
Economical:	Control over cost overrun	**Legal :**	Low contractual ambiguity
Social:	Satisfaction of community	**Operational:**	Quality standards maintained
Technological:	Use of latest Technology	**Stakeholder:**	Conflicts management

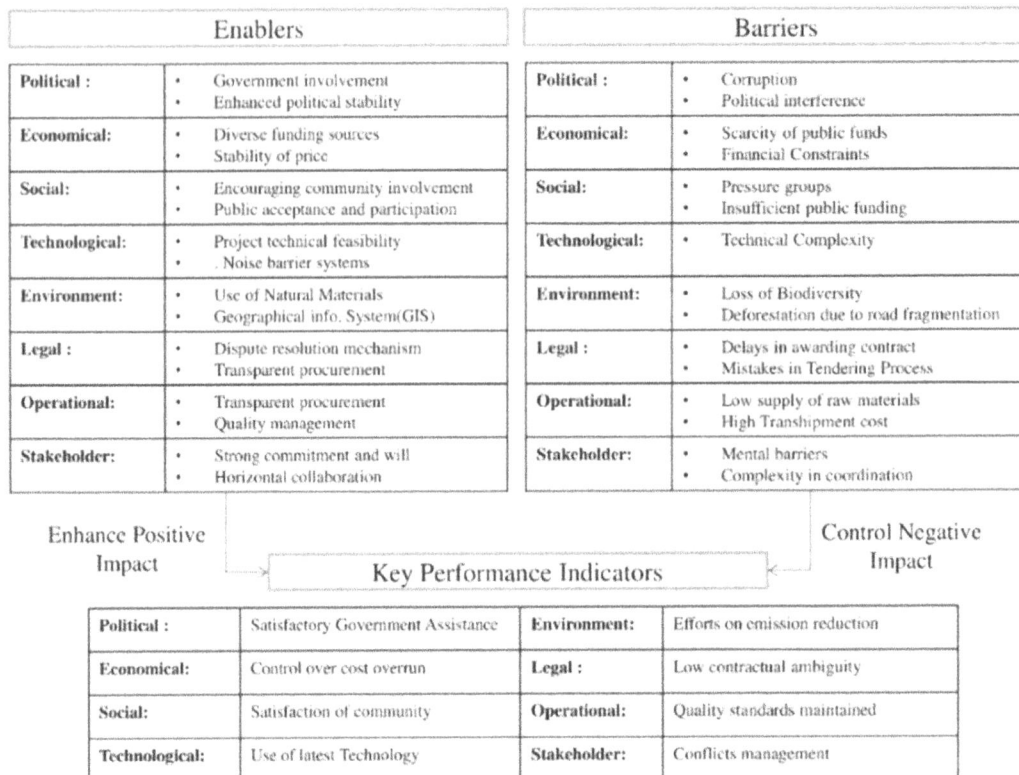

Figure 26.8 Inter-relationship identified framework representing success in TCM

Source: Compiled by author

questionnaires to do empirical study on the same. This will help in better decision-making and will eventually help in the economic betterment of TCM.

ACKNOWLEDGMENT

We sincerely thank NICMAR University and the ICCRIP Conference 2024 for their invaluable support in facilitating this research. The academic resources and intellectual environment provided by NICMAR University, along with the platform for research dissemination offered by the ICCRIP Conference, were crucial to the completion and presentation of this study. We also extend our gratitude to our mentors, peers, and families for their unwavering encouragement and support.

REFERENCES

[1] Alexander, S. M., & Waters, N. M. (2000). "The effects of highway transportation corridors on wildlife: a case study of banff national park." *Transportation Research Part C: Emerging Technologies,* 8(1), 307–20. https://doi.org/https://doi.org/10.1016/S0968-090X(00)00014-0.
[2] Behrends, S., & Flodén, J. (2012). "The effect of transhipment costs on the performance of intermodal line-trains." *Logistics Research,* 4(3), 127–36. https://doi.org/10.1007/s12159-012-0066-0.

[3]   Bezpalov, V., Bezpalov, V. V., Gukasyan, G., Gukasyan, G., Okhrimenko, I., & Okhrimenko, I. (2022). "Economic corridors in the context of the development of macroregions." *Innovative Infrastructure Solutions.* https://doi.org/10.1007/s41062-022-00848-2.

[4]   Bezpalov, V., Gukasyan, G., & Okhrimenko, I. (2022). "Economic corridors in the context of the development of macroregions." *Innovative Infrastructure Solutions,* 7(4), 1–9. https://doi.org/10.1007/s41062-022-00848-2.

[5]   Claire, K., & Njenga, G. (2021). "Decision making processes and road construction projects implementation in rwanda. a case of zindiro-birembo-gasanze road construction project in Gasabo District, Rwanda." *International Journal of Scientific and Research Publications (IJSRP),* 11(11), 90–110. https://doi.org/10.29322/ijsrp.11.11.2021.p11912.

[6]   Crainic, T. G., Gendreau, M., & Potvin. J-Y. (2009). "Intelligent fFreight-transportation systems: assessment and the contribution of operations research." *Transportation Research Part C: Emerging Technologies,* 17(6), 541–57. https://doi.org/10.1016/j.trc.2008.07.002.

[7]   Dablanc, L. (2007). "Goods transport in large European Cities: difficult to organize, difficult to modernize." *Transportation Research Part A: Policy and Practice,* 41(3), 280–85. https://doi.org/10.1016/j.tra.2006.05.005.

[8]   Djordjević, B., Ståhlberg, A., Krmac, E., Mane, A. S., & Kordnejad, B. (2024). "Efficient Use use of European rail freight corridors: current status and potential enablers." *Transportation Planning and Technology,* 47(1), 62–88. https://doi.org/10.1080/03081060.2023.2294344.

[9]   Fikreyesus, D. C., and & Jha, K. N. (2020). "Enabling successful application of PPPs in New new (iInexperienced) Marketsmarkets: implications of PPPs' success and failure in toll roads." *Journal of Legal Affairs and Dispute Resolution in Engineering and Construction,* 12(4), 5020015. https://doi.org/10.1061/(ASCE)LA.1943-4170.0000434.

[10] Ghani, E., Goswami, A., G., & Kerr, W. R. (2016). "Highway to success: the impact of the golden quadrilateral project for the location and performance of Indian Manufacturingmanufacturing." *Economic Journal,* 126(591), 317–57. https://doi.org/10.1111/ecoj.12207.

[11] Grisolía, J. M., López, F., & Ortúzar, J. D. (2015). "Burying the highway: the social valuation of community severance and amenity." *International Journal of Sustainable Transportation,* 9(4), 298–309. https://doi.org/10.1080/15568318.2013.769038.

[12] Islam, D. M. Z., & Zunder, T. H. (2018). "Experiences of rail intermodal freight transport for low-density high value (LDHV) Goods goods in Europe." *European Transport Research Review,* 10(2). https://doi.org/10.1186/s12544-018-0295-7.

[13] Karam, A., Hussein, M., & Reinau, K. H. (2021). "Analysis of the barriers to implementing horizontal collaborative transport using a hybrid fuzzy Delphi-AHP Aapproach." *Journal of Cleaner Production,* 321(April), 128943. https://doi.org/10.1016/j.jclepro.2021.128943.

[14] Kazaz, A., & Acikara, T. (2015). "Comparison of labor productivity perspectives of project managers and craft workers in turkish construction industry." *Procedia Computer Science,* 64, 491–96. https://doi.org/10.1016/j.procs.2015.08.548.

[15] Reza, R. K., Mavi, N. K., Olaru, D., Biermann, S., & Chi, S. (2022). "Innovations in Freight transport: a systematic literature evaluation and covid implications." *The International Journal of Logistics Management,* 33(4), 1157–95. https://doi.org/10.1108/IJLM-07-2021-0360.

[16] Kumar, A., & Anbanandam, R. (2022a). "Assessment of Environmental and social sustainability performance of the freight transportation industry: an index-based approach." *Transport Policy,* 124(40), 43–60. https://doi.org/10.1016/j.tranpol.2020.01.006.

[17] Kumar, A., & Anbanandam, R. (2022b). "Assessment of environmental and social sustainability performance of the freight transportation industry: an index-based approach." *Transport Policy,* 124(August), 43–60. https://doi.org/10.1016/J.TRANPOL.2020.01.006.

[18] Kurapati, S., Kourounioti, I., Lukosch, H., Tavasszy, L., & Verbraeck, A. (2018). "Fostering sustainable transportation operations through corridor management: a simulation gaming approach." *Sustainability (Switzerland),* 10(2). https://doi.org/10.3390/su10020455.

[19] Lee, G., You, S. I., Ritchie, S. G., Saphores, J. D., Jayakrishnan, R., & Ogunseitan, O. (2012). "Assessing air quality and health benefits of the clean truck program in the alameda corridor, CA." *Transportation Research Part A: Policy and Practice*, 46(8), 1177–93. https://doi.org/10.1016/j.tra.2012.05.005.

[20] Lopez, C., Shane, J., S., & Asce, A. M. (2014). "Keys to success in megaproject management in Mexico and the United States : Case study." *Journal of Construction Engineering and Management*, 140(4), 1–7. https://doi.org/10.1061/(ASCE)CO.1943-7862.0000476.

[21] Ludovico, D. D., Giacobbe, B. D., & Ovidio, G. D. (2021). "Analysis of European land transport network, MEGAs and socio-economic setting through territorial frames model.," 8(81).

[22] Ortega, E. (2015). "Ecological connectivity analysis to reduce the barrier effect of roads." *Core.Ac.Uk*. https://core.ac.uk/download/pdf/78495818.pdf.

[23] Robert, O. K., & Chan, A. P. C. (2015). "Review of studies on the critical success factors for public-private partnership (PPP) pProjects from 1990 to 2013." *International Journal of Project Management* 33(6), 1335–46. https://doi.org/10.1016/j.ijproman.2015.02.008.

[24] Padjen, J. (1998). "Eastern Adriatic and Ionic Corridorcorridor: new link between Central and South-eastern Europe." *Transport Reviews*, 18(4), 311–19. https://doi.org/10.1080/01441649808717021.

[25] Palomba, M., M Palomba, Russo, G., Russo, Federico Amadini, F., Amadini, Giampiero Carrieri, G., & Carrieri, Aanand Jain, and Jain, A. R. (2013). "Chenani-Nashri Tunneltunnel, the Longest road tunnel in india: a challenging case for design-optimization during construction." *Null*. https://doi.org/10.1201/b14769-132.

[26] Paul, J., Rosado-serrano, A., & Rosado-serrano, A. (2019). "Gradual internationalization vs born-global / international new venture models a review and research agenda." https://doi.org/10.1108/IMR-10-2018-0280.

[27] Purohit, R. R., Chaudhari, R. S., & Saraswat, A. (2023). "A study of land acquisition and compensation process for national highway in Maharashtra." *World Journal of Advanced Engineering Technology and Sciences*, 9(1), 307–15. https://doi.org/10.30574/wjaets.2023.9.1.0122.

[28] Rizopoulos, D., Laskari, M., Kouloumbis, G., Fergadiotou, I., Durkin, P., Kaare, K. K., and Muhammad Mahtab Alam. (2022). "5G as an enabler of connected-and-automated mobility in european cross-border corridors—a market assessment." *Sustainability (Switzerland)*, 14(21). https://doi.org/10.3390/su142114411.

[29] Roberts, M., Melecky, M., Bougna, T., & Xu, Y. (2020). "Transport Corridors and their wider economic benefits: a quantitative review of the literature." *Journal of Regional Science*, 60(2), 207–48. https://doi.org/10.1111/jors.12467.

[30] Sadullah, M., Ghazali, M., & Rashid, A. (2018). "Critical success factors in a public–private partnership highway project in Malaysia: Ampang–Kuala Lumpur elevated highway." *Proceedings of Institution of Civil Engineers: Management, Procurement and Law*, 170(6), 234–42. https://doi.org/10.1680/jmapl.16.00034.

[31] Sang, C., Olago, D. O., Nyumba, T. O., Marchant, R., & Thorn, J. P. R. (2022). "Assessing the Underlying drivers of change over two decades of land use and land cover dynamics along the standard gauge Gauge railway corridor, K, kenya." *Sustainability*, 14(10).

[32] Shankar, R., Pathak, D. K., & Choudhary, D. (2019). "Decarbonizing freight transportation: an integrated EFA-TISM approach to model enablers of dedicated freight corridors." *Technological Forecasting and Social Change*, 143(March), 85–100. https://doi.org/10.1016/j.techfore.2019.03.010.

[33] Shokat, D. M., & Jameel, A. K. (2023). "Identifying the cCrash worthiness and crash avoidance factors based on road features." *E3S Web of Conferences*, 427, 1–7. https://doi.org/10.1051/e3sconf/202342703010.

[34] Sinha, A. K., & Jha, K. N. (2021). "Financing Constraints of public–private partnership projects in India." *Engineering, Construction and Architectural Management*, 28(1), 246–69. https://doi.org/10.1108/ECAM-06-2018-0237.

[35] Stefanović, N., Milijić, S., & Hristić, N. D. (2020). "System approach in process of planning and project documentation preparation for highway corridors as an instrument for establishing the trans-european transport network." *Transportation Research Procedia*, 45(2019), 491–98. https://doi.org/10.1016/j.trpro.2020.03.043.

[36] Thomas, A. V., Kalidindi, S. N., & Ganesh, L. S. (2006). "Modelling and Assessment of critical risks in BOT road projects." *Construction Management and Economics,* 24(4), 407–24. https://doi.org/10.1080/01446190500435275.

[37] Wang, Z., Zhang, Y., Zheng, K., Zeng, R., Yuan, H., & Liu, J. (2023). "A review of mega-project management research from an organization science perspective: current status and future directions." *Developments in the Built Environment*, 16(May), 100254. https://doi.org/10.1016/j.dibe.2023.100254.

[38] Ansari, R, S A Banihashemi, R Taherkhani, and S Moradi. 2022. "Decision Support System for Analyzing Key Performance Indicators in Construction Projects Management." *International Journal of Engineering,* 35(5), 865–74.

[39] Flyvbjerg, Bent. 2008. "Public Planning of Mega-Projects: Overestimation of Demand and Underestimation of Costs." *Decision-Making on Mega-Projects: Cost-Benefit Analysis, Planning, and Innovation*, 120–44.

[40] Kiani Mavi, Reza, Neda Kiani Mavi, Doina Olaru, Sharon Biermann, and Sae Chi. 2022. "Innovations in Freight Transport: A Systematic Literature Evaluation and COVID Implications." *The International Journal of Logistics Management,* 33(4), 1157–95.

[41] Ludovico, Donato Di, Benedetta Di Giacobbe, and Gino D Ovidio. 2021. "Analysis of European Land Transport Network, MEGAs and Socio-Economic Setting through Territorial Frames Model," no. 81.

[42] Osei-Kyei, Robert, and Albert P C Chan. 2015. "Review of Studies on the Critical Success Factors for Public--Private Partnership (PPP) Projects from 1990 to 2013." *International Journal of Project Management,* 33(6), 1335–46.

[43] Rashid, Harunur. 2023. "Costs and Benefits of Regional Connectivity: An Analysis from Bangladesh Perspective." *South Asian Journal of Social Studies and Economics,* 20(3), 138–54.

[44] Roberts, Mark, Martin Melecky, Théophile Bougna, and Yan Xu. 2020. "Transport Corridors and Their Wider Economic Benefits: A Quantitative Review of the Literature." *Journal of Regional Science,* 60(2), 207–48.

[45] Schubert, G. 1990. "Noise Attenuation Developments for Major Transport Corridor Projects. 15th Arrb Conference, Darwin, Northern Territory, 26–31 August 1990; Proceedings parts 1 to 7." *Publication of Australian Road Research Board* 15(7).

[46] Wolfswinkel, Joost F, Elfi Furtmueller, and Celeste P M Wilderom. 2013. "Using Grounded Theory as a Method for Rigorously Reviewing Literature." *European Journal of Information Systems,* 22(1), 45–55.

[47] Yuewen, Li, and Wu Zhong. 2012. "Research of Intelligent Model of Optimal Route for the Urban Public Transport." In *2012 Second International Conference on Business Computing and Global Informatization*, 695–98.

Advances in Construction, Real Estate, Infrastructure and Project Management – Anil Kashyap et al. (eds)
© 2026 Taylor & Francis Group, London, ISBN 978-1-041-13433-6

27 Role of Labour Productivity, Operating Expenses and Net-Worth in Profitability of Construction Industry: A Study in India

J. C. Edison[1]
Professor,
NICMAR Business School NICMAR University,
Pune, India

Pradeepta Kumar Samanta[2]
Associate Professor,
School of Construction NICMAR University,
Pune, India

■ **ABSTRACT:** The construction industry in India plays a critical role in the nation's economic development by providing job opportunities and supporting infrastructure development. Understanding the factors influencing profitability in this industry is essential for optimizing resource allocation and fostering sustainable growth. This paper attempts to discuss factors influencing the profitability of the construction sector. The identification of the profitability determinants can help understand the variables which are to be deployed judiciously to improve the profitability of the construction industry. A literature review was undertaken taking into consideration with a view to identifying major factors affecting profitability in the sector.

Subsequently, aggregate financial data of construction industry were collected from the CME database. The study has taken labour productivity, net worth, operating expenses as the endogenous dependent variables, inflation and GDP as exogenous variables and net profit of the construction industry as independent variable for analysing the factors in profitability in the construction industry in India. SPSS was used to analyze the quantitative data, regression analysis to test the association and strength and direction of variables, and to measure their significance by correlation test.

The results of this research show that net-worth, operating expenses and inflation have significant effects on construction industry's profitability. Overall, this study helps identify the determinants of profitability in Indian construction industry. By understanding the interplay between labour productivity, operating expenses, and net-worth, construction firms can devise informed strategies to optimize performance, enhance competitiveness, and sustain profitability in an evolving market landscape.

■ **KEYWORDS:** Labour productivity, Operating expenses, Net-worth, Profitability, Construction industry

[1]edisonjolly@gmail.com, jcedison@nicmar.ac.in; [2]samanta.pk@gmail.com, pksamanta@nicmar.ac.in

DOI: 10.1201/9781003669814-27

1. INTRODUCTION

The construction industry and the infrastructure development are extremely interconnected, influencing various sectors and aspects of the economy and society. The construction industry is the foundation of an economy since it carries out the development of the necessary infrastructure for the other sectors and industry [14]. The linkages (backward and forward) between construction industry and the infrastructure development create a web of interdependencies that influence economic growth, social progress, environmental sustainability, and technological advancement. Consequently, construction industry and its profitability are essential for an economy. The political, economic, labour, cost, cultural, and legal aspects can negatively impact the level of profitability of construction projects [25]. This study is an attempt at role of labour productivity, operating expenses and net-worth in profitability of construction industry in India. Understanding these determinants will enable stakeholders to strategically utilize identified variables to enhance profitability in the construction sector.

2. CONSTRUCTION INDUSTRY PROFITABILITY

The profitability of the construction industry can be influenced by several factors, including economic conditions, project types, geographic location, business environment, trade cycles, labour costs, material availability and its cost variations, regulatory environments, market demand, adoption of advanced technologies, and effectiveness of management practices. Effective management of costs, including labour, materials, and overhead, can be crucial for maintaining profitability. The construction business world started adopting lean methodologies focussed on reducing waste and improving efficiency in construction processes, for better productivity. Numerous studies on construction industry profitability was carried out in the past including Isaac Buhamizo et al. [31], Edison and Singla [19], Edison and Singla [18], Bilal, et al. [10], Wahana Halian, et al. [47], Iyer, and Kumar [32], Akintoye and Skitmore [1], etc. The study of Jahan, et al. [33] have identified the following as most crucial risk factors influencing profitability: increasing material costs; supply chain processes; payment; planning and scheduling problems; financial difficulties; and managing resources efficiently.

Research studies in different countries reveals that low profitability is a major cause of increased risk of failure in the construction industry [2, 13, 16, 21, 29, 36, 45, 47].

The factors affecting profitability can be sub divided into firm level factors, industry level factors and economy level factors. Ammar et al. [4] use interest rate, GDP, and new business activities as economic variables and used growth rate as in industry variable and firm size as a firm specific variable to find the impact of these on profitability of firm. Zouaghi et al. [49] use a multilevel approach to study the impact on firm, industry and region level factors on profitability of firms. The factors that affect the productivity are primarily firm specific, through Jarkas and Bitar [35] based on a primary investigation divides them into four categories namely management; technological; human/labour; and external. At the same time, the factors affecting profitability and productivity can be overlapping.

3. LITERATURE REVIEW

Studies identified variables that affect the profitability like timely payments, delays in project completion, risks and its management, cost of finance, price variations, competition, corruption etc. [5, 8, 14, 15, 22, 28, 37, 38, 40]. Other variables like labour productivity, operating expenses, net worth, economic variables etc. affect the profitability of the construction industry.

The construction industry in India is one of the important sectors for the economic prosperity of the country [9]. Van Tam, et al. [44] in their study identified the critical factors affecting construction labour productivity from the perception of project managers compared to contractors' viewpoint. Being a job provider and offering infrastructure development, this area has always been in the centre of focus. This article aims to identify the drivers of profitability in the sector [44]. Through an extensive survey of existing literature, main elements affecting the profitability have been found. Knowing all these determinants will guide stakeholders on how to exploit the variables strategically to be able to increase the construction sector's profitability.

Hu and Liu [26] studied the overall efficiency of construction system. The study observed that the profitability performance and efficiency of construction show underperformance and a slight imbalance in Australia though the measures obtained for the efficiency factor indicate better achievement [26].

This study feels that this approach can be replicated worldwide for promoting sustainable development by construction projects, organizations to identify internal inefficiency constituents to recognize competitive advantages.

The study by Jarkas, et al. [34] ranked the relative position of the factors determining construction labour productivity in Oman. The Jarkas, et al. [34] study found 10 significant factors like "errors and omission" in design drawings, change orders throughout execution, working overtime, lack of labour supervision, rework etc. the study outcomes suggest methods to optimise productivity.

Thomas and Sudhakumar [6] in their study identified the factors influencing construction labour productivity in Kerala, India. It found "material related problems" as the chief factors of labour productivity. The political strikes, extreme climatic situations, changes in drawings, craftsmen turnover and absenteeism are "frequently occurring factors" in the projects in the state with significant negative impacts on productivity. Further, poor project planning and scheduling and lack of project coordination are among the top ten severe factors.

Azman et al. [43] studied the impact of institutional regulation on construction productivity of 55 public-listed construction Malaysian firms. The study observed that institutional regulation framework influences construction firms' total factor productivity in the long-term and must be considered by policymakers in formulating supporting incentives and policies [43].

3.1 Labour Productivity

Skill level of labour force and their productivity in construction sector is a critical issue affecting project timelines, costs, and overall profitability. Shortage of skilled labour can lead to project delays and increased costs, adversely affecting productivity. Better labour utilization by optimization of labour schedules and use subcontractors effectively to match labour supply with project demands may lead to improvement in labour productivity. It appears that well-organized and safe work sites enable workers to perform their tasks more efficiently. By addressing these factors and leveraging modern innovations, construction companies can improve labour productivity, resulting in more efficient project delivery and increased profitability.

3.2 Role of Operating Expenses in Profitability

Operating expenses are necessary for analyzing the profitability of construction companies. All costs related to daily operations necessary to manage the business are included in operating expenses. Managing these costs effectively can significantly impact the bottom line. The operating expenses directly affect the gross profit margin. High direct costs can reduce the margin unless managed effectively. Efficient procurement and cost control measures can help maintain healthy margins. Soaring operating expenses can strain cash flow, which makes it hard to meet immediate financial obligations. Maintaining a balanced cash flow is crucial for sustaining operations and taking on new projects.

4. METHODOLOGY

This study involves a number of chronological phases. The major factors which affect profitability of construction industry were identified from an analysis of literature reviewed. Subsequently, aggregate financial data of construction industry were collected from the CMIE database. The details are given below.

4.1 Data Source

The study has used data drawn from the Centre for Monitoring Indian Economy (CMIE) Industry Analysis database. Aggregate data of 29 years are used for the study. Period of the study ranges from 1993–94 to 2021–22. Population of the study include on an average 1300 firms from which the aggregate data was provided by the CMIE Industry Analysis database.

The analytical methods employed in this study are discussed as follows:

4.2 Variables

In order to carry out analysis the following are utilized. Industry-specific variables applied to perform the analysis for the construction industry include the construction labour productivity, net-worth, operating expenses and variables related to the economy such as economic growth and inflation. Table 27.1 presents the variables and its measure used for the construction industry analysis.

4.3 Regression Model

Developed a multiple regression model validating factors that affect construction industry's profitability. Among them, labour productivity, net worth, operating expenses, economic growth and inflation were used in the model to confirm and validate the model to forecast.

The model is as follows:

$$\text{Profitability} = \alpha + \beta_1 LP + \beta_2 NW + \beta_3 OpEx + \beta_4 GDPG + \beta_4 WPI + \varepsilon$$

where, LP = Labour Productivity, NW = Net worth, OpEx = Operating expenses, GDPG = Economic Growth and WPI = Inflation.

4.4 Data Analysis

The quantitative data was analysed using SPSS. Furthermore, regression analysis was also employed to analyze the relationship, strength and direction of the variables under study. Moreover, correlation analysis was carried out to establish the significance of the variables.

Table 27.1 Variables and its measure used for the construction industry analysis

Variable	Measure	References
Dependent Variable Profitability	Profit after tax (Total revenue minus Total expenses)	Isaac Buhamizo et al. [31], Edison and Singla [19], Edison and Singla [18], Bilal, et al. [10], Wahana Halian, et al. [48], Akintoye and Skitmore [1]
Independent Variables		
Industry Specific Variables Labour productivity	Net revenue ÷ Labour cost at organization level	Mohammed Hamza, et al. [42], Gopinath Selvam, et al. [24], Edison and Singla [20], Abdel-Hamid and Abdelhaleem [41], Payam Goodarzizad, et al. [46]
Net worth	Assets minus Liabilities	
Operating expenses	Direct costs incurred by the industry	
Economic Variables		
Economic Growth Inflation	Growth in GDP (GDPG) Wholesale Price Index (WPI)	

Source: Authors compilation from literature

5. RESULTS AND DISCUSSION

The financial soundness of construction sector can vary considerably depending on various factors including economic situations, market demand, regulatory environment, technological advancements, and industry dynamics. This is determined by a complex interplay of factors like revenue trends, profitability metrics, cost management, cash flow, risk management, market competition, regulatory environment, technological advancements, and economic conditions. Identifying determinants for assessing and projecting an industry's financial performance has got tremendous significance due to its applications in improving the profitability.

Profit is one of the key measures of the performance of firms, and a lot of research has gone into this over the years. Smith, 1980 indicated "the effect of working capital management" on the value, risk and profitability of companies. Smith's studies, referring the previous works, examined factors that influence a firm's financial performance, including leverage, economic activity, company growth, cash flow (operating), size, industry nature, operating cycle, and ROA. Pinches, 1991 specified that, different working capital policies are subject to long-standing debates about risk-return trade-offs. In total, conservative working capital approaches have a lower return and higher risk, while aggressive approaches have a higher return and lower risk (Pinches G.E, 1991).

Earlier studies of Hvide and Møen [27], Bukit et al. [12], and Flamini et al. [23] have largely focused on examining the influence of determinants, like financial ratios, of a group of firms on financial performance. Through this the decision-makers can achieve an overall view of its financial condition and develop strategies for performance enhancement.

5.1 Measure of Variables

Dependent Variable

Net profit plays a central role in the financial performance of the industry, especially construction, as it reflects the bottom-line profitability of construction companies after accounting for all expenses, taxes, and other financial obligations. Construction firms that focus on maximizing net profit through effective cost management, revenue growth, and operational efficiency are better positioned to achieve sustainable success and long-term competitiveness. Firms with higher net profit margins demonstrate skilful cost control and have strong potential to make considerable profits from their business activities [30].

The construction industry's overall net profit (Dependent variable), for 29 years, is examined and forecasted in terms of profit after tax.

Independent Variables

Labour productivity plays a crucial role in the financial performance of the construction industry. It directly impacts project costs, timelines, quality, and ultimately, the profitability of construction companies. Labour productivity one of the major factors affecting construction project delay and profitability [39].

Importance of net worth in financial strength and profitability of firms are widely discussed [3]. Net worth represents the residual value of a construction firm's assets after deducting its liabilities. It has an important part in the financial performance and stability of construction firms. It has a crucial role in shaping financial performance, development prospects, and resilience of firms. It is considered that by maintaining a healthy net worth, construction firms can enhance their financial stability, investment attractiveness, and stakeholder confidence, driving sustainable success and value creation in the dynamic construction industry.

Construction firms that effectively manage costs, mitigate risks, innovate, and deliver value to clients are better positioned to achieve sustained profitability and success in the dynamic construction market. Operating expenses are a critical component of the financial performance of construction companies. It has a crucial role in shaping the financial performance, competitiveness & sustainability of construction companies. It appears that by managing operating expenses effectively, construction firms can enhance profitability, cash flow, and operational efficiency, positioning themselves for long-term success in the dynamic construction industry.

Considering vital nature of the above three variables, i.e. labour productivity, net-worth, operating expenses were taken as the endogenous dependent variables. The profitability of the construction industry can vary based on numerous factors, including economic conditions, market demand, industry competition, regulatory environment, and company-specific factors. In order to include the exogenous factors two factors were considered, i.e. inflation and GDP as exogenous variables.

Using a linear regression model, this study analyzed the relationship between labour productivity, operating expenses and net-worth in profitability of construction industry. The results reveal that all these parameters are significantly contributing to the profit of construction industry in India.

Labour productivity showed the maximum significant influence on profitability of construction industry, which means that while keeping everything else constant, a change of one unit in labour productivity leads to increasing profitability by 44668.728 units from the construction industry (refer Table 27.1).

The other most significant variables are net-worth, operating expenses and inflation. Operating expenses and inflation are negatively related.

The study showed that the labour productivity has the maximum impact on profitability. Financial strength (net-worth) and operating expenses also impacts the profitability of the construction industry.

Table 27.2 shows the coefficients of regression analysis and the linear regression equation.

Table 27.2 Coefficients[a]

Model	B	Sig.
(Constant)	–90876.775	.084
LP	44668.728	.000
NW	.164	.000
OpEx	–.096	.000
GDPG	–3618.519	.205
WPI	–5361.437	.001

a. Dependent Variable: PAT
Profitability = (90876.775) + 44668.728 LP + 0.164 NW + (0.096)OpEx + (3618.519)GDPG + –5361.437WPI
Source: Author

Model Verification

The model was verified using the F-test. In this study, similar to Archer and Lemeshow [7], we calculated goodness of fit by comparing the table of F-values with the calculated F-value of SPSS.

An appropriate model has a computed "F-value higher than the critical F-value". With a "degree of freedom (df) of 5' and a selected "alpha p-value of 0.05", the critical F-value is calculated to be 5.69.

As presented in Table 27.3, the calculated F-value is 34.542 that is exceeding the critical value at 5.69; therefore, the model is appropriate and closely fits the ideal model.

Table 27.3 ANOVA[a] table for the model verification

	Model	Sum of Squares	df	Mean Square	F	Sig.
1	Regression	2.212E+11	5	44248118955	34.52	.000[b]
	Residual	28181970791	22	1280998672		
	Total	2.494E+11	27			

a. Dependent Variable: PAT
b. Predictors: (Constant), WPI, GDPG, LP, NW, OpEx

Source: Author

Model Fitness Test

A regression model was developed. The coefficient of determination, R^2, was evaluated. As per our assumptions, the forecasted and adjusted R^2 value are reasonably close. The greater the value of R^2, the better and acceptable. "The closer the forecasted and adjusted R^2 values are, the stronger the model and the better it predicts the response" [11]. Using the model (Table 27.4), the R^2 value was 0.88, meaning that 89% of the variance in profitability of the Indian construction industry is explained by the factors chosen for this study, which means that only 11% of the total variance unexplained by the model. Other factors might have been neglected in the study, which may explain the remaining 11%. The adjusted R^2 value is 0.861, which thus adds further strength to the model developed.

Table 27.4 Model summary[b]

Model	R	R Square	Adjusted R Square	Std. Error of the Estimate
1	.942[a]	.887	.861	35791.04179
a. Predictors: (Constant), WPI, GDPG, LP, NW, OpEx				
b. Dependent Variable: PAT				

Source: Author

6. CONCLUSION

The study has taken labour productivity, net worth, operating expenses as the endogenous dependent variables, inflation and GDP as exogenous variables and net profit of the construction industry as independent variable for analysing the factors in profitability of the construction industry in India. The results of this study show that net-worth, operating expenses and inflation have significant effects on construction industry's profitability. GDP have barely any significant effect on construction industry's profitability. Understanding and controlling operating expenses is essential for assessing the viability and profitability of firms. Conscientious efforts in these areas are required to improve profitability of the construction industry. Effective management of operating expenses through strategic planning, efficient resource utilization, technology adoption, and continuous improvement is able to significantly enhance construction industry profitability.

Future research is needed to compile firm-wise evidence to prove how these factors are effective at the firm level since the firms constitute the industry and all the firms in the industry data may not be same for all the years.

This research study acknowledges limitations that have to be considered like; the analysis concentrates precisely on macro-data of the industry. Both the exogenous variables and the PAT are industry aggregates. Therefore, the findings may not be generalized to other countries.

REFERENCES

[1] Akintoye, A. S., & Skitmore, M. (1991). Profitability of UK construction contractors. *Construction Management and Economics*, 9, 311–325. DOI: 10.1080/01446199100000025 Accessed from: http://eprints.qut.edu.au/archive/00004434/.

[2] Alavipour, S. M. R., & Arditi, D. (2019). Maximizing expected contractor profit using an integrated model. *Engineering, Construction and Architectural Management*, 26(1), 118–138. https://doi.org/10.1108/ECAM-04-2018-0149.

[3] Aldaas, A. (2021). The effect of firm life cycle on profitability: evidence from Jordanian firms. *Management Science Letters*, 11, 1919–1926. DOI:10.5267/j.msl.2021.1.009.

[4] Ammar, A., Hanna, A. S., Nordheim, E. V., & Russell, J. S. (2003). Indicator variables model of firm's size-profitability relationship of electrical contractors using financial and economic data. *Journal of Construction Engineering and Management*, 129(2), 192–197.

[5] Ansah, S. K. (2011). Causes and effects of delayed payments by clients on construction projects in Ghana. *Journal of Construction Project Management and Innovation*, 1(1), 27–45.

[6] Thomas, A. V., & Sudhakumar, J. (2013). Critical Analysis of the key factors affecting construction labour productivity – an Indian perspective. *International Journal of Construction Management*, 13 (4), 103–125.

[7] Archer, K. J., & Lemeshow, S. (2006). Goodness-of-fit test for a logistic regression model fitted using survey sample data. *The Stata Journal*, 6(1), 97–105. https:// doi.org/10.1177/1536867X0600600106.

[8] Balimwezo, R. (2009). A comparative study of local and foreign construction firms' participation in donor funded road construction projects: a case study of Uganda. Available from: http://makir.mak.ac.ug/handle/10570/3427 [Accessed on 7 July 2024].

[9] Bhattarai, K., & Negi, V. (2020). FDI and economic performance of firms in India. *Studies in Microeconomics*, 8(1), 44–74. https://doi.org/10.1177/2321022220918684.

[10] Bilal, M., Oyedele, L. O., Kusimo, H. O., Owolabi, H. A., Akanbi, L. A., Ajayi, A. O., et al. (2019). Investigating profitability performance of construction projects using big data: a project analytics approach. *Journal of Building Engineering*, 26, 100850. https://doi.org/10.1016/j.jobe.2019.100850.

[11] Blaikie, N. (2003). Analyzing Quantitative Data: From Description to Explanation. (1 st edn.), London: SAGE Publications Ltd.

[12] Bukit, R. B., Haryanto, B., & Ginting, P. (2018). Environmental performance, profitability, asset utilization, debt monitoring and company value. In IOP Conference Series: Earth and Environmental Science (Vol. 122, Article 012137). doi:10.1088/1755-1315/122/1/012137.

[13] Chan, T. K., & Martek, I. (2017). Profitability of large commercial construction companies in Australia. In Lamb, M. (Ed.), AUBEA 2017: Australasian Universities Building Education Association Conference 2017. (Vol. 1, pp. 139–146). Sydney: AUBEA. https://doi.org/10.29007/25c2.

[14] Colonnelli, E., & Ntungire, N. (2018). Construction and Public Procurement in Uganda. Helsinki: UNU-WIDER. WIDER Working Paper 2018/180. https://doi. org/10.35188/UNU-WIDER/2018/622-7.

[15] Colonnelli, E., & Ntungire, N. (2018). Construction and public procurement in Uganda. *Mining for Change*, 10.

[16] Creditsafe (2018). Watch Dog Report: A Statistical Analysis of the UK Economy. Q2 2018. Caerphilly, UK: Creditsafe. Available from: https://www.bing.eom/ck/a? !&&p=e1e89ccb758589c6JmltdHM9 MTY4NTA1OTIwMCZpZ3VpZD0yNzViMmR kNS0wOWQxLTY0OWMtMWFlOC0zZjRiMDgyYzY1N jEmaW5zaWQ9NTE5Ng&ptn=3&hsh=3&fclid=275b2dd5-09d1-649c-1ae8-3f4b082c6561&psq=creditsafe-watchdog-report-q1-2018&u = a1aHR0cHM6Ly93d3cuY3JlZGl0c2 FmZS5jb20vY29udGVudC9kYW0vdW svcGRmcy93YXRjaGRvZy1y-ZXBvcnRzL1dhdGGNoZG9nLVEyLTIwMTgtRnVsbC1SZXBvcnQucGGRm&n tb=1 [Accessed on 2 July 2024].

[17] Dambra, M., Gustafson, M. T., & Pisciotta, K. (2021). What is the effect of an additional dollar of IPO proceeds? *Journal of Corporate Finance*, 66, 101795, 10.1016/j.jcorpfin.2020.101795.

[18] Edison, J. C., & Singla, H. (2020). Comparative analysis of profitability of real estate, industrial construction and infrastructure firms: evidence from India. *Journal of Financial Management of Property and Construction*, 25(2), 273–291. doi.org/10.1108/JFMPC-08-2019-0069.

[19] Edison, J. C., & Singla, H. (2021). The mediating effect of productivity on profitability in Indian construction firms. *Journal of Advances in Management Research*, 18(1), 152–169. https://doi. org/10.1108/JAMR-05-2020-0092.

[20] Edison, J. C., & Singla, H. K. (2022). Examining the productivity and technical efficiency of industrial sector using stochastic frontier analysis in India. *Orissa Journal of Commerce*, 43(4), 28–45.

[21] El-Kholy, A. M., & Akal, A. Y. (2019). Determining the stationary financial cause of contracting firms failure. *International Journal of Construction Management*, 21(8), 818–833. https://doi.org/10.1080/156 23599.2019.1584836.

[22] Enshassi, A., Al-Hallaq, K., & Mohamed, S. (2006). Causes of contractor's business failure in developing countries: the case of Palestine. *Journal of Construction in Developing Countries*, 11(2), 1–14.

[23] Flamini, V., McDonald, C. A., & Schumacher, L. B. (2009). The Determinants of Commercial Bank Profitability in Sub-Saharan Africa (No. 2009/015). IMF.

[24] Selvam, G., Madhavi, T. C., Naazeema Begum, T. P., & Sudheesh, M. (2022). Impact of labour productivity in estimating the duration of construction projects. *International Journal of Construction Management*, 22 (12), 2398–2404.

[25] Han, S. H., Park, S. H., Kim, D. Y., Kim, H., & Kang, Y. W. (2007). Causes of bad profit in overseas construction projects. *Journal of Construction Engineering and Management*, 133, 932–943. http://dx.doi.org/10.1080/01446193.2016.1180415.

[26] Hu, X., & Liu, C. (2016). Profitability performance assessment in the Australian construction industry: a global relational two-stage DEA method. *Construction Management and Economics*, 34 (3), 147–159.

[27] Hvide, H., & Møen, J. (2007). Liquidity constraints and entrepreneurial performance. department of business and management science, norwegian school of economics. CEPR Discussion Paper No. DP6495. https://papers.ssrn.com/sol3/papers.cfm?abstract_id=1138959.

[28] Hwee, N. G., & Tiong, R. L. K. (2002). Model on cash flow forecasting and risk analysis for contracting firms. *International Journal of Project Management*, 20(5), 351–363. https://doi.org/10.1016/S0263-7863(01)00037-0.

[29] Ibn-Homaid, N. T., & Tijani, I. A. (2015). Financial analysis of a construction company in Saudi Arabia. *International Journal of Construction Engineering and Management*, 4(3), 80– 86.

[30] Imtiaz, F., Mahmud K., & Faisal, S. (2019). The determinants of profitability of non-bank financial institutions in Bangladesh. *International Journal of Economics and Finance*, 11(6), 25. 10.5539/ijef. v11n6p25.

[31] Buhamizo, I., Muhwezi, L., & Sengonzi, R. (2023). A regression model to enhance the profitability of local construction contractors in Uganda. *Journal of Construction in Developing Countries*, 28(1), 221–241. https://doi.org/10.21315/jcdc-08-21-0137.

[32] Iyer, K. C., & Kumar, R. (2016). Impact of delay and escalation on cash flow and profitability in a real estate project. *Procedia Engineering*, 145, 388–395. https://doi.org/10.1016/j.proeng.2016.04.097.

[33] Jahan, S., Khan, K. I. A., Thaheem, M. J., Ullah, F., Alqurashi, M., & Alsulami, B. T. (2022). Modeling profitability-influencing risk factors for construction projects: a system dynamics approach. *Buildings*, 12, 701. https://doi.org/10.3390/ buildings12060701.

[34] Jarkas, A. M., Al Balushi, R. A., & Raveendranath, P. K. (2015). Determinants of construction labour productivity in Oman. *International Journal of Construction Management*, 15(4), 332–344.

[35] Jarkas, A. M., & Bitar, C. G. (2012). Factors affecting construction labor productivity in Kuwait. *Journal of Construction Engineering and Management*, 138(7), 811–820.

[36] John, K., Gwaya, A. O., & Wanyona, G. (2019). Causes of contractors' failure in the construction industry in Rwanda. *International Journal of Innovative Science and Modern Engineering*, 5(12), 75–82.

[37] Kikwasi, G. J. (2012). Causes and effects of delays and disruptions in construction projects in Tanzania. In Australasian Journal of Construction Economics and Building Conference Series (Vol. 1, no. 2, pp. 52–59). https://doi.org/10.5130/ajceb-cs. v1i2.3166.

[38] Laryea, S., & Hughes, W. (2006). The price of risk in construction projects. In Boyd, D. (Ed.), Proceedings: 22nd Annual ARCOM Conference. Birmingham, UK: Association of Researchers in Construction Management (pp. 553–561).

[39] Mahamid, I. (2022). Relationship between delay and productivity in construction projects. *International Journal of Advanced and Applied Sciences*, 9(2), 160–166. https://doi.org/10.21833/ijaas.2022.02.018 C.

[40] Mishra, A. K., & Regmi, U. (2017). Effects of price fluctuation on the financial capacity of "Class A" contractors. *International Journal of Creative Research Thoughts*, 5(4), 1920–1938.

[41] Abdel-Hamid, M., & Abdelhaleem, H. M. (2022). Impact of poor labor productivity on construction project cost. *International Journal of Construction Management*, 22(12), 2356–2363. DOI: 10.1080/15623599.2020.1788757.

[42] Hamza, M., Shahid, S., Bin Hainin, M. R., & Nashwan, M. S. (2022). Construction labour productivity: review of factors identified. *International Journal of Construction Management*, 22(3), 413–425, DOI: 10.1080/15623599.2019.1627503.

[43] Azman, M. A., Chuweni, N. N., Muhamad Halil, F., Ku Azir, K. M. A., Lee, B. L., Juhari, F. N., et al. (2024). The impact of the change in institutional regulation on construction productivity: firm-level evidence in a developing economy. *Construction Management and Economics*, 42(3), 199–214.

[44] Van Tam, N., Quoc Toan, N., Tuan Hai, D., & Le Dinh Quy, N. (2021). Critical factors affecting construction labor productivity: A comparison between perceptions of project managers and contractors. *Cogent Business and Management*, 8(1), 1863303.

[45] Oladimeji, O., & Aina, O. O. (2018). Cash flow management techniques practices of local firms in Nigeria. *International Journal of Construction Management*, 21(4), 395–403. https://doi.org/10.1080/15 623599.2018.1541705.

[46] Goodarzizad, P., Mohammadi Golafshani, E., & Arashpour, M. (2023). Predicting the construction labour productivity using artificial neural network and grasshopper optimisation algorithm. *International Journal of Construction Management*, 23(5), 763–779. DOI: 10.1080/15623599.2021.1927363.

[47] Rajasekhar, R. (2017). Financial performance evaluation of construction industries. *International Journal of Scientific and Research Publications*, 7(1), 157–175. Available from: https://www.ijsrp.org/research-paper-0117/ijsrp-p6129.pdf.

[48] Halian, W., Sinaga, B. M., & Novianti, T. (2020). Factors affecting profitability of construction company sub-sector. *Indonesian Journal of Business and Entrepreneurship*, 6(2), 118–126. DOI: http://dx.doi.org/10.17358/IJBE.6.2.118.

[49] Zouaghi, F., Sánchez-García, M., & Hirsch, S. (2017), What drives firm profitability? a multilevel approach to the Spanish agri-food sector. *Spanish Journal of Agricultural Research*, 15(3), 1–15.

Advances in Construction, Real Estate, Infrastructure and Project Management – Anil Kashyap et al. (eds)
© *2026 Taylor & Francis Group, London, ISBN 978-1-041-13433-6*

28 Use of Environment Friendly Materials for Green Concrete Production—A Review

Vrushali V. Nalawade[1]

Research Student,
Civil Engg, MGM University, Chh. Sambhajinagar,
Maharashtra, India

Mohammed Sadique[2]

Professor,
Department of Civil Engg, University, Chh. Sambhajinagar,
Maharashtra, India

■ **ABSTRACT:** In the era of advanced concrete technology, sustainability must be a core principle, aiming to enhance living standards while preserving resources for future generations. Engineers must consider environmental, economic, and social aspects, ensuring responsible resource management, especially as global warming accelerates. The construction industry, as a significant contributor to CO2 emissions, must focus on adopting sustainable materials.

Industrial byproducts such as Ground Granulated Blast Furnace Slag (GGBS), fly-ash, micro silica and metakaolin, along with agricultural residues like rice husk ash, bagasse ash, and eggshell powder, can serve as cement replacements while maintaining concrete's strength and durability. This paper reviews studies on GGBS as an eco-friendly cement replacement and its effects on concrete's mechanical and durability properties. GGBS and fly ash have a proven global track record, particularly in demanding environments. The paper explores their use in India's concrete industry, highlighting their potential to drive sustainable construction practices.

■ **KEYWORDS:** GGBS, Sustainable concrete, Industrial waste, Mechanical properties, Durability properties

1. INTRODUCTION

Ordinary Portland Cement (OPC) is the most widely used artificial material globally, with annual production surpassing billions of tonnes. When combined with water, aggregates, and additives,

[1]nalawadevrushali22@gmail.com, [2]mdsadeq2013@gmail.com

DOI: 10.1201/9781003669814-28

cement becomes an essential ingredient in producing durable concrete. Concrete offers numerous benefits, including its widespread availability, cost-effectiveness compared to materials like timber and steel, and its ease of mixing and molding. These attributes have solidified its indispensability in large-scale infrastructure projects over the past two centuries (45). However, the widespread adoption of concrete carries significant environmental consequences. It accounts for 10% of the total global carbon dioxide (CO_2) emissions from man-made activities, with cement production alone contributing 70–80%. Each tonne of cement produced releases between 0.75 and 0.94 tonnes of CO_2, primarily from limestone decomposition and fossil fuel combustion. These emissions, amounting to billions of tonnes annually, contribute significantly to global warming.

Efforts are underway to mitigate these environmental impacts by focusing on enhancing energy efficiency, exploring alternative raw materials, improving manufacturing processes, developing technologies for CO_2 capture and storage, and advocating for the use of mineral additives such as fly ash (FA) and ground granulated blast furnace slag (GGBS) in blended cements. These strategies aim to reduce CO_2 emissions while ensuring that the performance remains comparable to traditional cement-based materials [14].

1.1 Ecofriendly Cementitious Materials

Using cementitious materials, such as GGBS, fly ash, metakaolin and micro silica, presents a practical strategy to significantly reduce greenhouse gas emissions. These materials, sourced from industrial by-products, present a promising opportunity to significantly reduce emissions associated with Ordinary Portland Cement (OPC). Previous studies, including Table 28.1, have detailed their chemical compositions, highlighting high concentrations of pozzolanic elements such as silica, calcium, alumina, magnesia, and iron, all meeting ASTM standards [35]. This positions materials like GGBS, micro silica, metakaolin, and fly ash as feasible alternatives to OPC in concrete, offering environmental advantages without compromising performance [19]. Recognized for their binding capabilities, these industrial wastes—GGBS, metakaolin, fly-ash ans micro silica are increasingly acknowledged by researchers as promising components for blended cements, aimed at reducing environmental carbon footprints.

Table 28.1 Chemical composition comparison of various green materials over OPC

Chemical	OPC	Fly Ash	Micro silica	Meta-kaolin	GGBS	UFS	RHA	CBWA	ESP	GP
SiO_2	23.85	65.9	93.74	47.64	32.7	33.61	93.4	37.90	0.58	68.1
Al_2O_3	3.26	22.1	0.19	50.22	12.6	22.5	0.05	24.12	0.06	0.9
Fe_2O_3	1.77	3.4	0.95	0.24	0.3	1.31	0.06	15.48	0.02	0.6
MgO	1.27	0.7	0.17	0.05	5.4	6.8	0.35	1.89	0.06	1.8
CaO	62.32	1.6	0.40	0.05	41.9	34.0	0.31	4.98	52.10	14.5
NaO_2	0	0.6	0	0.15	0.4	0	0.1	0.95	0.15	12.2
K_2O	0.82	1.8	0.62	0.15	0	0	1.4	0.83	0.25	0.8
Reference	[19]	[34]	[20]	[21]	[8]	[17]	[36]	[40]	[7]	[15]

Source: Compiled by authors

Other industrial byproducts, including waste marble, foundry sand, slag, and rubber, can also function as potential binding materials. These pozzolanic substances primarily arise from manufacturing processes, offering the dual benefits of waste reduction and environmental sustainability.

Why GGBS?

Blast-furnace slag and its derivative, GGBS, are industrial byproducts commonly used as supplementary cementitious materials in cement and concrete production. GGBS is particularly notable for its self-hydration properties and pozzolanic action, likely due to its calcium oxide content of 30–40%, which also influences its color [41]. As a sustainable alternative, GGBS offers significant environmental benefits in concrete production. Being a byproduct, its proper disposal is essential. Incorporating GGBS in concrete mixes reduces dependence on traditional materials such as OPC, fine aggregates, and coarse aggregates. Multiple studies have demonstrated that substituting GGBS in concrete maintains its structural integrity [3].

Availability of GGBS in Indian Market

India ranks as the world's second-largest steel-producing country. Figure 28.1 illustrates global steel production volumes [39]. The Iron and Steel industry holds a crucial position in India's economy. According to worldsteel, India's crude steel production reached 106.5 million tonnes (MT) in 2018, marking a 4.9% increase from 101.5 MT in 2017. This growth enabled India to surpass Japan as the world's second-largest steel producer. In comparison, Japan's crude steel production declined by 0.3%, dropping to 104.3 MT in 2018 from its 2017 levels. As reported by the Indian Steel Association (ISA), India's total installed steel-making capacity stood at 154 MT as of March 2023.

Figure 28.1 Steel production across the world

Source: World Steel Association

2. PHYSICAL PROPERTIES OF GGBS

The physical properties of GGBS, including specific gravity, absorption capacity, grain size distribution, fineness modulus, moisture content, bulk density, specific surface area, and unit weight, are crucial in determining its suitability for use in concrete [4]. Table 28.2 presents an overview of these properties.

Table 28.2 Physical Properties of GGBS by various author

Sr. No.	Physical Property	Referance Author					
		[4]	[12]	[11]	[6]	[38]	[35]
1	Specific Gravity	2.8	2.83	2.85	2.85	2.82	2.85
2	Bulk Density (kg/m^3)	1394	1200	1668			1200
3	Fineness (cm2/g)			6260	4000	5000	3500

Source: Compiled by authors

GGBS has a specific gravity ranging from 2.5 to 2.9, similar to that of cement. The variations in these values across different studies are shown in Figure 28.2. The material exhibits an absorption capacity of around 1.2%, which may adversely affect the workability of concrete. The grain size of GGBS varies between 1.18 mm and 0.10 mm, with 62% of the particles falling within this range [39].

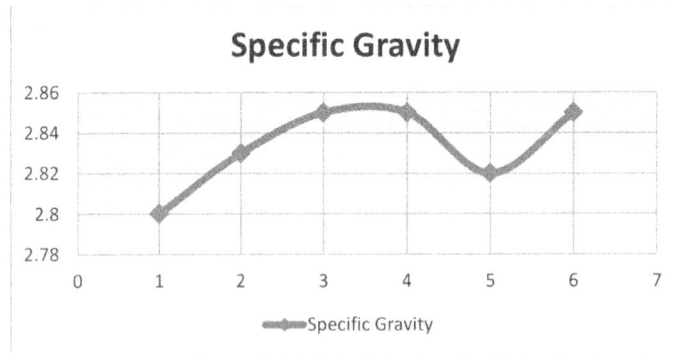

Figure 28.2 Comparative values of specific gravity from various literature
Source: Authors

The bulk density of GGBS ranges from 1200 to 1670 kg/m³, which is comparable to cement's density of 1440 kg/m³. GGBS also has a larger specific surface area, ranging from 3500 to 6000 cm²/g, significantly higher than cement's 3310 cm²/g [1]. This increased surface area requires more mortar for coverage, reducing the paste available for lubrication and affecting the flowability of concrete [10].

Overall, these physical properties highlight GGBS's potential for use in concrete, despite factors such as its absorption capacity and surface area that can influence its performance characteristics [13].

2.1 Chemical Composition of GGBS

The composition of slag in GGBS varies significantly depending on the raw materials used in iron production. It typically consists of silica, calcium, aluminum, magnesium, and oxygen, with silica accounting for more than 95% of its content [43].

Higher basicity enhances GGBS's hydraulic activity with alkaline activators, improving strength by boosting alumina content or compensating for calcium deficiencies with increased alumina or magnesia [16]. Minor fluctuations (up to 8–10%) in magnesia oxide content minimally affect strength, but concentrations exceeding 10% can impair it [11]. Most GGBS samples contain less than 10% magnesia oxide (Table 283). GGBS qualifies as a pozzolanic material under ASTM standards due to its accumulation of calcium, silica, magnesia, iron and alumina to over 70%, making it a viable substitute for OPC in concrete applications.

Table 28.3   Chemical composition of GGBS

Mass % Authors	[8]	[35]	[12]	[39]	[43]	[38]	[3]
SiO_2	32.7	31.65	35.32	35.34	36.9	34	35.85
Al_2O_3	12.6	12.40	15.63	11.59	12.3	14	13.39
Fe_2O_3	0.3	0.37	1.56	0.35	0.3	4	1.06
MgO	5.4	5.80	6.54	8.04	7.4	7	9.1
CaO	41.9	43.17	40.15	41.99	38.6	33	37.71
others							

Source: Authors

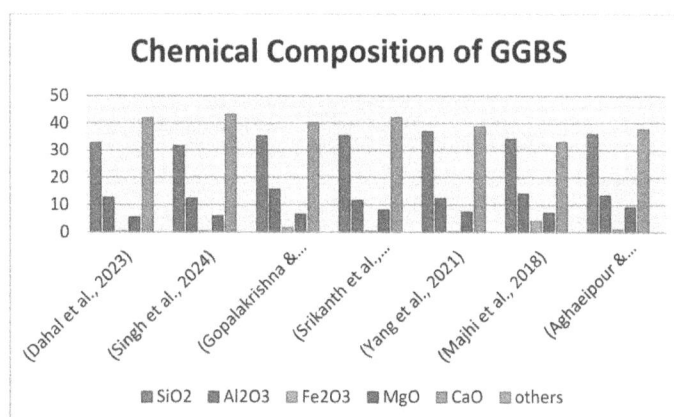

Figure 28.3   Chemical composition of GGBS

Source: Authors

2.2 Mechanical Properties of GGBS Concrete

Compressive Strength

Compressive strength is a crucial property in assessing the structural integrity and durability of concrete [9]. It measures the material's ability to withstand axial forces that compress or crush it, ensuring concrete can support applied loads without failure [2]. Table 28.4 and Figure 28.4 summarizes past research on the compressive strength of GGBS-incorporated concrete.

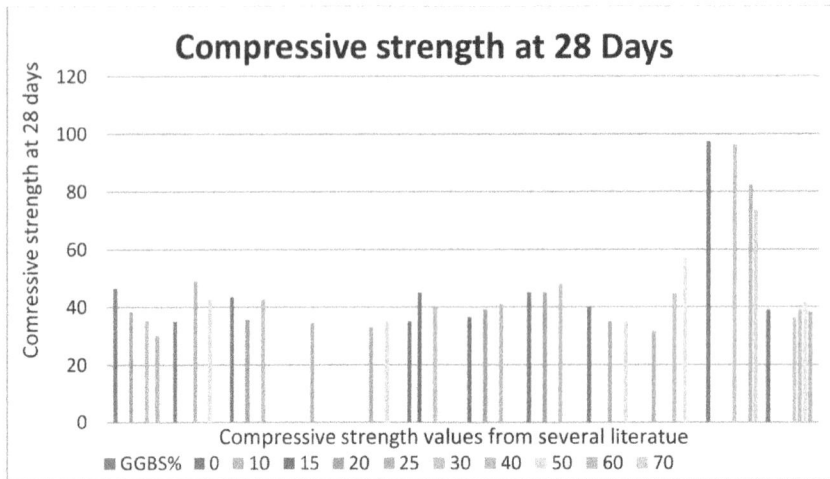

Figure 28.4 Compressive strength values from various literature

Source: Authors

Table 28.4 Values of compressive strength of GGBS concrete

Author	GGBS Replacement %	Compressive strength (28 Days) MPA
[46]	0, 20, 40 & 60%	46.5, 38.4, 35.2, 30.1
[44]	0, 25 & 50%	35, 49 & 43
[30]	0, 20% & 40%	43.4, 35.6 & 42.5
[47]	25% Ambient, air, water cooled	31.41, 30.99, 31.08
[1]	25 & 50%	33, 35
[50]	0, 15 & 30%	35, 45 & 40
[49]	0, 20 & 40%	36.42, 39.1 & 41
[3]	0, 20 & 40%	45, 45 & 48
[38]	0, 25 & 50%	40, 32 & 35
[48]	10, 30 & 50%	31.59, 44.73, 57.20
[39]	0, 30, 60 & 70%	97.375, 96.4, 82.34, 73.5
[35]	0, 30, 40, 50 & 60%	38.9, 36.32, 39.15, 41.82, 38.24

Source: Authors

Tensile Strength: Tensile strength measures stress as force per unit area. For non-uniform materials or assembled components, it may be represented as force or force per unit width. In the International System of Units (SI), tensile strength is expressed in pascals (Pa), often reported in multiples like megapascals (MPa). Direct testing of concrete's tensile strength is uncommon due to additional stresses caused by sample holding methods, leading to the use of various indirect approaches to assess it. The tensile strength of concrete demonstrates variability across different ratios of GGBS replacement.

Flexural Strength: Flexural strength, also referred to as modulus of rupture, bend strength, or transverse rupture strength, is a material property that describes the stress a material experiences just before yielding in a flexure test. The most common method used to measure this is the transverse bending test, where a specimen with a circular or rectangular cross-section is bent until fracture or yielding occurs, typically using a three-point flexural test technique. The flexural strength represents the maximum stress the material undergoes at the point of yield and is measured in terms of stress.

2.3 Durability Properties of GGBS

Density: Density is a key parameter for evaluating the quality of concrete, especially its durability [3]. Denser concrete exhibits greater strength, fewer voids, and lower porosity compared to less dense concrete [35]. A reduction in void spaces decreases the permeability of concrete to water and soluble components, thereby lowering water absorption and enhancing its durability. Water often contains harmful chemicals that contribute to the degradation of concrete [9].

Permeability: The permeability of a concrete crucially affects its durability by controlling the rate of moisture and aggressive chemical penetration, thereby reducing susceptibility to cracking. Studies indicate that substituting GGBS decreases concrete permeability [5]. For instance, after 28 days, concrete without GGBS had a maximum permeability of 19.5 mm, whereas with 10% GGBS and 10% MK, permeability decreased to 9 mm.

Reduced permeability in concrete is essential for durability, as it signifies resistance to chemical attack. This decrease is attributed to GGBS's pozzolanic reaction with secondary cementitious materials (CSH) improves mortar binding properties and reducing permeability. Additionally, GGBS's micro-filling effect in aggregate voids results in a denser mass, further lowering permeability. However, excessive GGBS (over 20%) can decrease workability and increase voids in concrete, potentially elevating permeability [24].

Dry shrinkage: Dry shrinkage, crucial for cementitious composites' durability, results from capillary water loss, causing contraction and internal cracks. Increasing GGBS content in concrete reduces drying shrinkage by lowering cement paste volume [18]. Mineral additives also decrease heat of hydration, slowing water evaporation and crack formation. GGBS's pozzolanic reaction produces CSH, densifying concrete and further curbing shrinkage [45]. Fly ash aids by filling microspores, enhancing internal compactness. Shrinkage primarily affects mortar volume, while coarse aggregate stabilizes cement paste volume. Combined pozzolanic reactions and micro-filling enhance cement paste binding and density, diminishing concrete's dry shrinkage.

2.4 Advantages of using GGBS in Concrete

Concrete incorporating GGBS cement exhibits greater ultimate strength compared to concrete made solely with Portland cement [44]. This is due to its higher proportion of calcium silicate hydrates (CSH), which enhance strength, and a lower content of free lime, a component that does not contribute to concrete strength [17]. Additionally, concrete with GGBS continues to gain strength over time, offering several advantages, including:

```
┌─────────────────┐   ┌─────────────────┐   ┌─────────────────┐   ┌─────────────────┐
│ Increased       │   │ Enhanced        │   │ Reduces risk of │   │ Minimizes       │
│ strength and    │──▷│ workability     │──▷│ thermal cracking│──▷│ shrinkage cracks│
│ durability      │   │                 │   │                 │   │                 │
└─────────────────┘   └─────────────────┘   └─────────────────┘   └─────────────────┘
        │
        ▽
┌─────────────────┐   ┌─────────────────┐   ┌─────────────────┐   ┌─────────────────┐
│ Lowers voids    │   │ Provides a      │   │ Improves        │   │ Facilitates     │
│ and permeability│──▷│ workable mix    │──▷│ flexural        │──▷│ pumping and     │
│                 │   │                 │   │ strength        │   │ compaction      │
└─────────────────┘   └─────────────────┘   └─────────────────┘   └─────────────────┘
        │
        ▽
┌─────────────────┐   ┌─────────────────┐   ┌─────────────────┐   ┌─────────────────┐
│ Increases       │   │ High resistance │   │ Low heat of     │   │ Highly resistant│
│ resistance to   │──▷│ to chloride     │──▷│ hydration       │──▷│ to alkali-silica│
│ sulphate attack │   │ ingress         │   │                 │   │ reaction        │
└─────────────────┘   └─────────────────┘   └─────────────────┘   └─────────────────┘
        │
        ▽
┌─────────────────┐   ┌─────────────────┐   ┌─────────────────┐   ┌─────────────────┐
│ Improves surface│   │ Offers          │   │ Gradual         │   │ Environmentally │
│ finish and      │──▷│ sustainability  │──▷│ hydration       │──▷│ friendly (low   │
│ aesthetics      │   │ benefits        │   │ reduces heat    │   │ emissions)      │
│                 │   │                 │   │ generation      │   │                 │
└─────────────────┘   └─────────────────┘   └─────────────────┘   └─────────────────┘
```

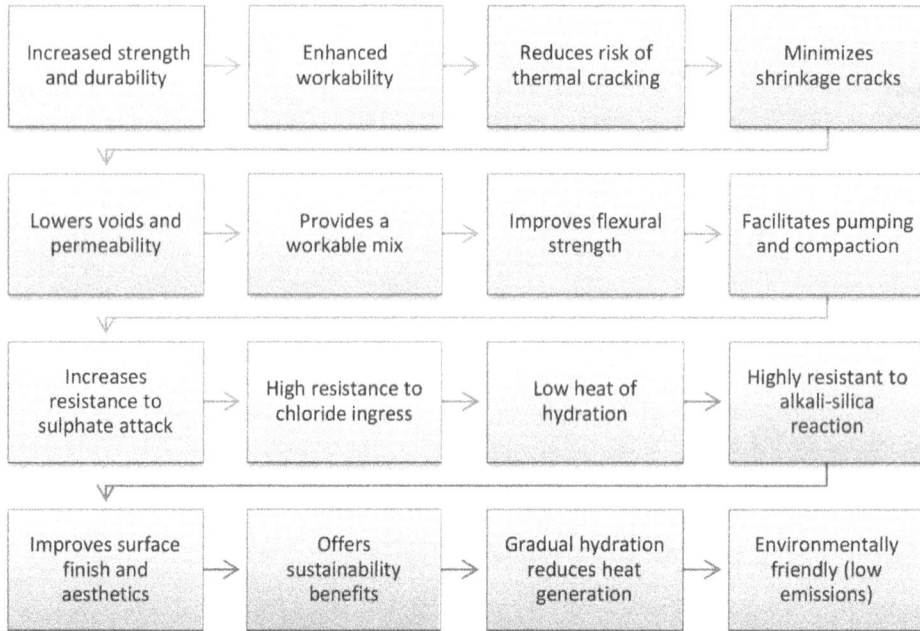

Figure 28.5   Advantage of GGBS based concrete as a partial replacement of cement

Source: Authors

2.5 Disadvantages of using GGBS in Concrete

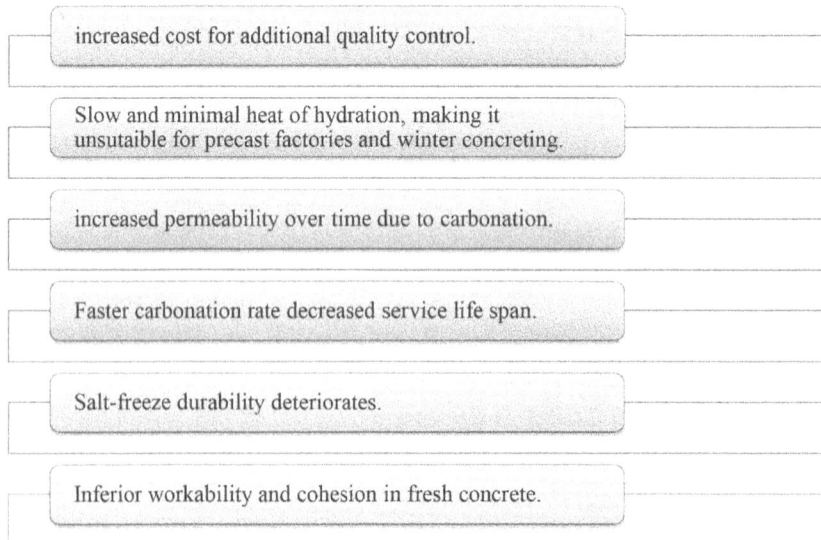

```
┌──────────────────────────────────────────────────────────┐
│ increased cost for additional quality control.            │
└──────────────────────────────────────────────────────────┘

┌──────────────────────────────────────────────────────────┐
│ Slow and minimal heat of hydration, making it             │
│ unsutaible for precast factories and winter concreting.   │
└──────────────────────────────────────────────────────────┘

┌──────────────────────────────────────────────────────────┐
│ increased permeability over time due to carbonation.      │
└──────────────────────────────────────────────────────────┘

┌──────────────────────────────────────────────────────────┐
│ Faster carbonation rate decreased service life span.      │
└──────────────────────────────────────────────────────────┘

┌──────────────────────────────────────────────────────────┐
│ Salt-freeze durability deteriorates.                      │
└──────────────────────────────────────────────────────────┘

┌──────────────────────────────────────────────────────────┐
│ Inferior workability and cohesion in fresh concrete.      │
└──────────────────────────────────────────────────────────┘
```

Figure 28.6   Disadvantages of using GGBS

Source: Authors

Despite its numerous benefits, Ground Granulated Blast Furnace Slag (GGBS) based concrete has some disadvantages. One significant drawback is the slower rate of strength gain compared to conventional Portland cement, particularly in early stages, which can delay construction schedules. GGBS concrete may also require stricter quality control during mixing to ensure proper blending and performance. Additionally, its lighter color compared to traditional concrete can affect aesthetic requirements in certain applications. Transporting and sourcing GGBS can increase costs and carbon emissions if it is not locally available. Furthermore, the reliance on GGBS ties its availability to the steel industry, which could pose supply challenges if production scales down in the future. Following are the disadvantages of using GGBS as a partial of fully replacement of cement.

3. CONCLUSION

This review examines the latest literature on the utilization and efficacy of GGBS in concrete. It covers GGBS manufacturing methods, its physical as well as chemical composition, hydration reactions, and its impact on various concrete properties including fresh characteristics, mechanical strength, permeability, and durability. The seminar aimed to raise awareness about the environmental implications and technical advantages of using GGBS for sustainable construction. Overall benefits of GGBS in concrete include:

- GGBS exhibits physical properties such as specific gravity and bulk density similar to cement, but features a larger surface area and angular particle texture, as revealed by SEM analysis.
- Its chemical composition enables effective partial replacement of cement up to a specific percentage.
- Replacing cement with GGBS reduces the heat of hydration, attributed to its slower pozzolanic reaction compared to cement.
- Due to its surface characteristics, GGBS may decrease concrete workability, often requiring plasticizers, particularly at higher replacement levels.
- Mechanical properties, including compressive strength, split tensile strength, and flexural strength, are enhanced with GGBS through pozzolanic reactions and the micro-filling of voids. However, excessive replacement can lead to reduced workability, negatively impacting these properties and overall durability.
- Enhanced durability is observed in terms of reduced dry shrinkage, permeability, chloride penetration, and resistance to acid attacks, attributed to GGBS's micro-filling and pozzolanic reaction effects.
- Optimal dosage of GGBS is crucial for maximizing mechanical strength and durability. Different studies suggest varying optimum percentages, with many researchers recommending around 20% GGBS content.

REFERANCES

[1] Tamilarasan, A., & Suganya, O. M. (2024). Investigation on the properties of sustainable alkali-activated GGBS mortar with nano silica and dredged marine sand: Mechanical, microstructural and durability aspects. *Case Studies in Construction Materials*, 20, e03148.

[2] Abdullahi, M., Ojelade, G., & Auta, S. (2017). Modified water-cement ratio law for compressive strength of rice husk ash concrete. *Nigerian Journal of Technology*, 36, 373.

[3] Aghaeipour, A., & Madhkhan, M. (2017). Effect of ground granulated blast furnace slag (GGBFS) on RCCP durability. *Construction and Building Materials*, 141, 533–541.

[4] Amar, R., Devanand, R., Harsha, H. N., & Sachin, K. C. (2023). Experimental studies on GGBS-based geopolymer concrete. Materials Today: Proceedings.

[5] Amran, M., Al-Fakih, A., Chu, S. H., Fediuk, R., Haruna, S., Azevedo, A., & Vatin, N. (2021). Long-term durability properties of geopolymer concrete: an in-depth review. *Case Studies in Construction Materials*, 15, e00661.

[6] Anil Kumar Reddy, N., & Ramujee, K. (2022). Comparative study on mechanical properties of fly ash & GGBFS-based geopolymer concrete and OPC concrete using nano-alumina. Materials Today: Proceedings, 60, 399–404.

[7] Chong, B. W., Othman, R., Ramadhansyah, P. J., Doh, S. I., & Li, X. (2020). Properties of concrete with eggshell powder: a review. Physics and Chemistry of the Earth, Parts A/B/C, 120, 102951.

[8] Dahal, M., Oinam, Y., Vashistha, P., Oh, J.-E., & Pyo, S. (2023). Cementless ultra-high performance concrete (UHPC) using CaO-activated GGBFS and calcium formate as an accelerator. *Journal of Building Engineering*, 75, 107000.

[9] Divsholi, B. S., Lim, T. Y. D., & Teng, S. (2014). Durability properties and microstructure of ground granulated blast furnace slag cement concrete. *International Journal of Concrete Structures and Materials*, 8(2), Article 2.

[10] Ghasemi, Y., Emborg, M., & Cwirzen, A. (2019). Exploring the relation between the flow of mortar and specific surface area of its constituents. *Construction and Building Materials*, 211, 492–501.

[11] Ghosh, D., Ma, Z. J., & Hun, D. (2023). Effect of GGBFS slag on CSA-based ternary binder hydration, and concrete performance. *Construction and Building Materials*, 386, 131554.

[12] Gopalakrishna, B., & Dinakar, P. (2023). Mix design development of fly ash-GGBS-based recycled aggregate geopolymer concrete. *Journal of Building Engineering*, 63, 105551.

[13] Hakeem, I. Y., Zaid, O., Arbili, M. M., Alyami, M., Alhamami, A., & Alharthai, M. (2024). A state-of-the-art review of the physical and durability characteristics and microstructure behavior of ultra-high-performance geopolymer concrete. *Heliyon*, 10(2), e24263.

[14] Hanifa, M., Agarwal, R., Sharma, U., Thapliyal, P. C., & Singh, L. P. (2023). A review on CO2 capture and sequestration in the construction industry: Emerging approaches and commercialized technologies. *Journal of CO2 Utilization*, 67, 102292.

[15] Islam, G. M. S., Rahman, M. H., & Kazi, N. (2017). Waste glass powder as partial replacement of cement for sustainable concrete practice. *International Journal of Sustainable Built Environment*, 6(1), 37–44.

[16] Kumar, K. V., & Reddy, N. S. S. (2022). An experimental investigation on partial replacement of cement by GGBS & full replacement of river sand by Robo Sand in concrete. *Journal of Engineering and Technology*, 9(7).

[17] Kumar Sinha, A., & Talukdar, S. (2023). Mechanical and bond behaviour of high-volume ultrafine-slag blended fly ash-based alkali-activated concrete. *Construction and Building Materials*, 383, 131368.

[18] Li, Q., & Zhang, Q. (2019). Experimental study on the compressive strength and shrinkage of concrete containing fly ash and ground granulated blast-furnace slag. *Structural Concrete*, 20(5), 1551–1560.

[19] Raza, F., Pathan, D., Dangre, M., Deshmukh, N., Ali Khan, H., & Kumar Shaw, S. (2023). A state-of-the-art review on the mechanical behaviours of GGBFS concrete for sustainable development. Materials Today: Proceedings. https://doi.org/10.1016/j.matpr.2023.05.530

[20] Mahalakshmi, M., Kumaravel, T., & Sathishkumar, S. (2017). Utilization of GGBS in concrete with partial replacement of cement and fine aggregate. *International Journal of Civil Engineering and Technology*, 8(6), 825–832.

[21] Manjunatha, S. G., Sannibabu, T., Rao, M. N., & Kumar, G. S. (2022). Experimental study on GGBS concrete using nano-alumina and metakaolin as supplementary materials. Materials Today: Proceedings, 60, 387–393.

[22] Muthukumar, K., & Balasubramanian, V. (2021). Strength and durability characteristics of concrete incorporating GGBFS and fly ash. *Advances in Concrete Construction*, 11(1), 11–19.

[23] Nagaveni, M., Deepa, C., & Suganya, T. R. (2021). The effect of supplementary cementitious materials and nano-silica on GGBS geopolymer concrete. Materials Today: Proceedings, 46, 353–359.

[24] Nataraja, M. C., Dondapati, R., & Kumar, D. R. (2023). Utilization of ground granulated blast furnace slag and dolomite powder in concrete. *Case Studies in Construction Materials*, 19, e01426.

[25] Neville, A. M. (2011). Properties of Concrete (5th ed.). Pearson.

[26] Pawar, A. M., & Raut, S. P. (2022). A review on the performance of geopolymer concrete using GGBS and metakaolin. Materials Today: Proceedings, 59, 220–225.

[27] Rafiq, S., Ahmed, W., & Aslam, F. (2023). Mechanical and durability properties of sustainable geopolymer concrete incorporating GGBFS and nano-silica. *Case Studies in Construction Materials*, 19, e02002.

[28] Ranjith Kumar, R., & Priyadharshini, B. (2018). Study on mechanical and durability properties of GGBS-based self-compacting concrete. *International Journal of Engineering & Technology*, 7(2.34), 310–313.

[29] Rashid, A., & Nasir, K. (2021). Effect of GGBS on the performance of concrete under various curing conditions. *Construction and Building Materials*, 300, 124283.

[30] Siddique, R., & Bennacer, R. (2012). Use of iron and steel industry by-product (GGBS) in cement paste and mortar. *Resources, Conservation and Recycling*, 69, 29–34. https://doi.org/10.1016/j.resconrec.2012.09.002

[31] Srinivasan, M., & Sathya, R. (2022). The effect of fly ash and GGBS-based geopolymer concrete on strength and durability properties. *Journal of Building Engineering*, 45, 103333.

[32] Venkatesh, K., & Ramesh, B. (2019). Effect of partial replacement of cement by GGBS and fine aggregate by foundry sand on mechanical properties of concrete. *Advances in Concrete Construction*, 7(1), 1–7.

[33] Xiao, J., Qiang, C., & Zhang, C. (2018). Effect of ground granulated blast furnace slag on the durability performance of recycled aggregate concrete. *Materials and Structures*, 51(4), 97.

[34] Zhang, Y., Zheng, X., & Zhu, J. (2023). Impact of supplementary cementitious materials on ultra-high-performance concrete durability. *Cement and Concrete Research*, 165, 107096.

[35] Singh, R. P., Vanapalli, K. R., Jadda, K., & Mohanty, B. (2024). Durability assessment of fly ash, GGBS, and silica fume based geopolymer concrete with recycled aggregates against acid and sulfate attack. *Journal of Building Engineering*, 82, 108354. https://doi.org/10.1016/j.jobe.2023.108354

[36] Mustafa Mohamed, A., Osman Bashir, M., Dirar, S., & Beshir Osman, N. (2022). Compressive strength of efficient self-compacting concrete with rice husk ash, fly ash, and calcium carbide waste additives using multiple artificial intelligence methods. *Structural Concrete*, 23(4), Article 4. https://doi.org/10.1002/suco.202100286

[37] Praveen Kumar, V. V., Prasad, N., & Dey, S. (2020). Influence of metakaolin on strength and durability characteristics of ground granulated blast furnace slag based geopolymer concrete. *Structural Concrete*, 21(3), Article 3. https://doi.org/10.1002/suco.201900415

[38] Majhi, R. K., Nayak, A. N., & Mukharjee, B. B. (2018). Development of sustainable concrete using recycled coarse aggregate and ground granulated blast furnace slag. *Construction and Building Materials*, 159, 417–430. https://doi.org/10.1016/j.conbuildmat.2017.10.118

[39] Srikanth, S., Krishna, C. B. R., Srikanth, T., Sai Nitesh, K. J. N., Nadh, V. S., Kumar, S., & Thanappan, S. (2022). Effect of Nano Ground Granulated Blast Furnace Slag (GGBS) Volume % on Mechanical Behaviour of High-Performance Sustainable Concrete. *Journal of Nanomaterials*, 2022, e3742194. https://doi.org/10.1155/2022/3742194

[40] Satheesh, M., Pugazhvadivu, M., Prabu, B., Gunasegaran, V., & Manikandan, A. (2019). Synthesis and Characterization of Coconut Shell Ash. *Journal of Nanoscience and Nanotechnology*, 19(7), Article 7. https://doi.org/10.1166/jnn.2019.16299

[41] Lodhi, U., & Choudhary, K. K. (2021). *Determination of strength of concrete by using Ground-granulated blast-furnace slag (GGBS)* (10). 3(10), Article 10.

[42] Silva, L. H. P., Nehring, V., de Paiva, F. F. G., Tamashiro, J. R., Galvín, A. P., López-Uceda, A., & Kinoshita, A. (2023). Use of blast furnace slag in cementitious materials for pavements—Systematic literature review and eco-efficiency. *Sustainable Chemistry and Pharmacy*, 33, 101030. https://doi.org/10.1016/j.scp.2023.101030

[43] Yang, J., Zeng, J., He, X., Hu, H., Su, Y., Bai, H., & Tan, H. (2021). Eco-friendly UHPC prepared from high volume wet-grinded ultrafine GGBS slurry. *Construction and Building Materials*, 308, 125057. https://doi.org/10.1016/j.conbuildmat.2021.125057

[44] Topçu, İ. B., & Boğa, A. R. (2010). Effect of ground granulate blast-furnace slag on corrosion performance of steel embedded in concrete. *Materials & Design, 31*(7), Article 7. https://doi.org/10.1016/j.matdes.2010.01.057

[45] Neupane, K., Kidd, P., Chalmers, D., Baweja, D., & Shrestha, R. (2016). Investigation on compressive strength development and drying shrinkage of ambient cured powder-activated geopolymer concretes. Australian Journal of Civil Engineering, 14(1), Article 1.

[46] Shariq, M., Prasad, J., & Masood, A. (2010). Effect of GGBFS on time dependent compressive strength of concrete. *Construction and Building Materials, 24*(8), Article 8. https://doi.org/10.1016/j.conbuildmat.2010.01.007

[47] Mathew, G., & Paul, M. M. (2014). Influence of Fly Ash and GGBFS in Laterized Concrete Exposed to Elevated Temperatures. *Journal of Materials in Civil Engineering, 26*(3), Article 3. https://doi.org/10.1061/(ASCE)MT.1943-5533.0000830

[48] Nagajothi, S., & Elavenil, S. (2021). Effect of GGBS Addition on Reactivity and Microstructure Properties of Ambient Cured Fly Ash Based Geopolymer Concrete. *Silicon, 13*(2), Article 2. https://doi.org/10.1007/s12633-020-00470-w

[49] Patra, R. K., & Mukharjee, B. B. (2017). Influence of incorporation of granulated blast furnace slag as replacement of fine aggregate on properties of concrete. *Journal of Cleaner Production, 165*, 468–476. https://doi.org/10.1016/j.jclepro.2017.07.125

[50] Özbay, E., Erdemir, M., & Durmuş, H. İ. (2016). Utilization and efficiency of ground granulated blast furnace slag on concrete properties – A review. *Construction and Building Materials, 105*, 423–434. https://doi.org/10.1016/j.conbuildmat.2015.12.153

Advances in Construction, Real Estate, Infrastructure and Project Management – Anil Kashyap et al. (eds)
© 2026 Taylor & Francis Group, London, ISBN 978-1-041-13433-6

29 | Implementing Decarbonization Across Contexts: From Wampanoag to the Amazigh—Sustainable Design for Community

Nea Maloo[1]
Assistant Professor,
College of Engineering and Architecture (CEA),
Howard University, Washington

N. Jonathan Unaka[2]
Assistant Professor,
School of Architecture and Design,
Wentworth Institute of Technology, Boston

■ **ABSTRACT:** Students who experience context, are more detailed in the intervention of creative designs. This paper presents a multitude of approaches to teaching and learning about high-performance building designs. This paper presents two case studies to illustrate how varied approaches to teaching about sustainable design can occur. In both cases, learners were immersed in a specific context and were encouraged to seriously consider contextual conditions while attempting low-carbon design responses. In one case a conventional required architecture design studio course attempted a project with a site in a rural part of the Atlas Mountains. Learners are encouraged to be sensitive to the local context. As such, their design responses employ local knowledge of traditional, indigenous practices of passive design strategies, and available resources. The other case is an elective course taught as a workshop/ seminar involving students from multiple disciplines, which tackled a site in New Bedford, Massachusetts in the USA. This multi-disciplinary team entered their design proposal in the Solar Decathlon Design Competition, in which teams from around the world compete to produce buildings that achieve net-zero energy and net-zero carbon, typically through integrated passive and active design strategies and systems, with stringent building performance goals.

■ **KEYWORDS:** Sustainable design, Contextual learning, Passive design strategies

1. TIZYANE, AZZADEN VALLEY, HIGH ATLAS, MOROCCO

The Amazigh are a group of peoples scattered across North Africa, from Tunisia in the north to Mali in the south, from Egypt in the east to Morocco in the west. Their diaspora is found in every corner

[1]nea.maloo@howard.edu; [2]spiritcircle@gmail.com, unaka@wit.edu

DOI: 10.1201/9781003669814-29

of the globe, including a very old community in the Canary Islands. As such, there are Amazigh in lush mountain ecosystems and others in deserts. They can be found on coastlines and in hinterland areas [1].

The structure of Amazigh communities is quite egalitarian. Community members assert various forms of equality and claim freedom from the over-arching extant governmental or bureaucratic structures. Many of the locations where the Amazigh live would be considered relatively harsh environments for a modern city dweller. Modern infrastructure and systems are not readily available in many cases. For instance, pipe-borne water is not entirely rare but is not ubiquitous either. Sometimes there are good roads, however, they are subject to damage due to erosion, avalanches, and other stresses due to the harsh environment. Transporting building materials can be challenging. As such, the communities tend to be self-reliant and community members are often interdependent. There is greater reliance on local resources and knowledge [2].

1.1 Water and Freedom

This necessity to be self-reliant and to rely on local resources is heavily influenced by the belief that water is sacred. As such, in arid regions where water is scarce, vast infrastructure systems called Kheterra for moving water were used to bring water hundreds of kilometers. Oasis were constructed which in turn altered the ecosystem. These systems help to provide water for domestic use and irrigating fields.

In mountainous areas where water was abundant from ice melt, elaborate canal systems channeled water to multiple villages. In all cases, the people felt a sacred duty to protect the water and to share it. The Amazigh view water, and its purity, as sacred. Hence, the obstruction of water is seen as taboo. In villages and towns across the Atlas Mountain range, ice-melt water supply springs where water is sourced. Canals channel the water throughout the area. Community spirit prevents overdrawing water or blocking off canals. Villages at higher altitudes were responsible for letting water get to villages down-slope. The sacredness of water and its free access can be observed in contemporary times with public faucets dotted across the land where water from springs is accessible to all. Amazigh notions of sharing water extend to other useful items, including symbolic goods.

Various ecosystems inhabited by the Amazigh directly influence their culture and by extension how they build. And while many Amazigh groups have long settled in the places where they currently live, they all still seem comfortable with migration. As such, whether we speak of the Tuareg who still traverse the Sahara seasonally, others who live in the mountains, or those who dwell along the shore, they all engage in constant movement, often for work. These and other cultural aspects of life among the Amazigh have very interesting effects on how their communities are organized and can be seen in the way buildings are designed and built.

2. IMPLEMENTING DECARBONIZATION FOR TIZYANE, AZZADEN VALLEY, HIGH ATLAS, MOROCCO

The foregoing influenced the framing of an architecture design course. The course is one section in a graduate-level design studio, which occurs during the penultimate semester of the Master of Architecture program at Wentworth Institute of Technology. The following is the brief for a design-as-research case study in a village in the High Atlas Mountains which has spawned several other projects in the area, as well as down in the Draa River Valley. The subsequent remarks will discuss one such academic project which is an architectural design studio with graduate students [3].

2.1 The Azzaden Valley

Nestled in the Atlas Mountains, this valley is one of the homelands of the Amazigh. In each village along the slopes, rammed earth houses cluster around the minaret of a pink mosque. Tizyane is one such village. Tizyane has electricity, mostly to supply lights to the home. Due to the altitude, they receive a considerable amount of snow, and experience quite extreme temperatures. Ice-melt from the snow-capped mountains supply springs. As in most mountain villages, the inhabitants of Tizyane have built reservoirs on the slopes above their houses, near the springs. From the reservoir, water is piped and gravity-fed into about 75% of the homes. They have no hot water or indoor heating. Hence, bathing during the winter months is challenging [4].

In Moroccan culture, the Hamman is perhaps, almost as important a communal space as the mosque. People go to hammam once or twice a week. In the bathhouse, dead skin is vigorously scrubbed off. Afterward, they sip tea and catch up on the latest goings-on. The typical village bathhouse and its water are heated by a large wood-burning stove, typically using a small tree each day in a small village community hammam. Usually, there is a schedule that divides the day by gender for using the bathhouse. Young children go with the women.

2.2 Design of the Request for Proposals

In the village of Tizyane, there was communal no hammam. They were using hammams in neighboring villages, and therefore, tended to go less frequently than they normally would. These conditions constituted both medical and public health challenges as well as a negative social situation. Working with an NGO, the people of Tizyane decided to address their lack of a bathhouse while lowering their ecological footprint. As the conversation for this project evolved, the women of Tizyane requested ideas for a laundry and women's gathering space.

Properly addressing the villagers' needs should take advantage of readily sourced materials and local knowledge. Design drawings must be visually based, and not require words to facilitate clear communication between the clients and the designers.

2.3 Student Travel

The students traveled with faculty to Morocco and were given a wide-ranging introduction to southeastern Morocco. They spent some time in Marrakech learning about a complex ancient and contemporary, cosmopolitan African city. They went to the mountains and the desert, staying in tents and in a kasbah. While spending several days in Tizyane, the students met its inhabitants and conducted a typical community engagement activity as any urban designer or town planner would organize. After practicing with their peers at school, the students were able to organize this themselves. They learned about historical and contemporary strategies for accessing basic needs such as water, and studied the juxtapositions of modern with traditional construction techniques such as rammed earth, compressed earth bricks, tadelakt, stone, and others.

2.4 Student Proposals

Upon returning from the travel week, the students were allowed to interpret the proposal brief loosely. Design responses to the need for a community hammam, addressing the specific needs of the women, or any other design challenge should expect to sit right into the side of the mountain, like other buildings, for enhanced thermal comfort. If any proposed plan is successful in Tizyane, it may be adopted by other villages in the valley.

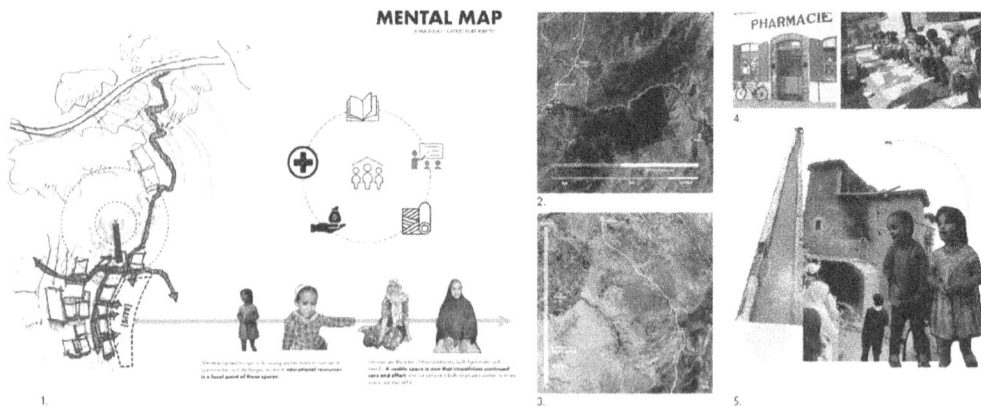

Figure 29.1 First design response: Community health center – Jona Sulaj and Catriel Kurtz Ribetto

Source: Authors

The first proposal was a women's clinic and birthing center. Responding directly to interviews and research done with the community about their emergent and preventative health needs, this team proposed a healthcare complex to provide for the basic needs of Tizyane and nearby villages.

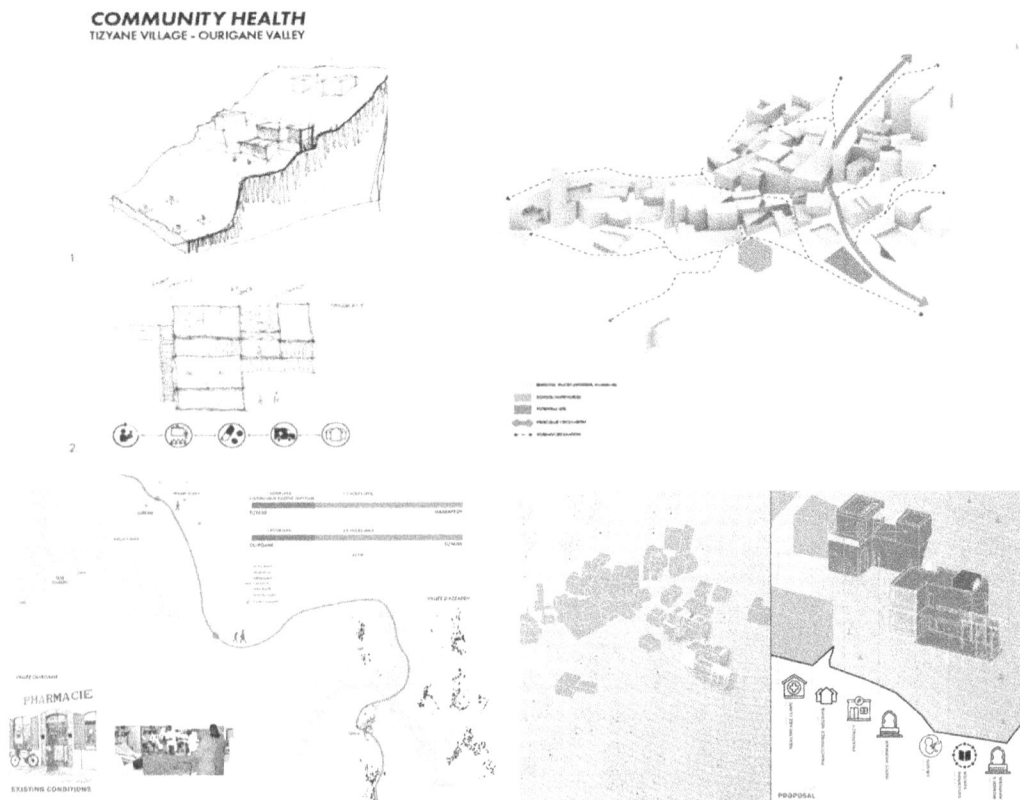

Figure 29.2 First design response: Community health center – Jona Sulaj and Catriel Kurtz Ribetto

Source: Authors

The proposal focuses on the needs of women, with OBGYN, and educational spaces, to be government-funded and run in collaboration with medical centers in the nearest urban centers. Included in the program is housing space for visiting health providers.

The second proposal sought to address the need for water by accessing it from the river for certain uses. The proposal addresses the ambivalence of community members towards the river in the valley. While it is a useful resource, it also constitutes a barrier to the outside. Their design is a modular bridge that allows for crossing the river as it uses water from the river.

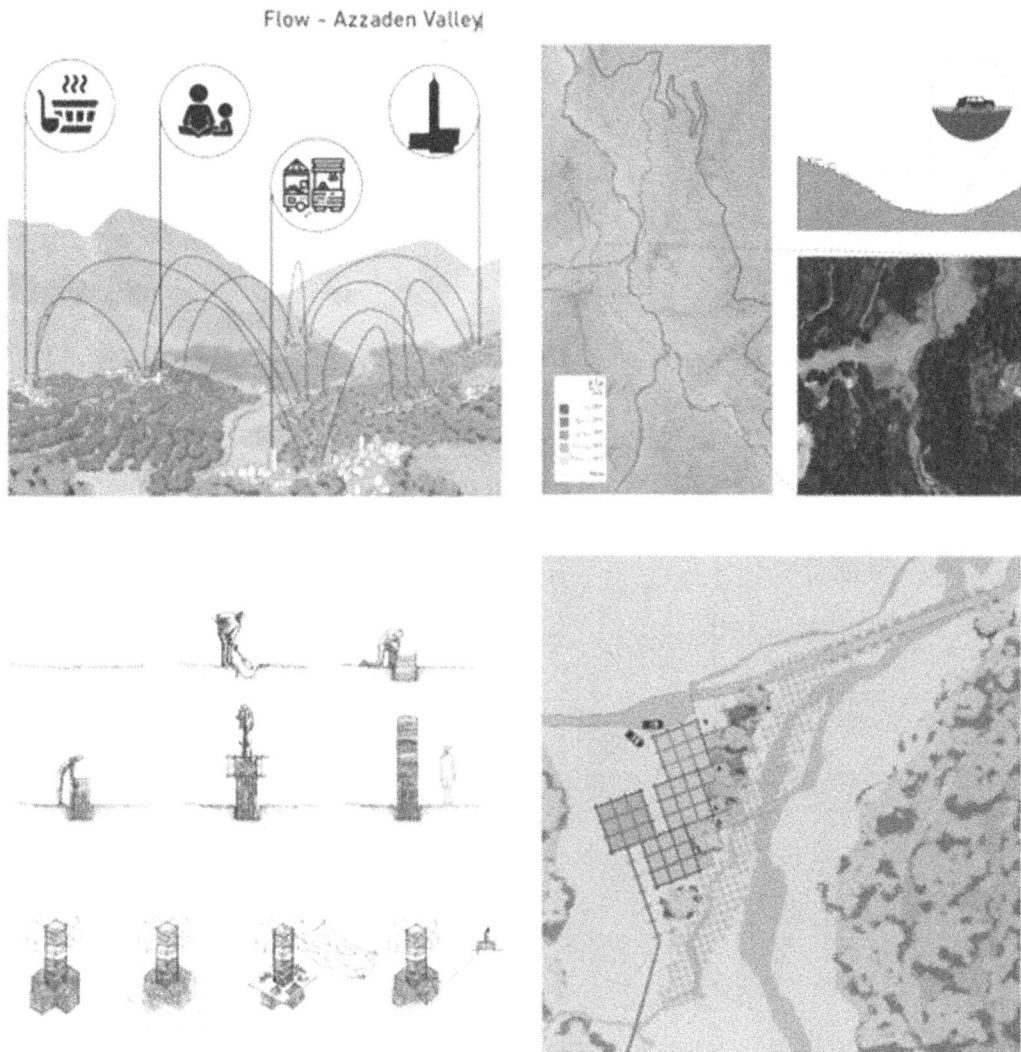

Figure 29.3 Second design response: FLOW – Bridge and water supply system – Andres Jimenez and Joseph Scheuermann

Source: Authors

The third proposal was dedicated mainly to the children of the village and involved play. The proposal was designed to be spread out across the entire village in alleyways, at corners, and other in-between spaces, such as on and between rooftops.

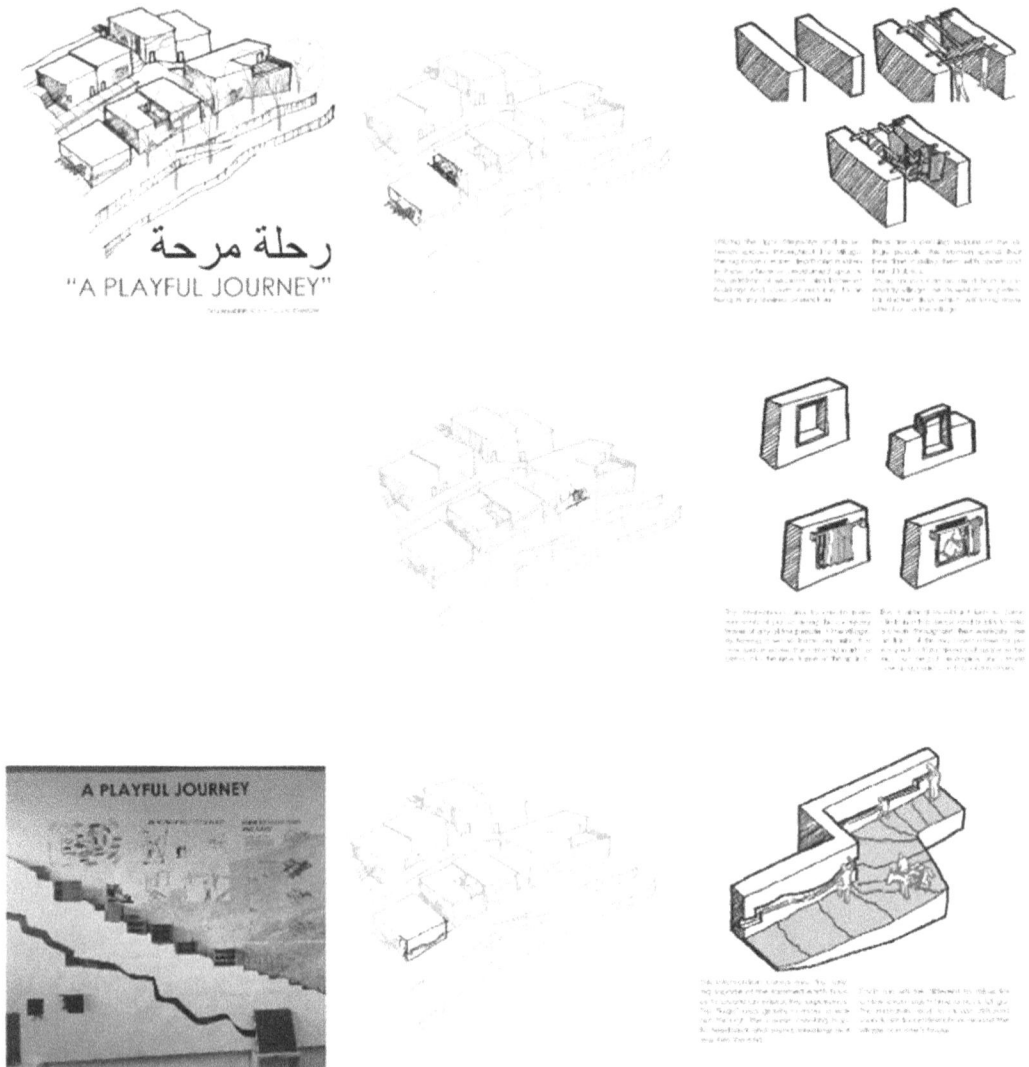

Figure 29.4 Third design response: A playful journey – Logan Middelton and Calvin Tucker

Source: Authors

Using rugs created by the women in the village and other local resources, the proposal turns the entire village into a play area for children and others. They also provide interesting moments of respite for the women.

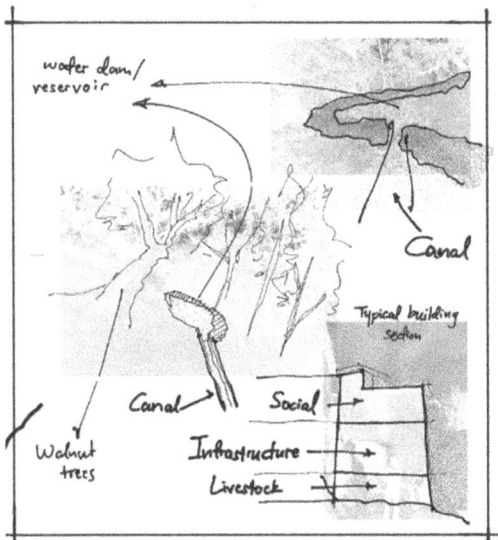

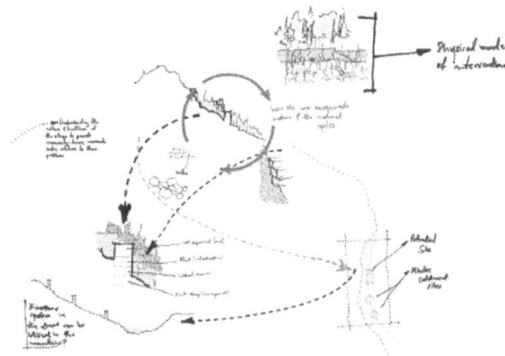

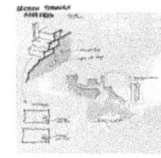

Figure 29.5   Fourth design response: Water commune – Joshua Ssebuwufu and Schuyler
Wilkerson

Source: Authors

The fourth proposal used the idea of the water elevation system as inspiration for providing a new
kind of housing. This proposal was designed in section to take full advantage of the steep mountain
slope. The design's form is in the familiar building language found in the village, with local materials
and techniques to address the need for water provision and processing at the scale of the building.

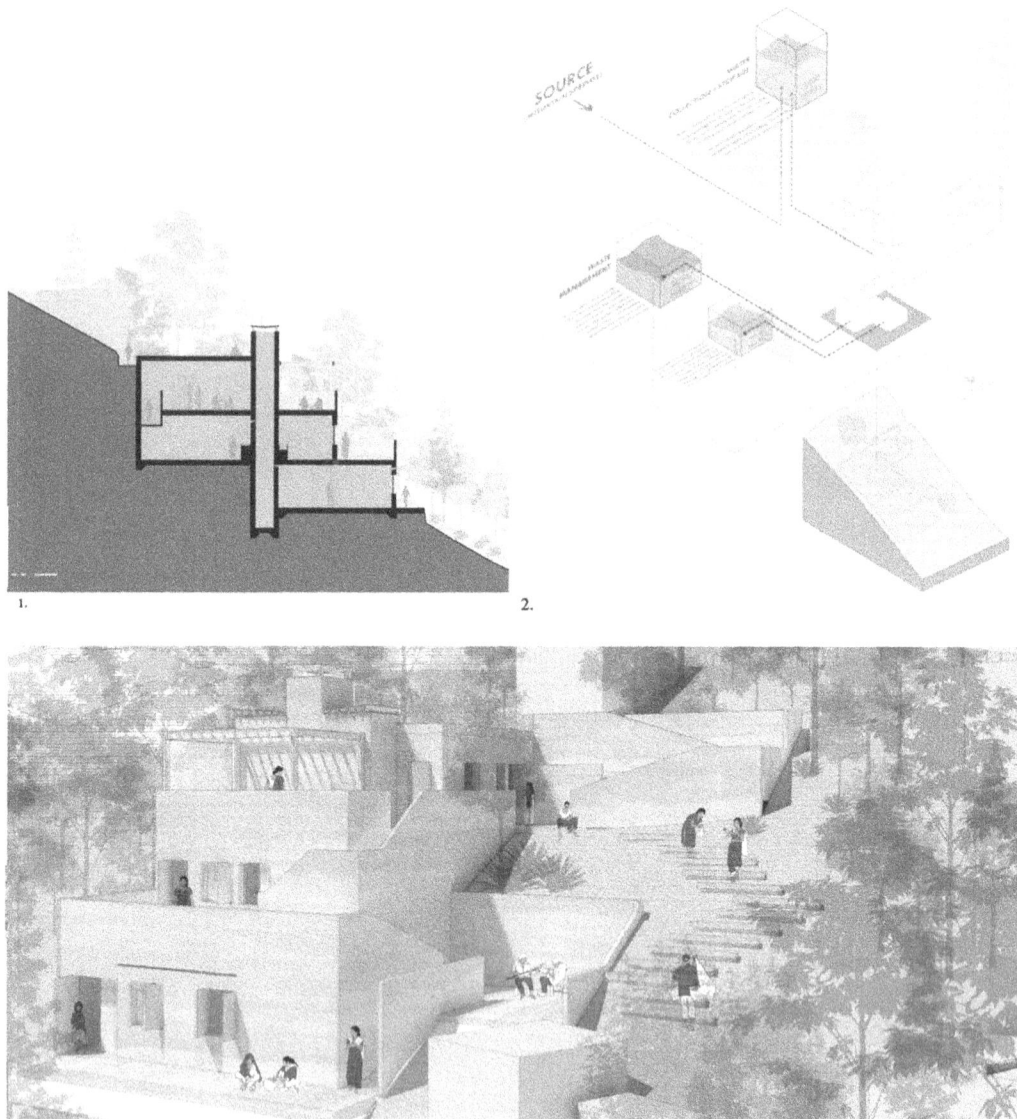

Figure 29.6 Fourth design response: Water commune – Joshua Ssebuwufu and Schuyler Wilkerson

Source: Authors

3. CONCLUSION

After being in Morocco, a main takeaway for the students was the possibility of a different way to be in the world. As such, their design responses were very different. Great care was taken to make the representation legible to experts and non-experts alike. The work of these student researchers is being built upon by successive students. While the design proposals addressed climate action, they also considered the specific social needs of the clients.

4. NEW BEDFORD, MA, USA

4.1 Context

New Bedford is a medium-sized town in the northeast United States known as the "City that Lit the World." It was first settled in 1600 by European settlers who seized the Wampanoag land and created the first booming whaling industry. Following New Bedford, several other whaling ports pop up throughout New England. The town of New Bedford built the first largest whaling ship and launched it to the world, paving the way for its largest fishing industry. During the 1800s, the city had the most prominent Quaker population, making it a haven for fugitive slaves, such as Frederick Douglass, to settle and provide successful business opportunities for all. This population influx and the growing whaling economy made New Bedford the richest per capita city during the 19th century. It also led to the opening of the first cotton mill in 1846 and the textile mill in 1848, which also contributed to the rising economy of the town. The multicultural population grew more diverse, with Portuguese and Cape Verdeans emigrating to the city. The economic growth, whaling, textile, and fishing industries created a rich atmosphere in art and history. Today, the town is the seat for the most significant offshore wind on the East Coast, creating new opportunities in clean energy and leading to more economic growth.

The town is in a temperate zone, with three months of summer and harsh winters with several inches of snow. Rising sea levels and other effects of climate change threaten the city, which has also faced slowing infrastructure and reduced federal investments. In the recent census, EPA deemed Bedford a Justice 40 location for various inequities and underrepresented population resources.

4.2 Problem Statement

The students were given the historic Hillman Firehouse, which was abandoned, as a case study to work on designing and reviving a public service building. The context of the town, with its deep history, had to be considered, and the existing building code, historic codes, and the sea level rise had to be incorporated. The need for affordable housing as a broader design goal was established as the program for the firehouse.

History of Firehouse

The historic new Bedford Hillman firehouse was constructed and put into commission in 1892. It served the region, saving lives and being part of the community as Engine No. 5 until 1950 the Civil Defense Department decommissioned it. It was abandoned, and the structure deteriorated until the city acquired it in 1994. In 2022, saving it from demolishing, WHALE (waterfront Historic Area League) acquired it and restored it.

4.3 The Howard Student Team

The students who worked on this project were part of a credit elective course, "Equitable performance buildings," and presented this in the United States Department of Energy Solar Decathlon Design Competition 2024. This paper will be the only section of their project discussing students' passive strategies and design decisions. Envision Resilience supported the travel and project—the students were interdisciplinary, from juniors and seniors in architecture and environmental studies majors.

Figure 29.7 Existing hillman fire house

Source: HU_Lumina Team

4.4 Site Context

The firehouse is located on the corner of Hillman & County streets; the Hillman Firehouse is in a prime location. The site is in the center of New Bedford and is well connected by street and public transportation. There is a public bus stop in front of the firehouses, and significant transportation route stops right in front of the firehouse,

The neighborhood recreation is only a five-minute walk, a seven-minute walk from the future Massachusetts South Coast railway station and pedestrian bridge connecting residents to Boston, and a ten-minute walk from downtown New Bedford's commercial, art, and waterfront properties.

4.5 Travel

Initially, Student designs and programs were partially framed before the organized travel to New Bedford to facilitate inquiries, verifications, and community engagement. Students met the housing authorities and toured the site, neighborhoods, community, and fishing port. The students also visited the whaling museum and the new offshore wind farm and learned much from the local architects and builders. The community partners gave the students the broader picture, the need for the people, and all the local resources. The students walked and researched the neighborhoods in the town's rich history and architecture. They felt the slight town feel of the city and the uniqueness of its people. This experience refined their goals to be better focused on the people. They came with three main goals: Regenerate, advocate, and Beacon. Regenerate, To design solutions to bring energy back into a building left abandoned and unused for decades, turning a blight on a prime street corner into high-performance affordable housing. Advocate for the use of the community and more renewable energy. Be a beacon to the community in design, energy, and resources.

4.6 Design Solutions

Hillman Lawn: To activate the street corner for the community of Hillman, the students designed a new outdoor program named the Hillman Lawn in this lawn space for an organic greenhouse, mural, and program to be used by the community. Vibrant art and a small-town feel were also engaged in the hillman lawn design by space for artist display. Students designed a Pergola for shade and created space for community events such as local AHA events and voter registration.

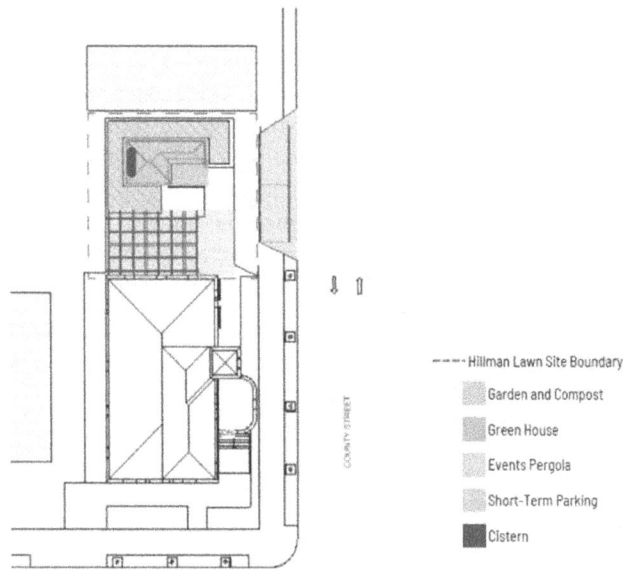

Figure 29.8 Site plan showing hillman lawn programming

Source: HU_Lumina Team

Passive Strategies

The use of thick brick walls and large windows, the addition of the historic balcony, which was abandoned to give more light on the second floor, and window wells and natural ventilation of the windows created more daylight in the basement. The addition of elevators to the firehouse made it universally accessible to all. The walls were insulated with materials not on the red list and procured locally to reduce embodied carbon. The passive strategies and tight thermal control reduced the mechanical loads on heating and cooling the buildings.

Figure 29.9 Existing monitor examples

Source: HU_Lumina

Designing New Monitor

The existing firehouse was Romanesque revival style with red brick and rich motifs. During the travel and research interpreting historical architecture, the students observed monitors in other neighborhood buildings. These monitors allowed whales to see if their husbands had returned to the sea from long travel, and these monitors provided light and direction for the ships approaching New Bedford. During the trip to Bedford and the analysis and synthesis of all research, students created an upper-level monitor for wellness to maximize the sunlight for the residents in the Hillman Firehouse housing. The Sun is a minimal resource, especially during the winter; this monitor provides maximum benefit to the occupant throughout the year.

Reusing of brick: The new design complements the existing historic firehouse brick in architecture and design. Students carefully matched the brick of the historic color of the firehouse to denote the bright orange-reddish flame.

Figure 29.10　Rendering of hillman lawn

Source: HU_Lumina Team

Stormwater Management

Given the increasing possibility of sea level rise, the design solution incorporated the ground level as a more permanent housing transient. Stormwater management and planting of native species improved the landscape.

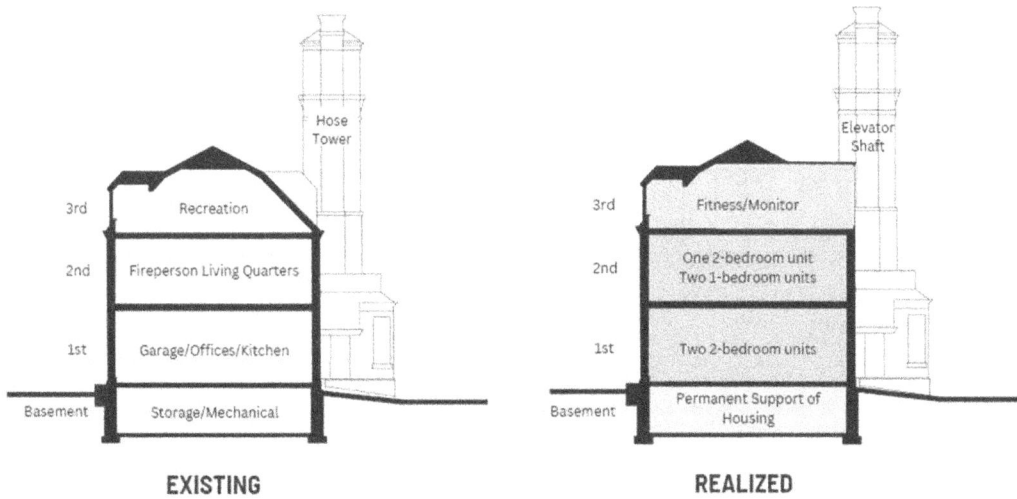

Figure 29.11   Programing of existing and realized

Source: HU _ Team Lumina

5. CONCLUSIONS

The project included an in-depth analysis of lifecycle, marketing, engineering, and embodied carbon, not presented in this paper. Field visits are essential in teaching architectural design. The interdisciplinary nature of the students provided in-depth discussions on topics of what the community needs and the local resources that can be tapped into. The students benefitted from the rich history of the firehouse and its iconic presence in the community. There was a sense of

Figure 29.12   Hillman firehouse student design rendereing

Source: HU_Lumina Team

pride when they talked to the neighbors about how their design and program would regenerate the street and the city. Another conclusion is learning to ask the right questions to the community partners, industry partners, and government officials. Interacting with all stakeholders, including contractors, enriched the learning and development of the design solutions. The purpose of taking existing buildings to reduce carbon emissions and reviving public service buildings is to generate affordable housing, and its program is a complete resource to society at large.

Notes:

All images and design work in Case 2 are produced by Team Lumina, Raeesah Amegankpoe, Jai Blyden, Alyse Dees, Bria Holton-Oswald, Julian Newnham, Inaya Samad, Journey O'Neal, Journey O'Neal, Clarissa Smith, Bruce Turner III, with guidance from faculty Prof. Nea maloo.

REFERENCES

[1] Hoffman, K. E. (2018). *We Share Walls: Language, Land, and Gender in Berber Morocco. Blackwell Studies in Discourse and Culture*. Malden, MA: Blackwell Publications.

[2] Maddy-Weitzman, B. (2011). *The Berber Identity Movement and the Challenge to North African States* (1st ed.). Austin, TX: University of Texas.

[3] Rayne, L., Brandolini, F., Makovics, J. L., Hayes-Rich, E., Levy, J., Irvine, H., . . . Bokbot, Y. (2023). Detecting Desertification in the Ancient Oases of Southern Morocco. *Scientific Reports*, 13(19424). Retrieved from https://doi.org/10.1038/s41598-023-46319-1

[4] Woodburn, J. (1982). Egalitarian Societies. *Man: The Journal of the Royal Anthropological Institute*, 17(3), 431–451.

For Product Safety Concerns and Information please contact our EU
representative GPSR@taylorandfrancis.com
Taylor & Francis Verlag GmbH, Kaufingerstraße 24, 80331 München, Germany